NanoScience and Technology

NanoScience and Technology

Series Editors:
P. Avouris B. Bhushan D. Bimberg K. von Klitzing H. Sakaki R. Wiesendanger

The series NanoScience and Technology is focused on the fascinating nano-world, mesoscopic physics, analysis with atomic resolution, nano and quantum-effect devices, nanomechanics and atomic-scale processes. All the basic aspects and technologyoriented developments in this emerging discipline are covered by comprehensive and timely books. The series constitutes a survey of the relevant special topics, which are presented by leading experts in the field. These books will appeal to researchers, engineers, and advanced students.

Multiscale Dissipative Mechanisms and Hierarchical Surfaces
Friction, Superhydrophobicity, and Biomimetics
Editors: M. Nosonovsky, B. Bhushan

Applied Scanning Probe Methods I
Editors: B. Bhushan, H. Fuchs,
S. Hosaka

Applied Scanning Probe Methods II
Scanning Probe Microscopy
Techniques
Editors: B. Bhushan, H. Fuchs

Applied Scanning Probe Methods III
Characterization
Editors: B. Bhushan, H. Fuchs

Applied Scanning Probe Methods IV
Industrial Application
Editors: B. Bhushan, H. Fuchs

Applied Scanning Probe Methods V
Scanning Probe Microscopy
Techniques
Editors: B. Bhushan, H. Fuchs, S. Kawata

Applied Scanning Probe Methods VI
Characterization
Editors: B. Bhushan, S. Kawata

Applied Scanning Probe Methods VII
Biomimetics and Industrial Applications
Editors: B. Bhushan, H. Fuchs

Applied Scanning Probe Methods VIII
Scanning Probe Microscopy Techniques
Editors: B. Bhushan, H. Fuchs, M. Tomitori

Applied Scanning Probe Methods IX
Characterization
Editors: B. Bhushan, H. Fuchs,
M. Tomitori

Applied Scanning Probe Methods X
Biomimetics and Industrial Applications
Editors: B. Bhushan, H. Fuchs, M. Tomitori

Applied Scanning Probe Methods XI
Scanning Probe Microscopy
Techniques
Editors: B. Bhushan, H. Fuchs

Applied Scanning Probe Methods XII
Characterization
Editors: B. Bhushan, H. Fuchs

Applied Scanning Probe Methods XIII
Biomimetics and Industrial Applications
Editors: B. Bhushan, H. Fuchs

Magnetic Microscopy of Nanostructures
Editors: H. Hopster and H.P. Oepen

The Physics of Nanotubes
Fundamentals of Theory, Optics
and Transport Devices
Editors: S.V. Rotkin and S. Subramoney

Atomic Force Microscopy, Scanning Nearfield Optical Microscopy and Nanoscratching
Application to Rough
and Natural Surfaces
By G. Kaupp

Nanocatalysis
Editors: U. Heiz, U. Landman

Roadmap of Scanning Probe Microscopy
Editors: S. Morita

Nanostructures – Fabrication and Analysis
Editor: H. Nejo

Bharat Bhushan
Harald Fuchs

Applied Scanning
Probe Methods XI

Scanning Probe Microscopy Techniques

With 113 Figures and 17 Tables
Including 22 Color Figures

 Springer

Editors

Prof. Dr. Bharat Bhushan
Ohio State University
Nanoprobe Laboratory for Bio- & Nanotechnology
& Biomimetics (NLB^2)
201 W. 19th Ave
Columbus, Ohio 43210-1142
USA
bhushan.2@osu.edu

Prof. Dr. Harald Fuchs
Universität Münster
FB 16
Physikalisches Institut
Wilhelm-Klemm-Str. 10
48149 Münster
Germany
fuchsh@uni-muenster.de

Series Editors

Professor Dr. Phaedon Avouris
IBM Research Division
Nanometer Scale Science & Technology
Thomas J.Watson Research Center, P.O. Box 218
Yorktown Heights, NY 10598, USA

Professor Bharat Bhushan
Nanoprobe Laboratory for Bio- & Nanotechnology
and Biomimetics (NLB^2)
201 W. 19th Avenue
The Ohio State University
Columbus, Ohio 43210-1142, USA

Professor Dr. Dieter Bimberg
TU Berlin, Fakutät Mathematik,
Naturwissenschaften,
Institut für Festkörperphysik
Hardenbergstr. 36, 10623 Berlin, Germany

Professor Dr., Dres. h. c. Klaus von Klitzing
Max-Planck-Institut für Festkörperforschung
Heisenbergstrasse 1, 70569 Stuttgart, Germany

Professor Hiroyuki Sakaki
University of Tokyo
Institute of Industrial Science,
4-6-1Komaba, Meguro-ku, Tokyo 153-8505
Japan

Professor Dr. RolandWiesendanger
Institut für Angewandte Physik
Universität Hamburg
Jungiusstrasse 11, 20355 Hamburg, Germany

ISBN: 978-3-540-85036-6   e-ISBN: 978-3-540-85037-3

NanoScience and Technology ISSN 1434-4904

Library of Congress Control Number: 2008933574

© Springer-Verlag Berlin Heidelberg 2009

Cover design: WMXDesign GmbH, Heidelberg

Printed on acid-free paper

9 8 7 6 5 4 3 2 1

springer.com

Preface for Applied Scanning Probe Methods Vol. XI–XIII

The extremely positive response by the advanced community to the Springer series on Applied Scanning Probe Methods I–X as well as intense engagement of the researchers working in the field of applied scanning probe techniques have led to three more volumes of this series. Following the previous concept, the chapters were focused on development of novel scanning probe microscopy techniques in Vol. XI, characterization, i.e. the application of scanning probes on various surfaces in Vol. XII, and the application of SPM probe to biomimetics and industrial applications in Vol. XIII. The three volumes will complement the previous volumes I–X, and this demonstrates the rapid development of the field since Vol. I was published in 2004. The purpose of the series is to provide scientific background to newcomers in the field as well as provide the expert in the field sound information about recent development on a worldwide basis.

Vol. XI contains contributions about recent developments in scanning probe microscopy techniques. The topics contain new concepts of high frequency dynamic SPM technique, the use of force microscope cantilever systems as sensors, ultrasonic force microscopy, nanomechanical and nanoindentation methods as well as dissipation effects in dynamic AFM, and mechanisms of atomic friction.

Vol. XII contains contributions of SPM applications on a variety of systems including biological systems for the measurement of receptor–ligand interaction, the imaging of chemical groups on living cells, and the imaging of chemical groups on live cells. These biological applications are complemented by nearfield optical microscopy in life science and adhesional friction measurements of polymers at the nanoscale using AFM. The probing of mechanical properties by indentation using AFM, as well as investigating the mechanical properties of nanocontacts, the measurement of viscous damping in confined liquids, and microtension tests using in situ AFM represent important contributions to the probing of mechanical properties of surfaces and materials. The atomic scale STM can be applied on heterogeneous semiconductor surfaces.

Vol. XIII, dealing with biomimetics and industrial applications, deals with a variety of unconventional applications such as the investigations of the epicuticular grease in potato beetle wings, mechanical properties of mollusc shells, electrooxidative lithography for bottom-up nanofabrication, and the characterization of mechanical properties of biotool materials. The application of nanomechanics as tools for the investigation of blood clotting disease, the study of piezo-electric polymers, quantitative surface characterization, nanotribological characterization of carbonaceous materials, and aging studies of lithium ion batteries are also presented in this volume.

We gratefully acknowledge the support of all authors representing leading scientists in academia and industry for the highly valuable contribution to Vols. XI–XIII. We also cordially thank the series editor Marion Hertel and her staff members Beate Siek and Joern Mohr from Springer for their continued support and the organizational work allowing us to get the contributions published in due time.

We sincerely hope that readers find these volumes to be scientifically stimulating and rewarding.

August 2008   Bharat Bhushan
                                                                    Harald Fuchs

Contents – Volume XI

Contents – Volume XII

Contents – Volume XIII

20   Electro-Oxidative Lithography and Self-Assembly Concepts for Bottom-Up Nanofabrication
           Stephanie Hoeppener, Ulrich S. Schubert

21   Application of SPM and Related Techniques to the Mechanical Properties of Biotool Materials
           Thomas Schöberl, Ingomar L. Jäger, Helga C. Lichtenegger

Contents – Volume I

Part III   Industrial Applications

Contents – Volume II

Contents – Volume III

Contents – Volume IV

Contents – Volume V

Contents – Volume VI

Contents – Volume VII

Contents – Volume VIII

Contents – Volume IX

Contents – Volume X

List of Contributors – Volume XI

Houssein Awada
Université Catholique de Louvain, Unité de chimie et de physique des hauts polymères (POLY), Croix du Sud 1 – 1348 Louvain-la-Neuve – Belgique (B)
e-mail: houssein.awada@uclouvain.be

Elmar Bonaccurso
Max-Planck-Institute for Polymer Research, Ackermannweg 10, D-55128 Mainz, Germany
e-mail: bonaccur@mpip-mainz.mpg.de

Paolo Bonanno
Department of Biophysical and Electronic Engineering, Unversity of Genova, Via all'Opera Pia 11a, I-16145 Genova, Italy
e-mail: paolo.bonanno@unige.it

Maurice Brogly
Université de Haute Alsace (UHA), Equipe Interfaces Sous Contraintes (ICSI - CNRS UPR 9069), 15 rue Jean Starcky – 68057 Mulhouse Cx – France (F)
e-mail: maurice.brogly@uha.fr

Hans-Jürgen Butt
Max-Planck-Institute for Polymer Research, Ackermannweg 10, D-55128 Mainz Germany
e-mail: butt@mpip-mainz.mpg.de

Lorenzo Calabri
CNR-INFM – National Research Center on nanoStructures and bioSystems at Surfaces (S3), Via Campi 213/a, 41100 Modena, Italy
e-mail: calabri.lorenzo@unimore.it.

M. Teresa Cuberes
Laboratorio de Nanotécnicas, UCLM, Plaza Manuel de Meca 1, 13400 Almadén, Spain
e-mail: teresa.cuberes@uclm.es

Dmytro S. Golovko
Max-Planck-Institute for Polymer Research, Ackermannweg 10, D-55128 Mainz,
Germany
e-mail: golovkod@mpip-mainz.mpg.de

Mykhaylo Evstigneev
Fakultät für Physik, Universität Bielefeld, Universitätsstr. 25, 33615 Bielefeld,
Germany
e-mail: Mykhaylo@Physik.Uni-Bielefeld.De

Harald Fuchs
Physikalisches Institut and Center for Nanotechnology (CeNTech), Universität
Münster, Wilhelm-Klemm-Str. 10, Münster D48149, Germany
e-mail: fuchsh@uni-muenster.de

Thomas Haschke
University of Siegen, Faculty 11, Department of Simulation, Am Eichenhang 50,
D-57076 Siegen, Germany
e-mail: haschke@simtec.mb.uni-siegen.de

Donna C. Hurley
National Institute of Standards & Technology, 325 Broadway, Boulder, Colorado
80305 USA
e-mail: hurley@boulder.nist.gov

Johann Jersch
Physikalisches Institut, Universität Münster, Wilhelm-Klemm-Str. 10, Münster
D48149, Germany
e-mail: jersch@uni-muenster.de

Olivier Noel
Université du Maine, Laboratoire de Phjysique de l'Etat Condensé (CNRS UMR
6087), Avenue Olivier Messiaen – 72085 Le Mans Cx 9 – France (F)
e-mail: olivier.noel@univ-lemans.fr

Stefano Piccarolo
Dipartimento di Ingegneria Chimica dei Processi e dei Materiali, Università di
Palermo, Viale delle Scienze, 90128 Palermo, Italy and INSTM Udr Palermo
e-mail: piccarolo@unipa.it

Nicola Pugno
Department of Structural Engineering, Politecnico di Torino, Corso Duca degli
Abruzzi 24, 10129 Torino, Italy, National Institute of Nuclear Physics, National
Laboratories of Frascati, Via E. Fermi 40, 00044, Frascati, Italy
e-mail: nicola.pugno@polito.it

Roberto Raiteri
Department of Biophysical and Electronic Engineering, Unversity of Genova, Via all'Opera Pia 11a I-16145 Genova, Italy
e-mail: rr@unige.it

Davide Tranchida
Dipartimento di Ingegneria Chimica dei Processi e dei Materiali, Università di Palermo, Viale delle Scienze, 90128 Palermo, Italy and INSTM Udr Palermo

Sergio Valeri
CNR-INFM – National Research Center on nanoStructures and bioSystems at Surfaces (S3), Via Campi 213/a, 41100 Modena, Italy. Department of Physics, University of Modena and Reggio Emilia, via Campi 213/a 41100 Modena, Italy
e-mail: sergio.valeri@unimo.it

Wolfgang Wiechert
University of Siegen, Faculty 11, Department of Simulation, Am Eichenhang 50, D-57076 Siegen, Germany
e-mail: wolfgang.wiechert@uni-siegen.de

List of Contributors – Volume XII

N. Agraït
Departamento de Física de la Materia Condensada C-III, Universidad Autonoma de Madrid, Madrid 28049, Spain
e-mail: nicolas.agrait@uam.es

Jean-Pierre Aimé
CPMOH, Université Bordeaux1, 351 Cours de la Libération, 33405 Talence Cedex, France
e-mail: jp.aime@cpmoh.u-bordeaux1.fr

David Alsteens
Unité de Chimie des Interfaces, Université Catholique de Louvain, Croix du Sud 2/18, B-1348 Louvain-la-Neuve, Belgium
e-mail: david.alsteens@uclouvain.be

Guillaume André
Unité de Chimie des Interfaces, Université Catholique de Louvain, Croix du Sud 2/18, B-1348 Louvain-la-Neuve, Belgium
e-mail: guillaume.andre@uclouvain.be

Sophie Bistac
Université de Haute-Alsace, CNRS, 15, rue Jean Starcky, BP 2488, 68057 Mulhouse Cedex, France
e-mail: sophie.bistac-brogly@uha.fr

Massimiliano Bocciarelli
Politecnico di Milano, Dipartimento di Ingegneria Strutturale, piazza Leonardo da Vinci 32, 20133 Milano, Italy
e-mail: bocciarelli@stru.polimi.it

Rodolphe Boisgard
CPMOH, Université Bordeaux1, 351 Cours de la Libération, 33405 Talence Cedex, France
e-mail: r.boisgard@cpmoh.u-bordeaux1.fr

Gabriella Bolzon
Politecnico di Milano, Dipartimento di Ingegneria Strutturale, piazza Leonardo da Vinci 32, 20133 Milano, Italy
e-mail: gabriella.bolzon@polimi.it, bolzon@stru.polimi.it

Enzo J. Chiarullo
Politecnico di Milano, Dipartimento di Ingegneria Strutturale, piazza Leonardo da Vinci 32, 20133 Milano Italy
e-mail: chiarullo@stru.polimi.it

Touria Cohen-Bouhacina
CPMOH, Université Bordeaux1, 351 Cours de la Libération, 33405 Talence Cedex, France
e-mail: t.bouhacina@cpmoh.u-bordeaux1.fr

Etienne Dague
Unité de Chimie des Interfaces, Université Catholique de Louvain, Croix du Sud 2/18, B-1348 Louvain-la-Neuve, Belgium
e-mail: etienne.dague@laas.fr

Jurg Dual
ETH Zentrum, IMES – Institute of Mechanical Systems, CLA J23.2, Department of Mechanical and Process Engineering, 8092 Zürich, Switzerland
e-mail: juerg.dual@imes.mavt.ethz.ch

Yves F. Dufrêne
Unité de Chimie des Interfaces, Université Catholique de Louvain, Croix du Sud 2/18, B-1348 Louvain-la-Neuve, Belgium
e-mail: dufrene@cifa.ucl.ac.be

Vincent Dupres
Unité de Chimie des Interfaces, Université Catholique de Louvain, Croix du Sud 2/18, B-1348 Louvain-la-Neuve, Belgium
e-mail: vincent.dupres@uclouvain.be

Robert H. Eibl
Plainburgstr. 8, 83457 Bayerisch Gmain, Germany
e-mail: robert_eibl@yahoo.com

Grégory Francius
Unité de Chimie des Interfaces, Université Catholique de Louvain, Croix du Sud 2/18, B-1348 Louvain-la-Neuve,
Belgium
e-mail: gregory.francius@uclouvain.be

Hongjun Gao
Nanoscale Physics & Devices Laboratory, Institute of Physics, Chinese Academy of
Sciences, P. O. Box 603, Beijing 100080, China
e-mail: hjgao@aphy.iphy.ac.cn

Pietro Giuseppe Gucciardi
CNR-Istituto per i Processi Chimico-Fisici, Salita Sperone c.da Papardo, I-98158
Messina, Italy
e-mail: gucciardi@me.cnr.it

Haiming Guo
Nanoscale Physics & Devices Laboratory, Institute of Physics, Chinese Academy of
Sciences, P. O. Box 603, Beijing 100080, China
e-mail: hmguo@aphy.iphy.ac.cn

Cedric Hurth
CPMOH, Université Bordeaux1, 351 Cours de la Libération, 33405 Talence Cedex,
France
e-mail: cedric.hurth@asu.edu

Cédric Jai
CPMOH, Université Bordeaux1, 351 Cours de la Libération, 33405 Talence Cedex,
France
e-mail: c.jai@free.fr

Udo Lang
ETH Zentrum, IMES – Institute of Mechanical Systems, Department of Mechanical
and Process Engineering, 8092 Zürich, Switzerland
e-mail: udo.lang@imes.mavt.ethz.ch

Abdelhamid Maali
CPMOH, Université Bordeaux1, 351 Cours de la Libération, 33405 Talence Cedex,
France
e-mail: a.maali@cpmoh.u-bordeaux1.fr

J.J. Riquelme
Departamento de Física de la Materia Condensada C-III, Universidad Autonoma de
Madrid, Madrid 28049, Spain
e-mail: juanjo.riquelme@uam.es

Gabino Rubio-Bollinger
Departamento de Física de la Materia Condensada C-III, Universidad Autonoma de
Madrid, Madrid 28049, Spain
e-mail: gabino.rubio@uam.es

Marjorie Schmitt
Université de Haute-Alsace, CNRS, 15, rue Jean Starcky, BP 2488, 68057 Mulhouse
Cedex, France
e-mail: Marjorie.Schmitt@uha.fr

Claire Verbelen
Unité de Chimie des Interfaces, Université Catholique de Louvain, Croix du Sud
2/18, B-1348 Louvain-la-Neuve, Belgium
e-mail: claire.verbelen@uclouvain.be

S. Vieira
Departamento de Física de la Materia Condensada C-III, Universidad Autonoma de
Madrid, Madrid 28049, Spain
e-mail: sebastian.vieira@uam.es

Yeliang Wang
Nanoscale Physics & Devices Laboratory, Institute of Physics, Chinese Academy of
Sciences, P. O. Box 603, Beijing 100080, China
e-mail: ylwang@aphy.iphy.ac.cn

List of Contributors – Volume XIII

Francois Barthelat

Department of Mechanical Engineering, McGill University, Macdonald Engineering Building, Rm 351, 817 Sherbrooke Street West, Montreal, Quebec H3A 2K6
e-mail: francois.barthelat@mcgill.ca

Bharat Bhushan

Nanotribology Laboratory for Information Storage and MEMS/NEMS (NLIM), Ohio State University, Columbus, OH 43210, USA

Sophie Bistac

Université de Haute-Alsace, 15, rue Jean Starcky, BP 2488, 68057, Mulhouse Cedex, France
e-mail: Sophie.Bistac-Brogly@uha.fr

Horacio D. Espinosa

Department of Mechanical Engineering, Northwestern University, 2145 Sheridan Rd., Evanston, IL 60208-3111, USA
e-mail: espinosa@northwestern.edu

Stanislav Gorb

Evolutionary Biomaterials Group, Max-Planck-Institut für Metallforschung, Heisenbergstrasse 3, 70569 Stuttgart, Germany
e-mail: s.gorb@mf.mpg.de

Stephanie Hoeppener

Laboratory of Macromolecular Chemistry and Nanoscience, Eindhoven University of Technology, P.O. Box 513, 5600 MB Eindhoven, The Netherlands Center for NanoScience, Lehrstuhl für photonik und Optoelektronik Luduig - Maximilians - Universität München, Geschwister-Scholl platz 1, 80333 München, Germany
e-mail: s.hoeppener@tue.nl

Ingomar L. Jäger

Department of Materials Science, University of Leoben, Jahnstrasse 12, 8700
Leoben, Austria
e-mail: ingomar@unileoben.ac.at

Taekwon Jee

Mechanical Engineering, Texas A&M University, College Station, TX 77843-3123
e-mail: taekwonjee@gmail.com

Hyungoo Lee

Department of Mechanical Engineering, Texas A&M University,
College Station, TX 77843, USA
e-mail: thanku7@gmail.com

Hong Liang

Department of Mechanical Engineering, Texas A&M University,
College Station, TX 77843-3123, USA
e-mail: hliang@tamu.edu

Helga C. Lichtenegger

Institute of Materials Science and Technology E308, Vienna University of
Technology, Favoritenstrasse 9-11, 1040 Wien, Austria
e-mail: helga.lichtenegger@tuwien.ac.at

Shrikant C. Nagpure

Nanotribology Laboratory for Information Storage and MEMS/NEMS (NLIM),
Ohio State University, Columbus, OH 43210, USA
e-mail: nagpure.1@osu.edu

H. Peisker

Evolutionary Biomaterial Group, Max-Plank-Institut für Metallforschung,
Heisenbergstrasse 3, 70569 Stuttgart, Germany

Jee E. Rim

Mechanical Engineering, Northwestern University, 2145 Sheridal Road,
Technological Institute B224, Evanton, IL 60208
e-mail: j-rim@northwestern.edu

Maria Cecília Salvadori

Institute of Physics, University of São Paulo, C.P. 66318, CEP 05315-970, São
Paulo, SP, Brazil
e-mail: mcsalvadori@if.usp.br

Marjorie Schmitt
Université de Haute-Alsace, 15, rue Jean Starcky, BP 2488, 68057, Mulhouse
Cedex, France
e-mail: Marjorie.Schmitt@uha.fr

Matthias Schneider
University of Augsburg, Experimental Physics I, Universitätsstr. 1, 86159 Augsburg,
Germany
e-mail: matthias.schneider@physik.uni-augsburg.de

Thomas Schöberl
Erich Schmid Institute of Materials Science of the Austrian Academy of Sciences,
Jahnstrasse 12, 8700 Leoben, Austria
e-mail: schoeber@unileoben.ac.at

Ulrich S. Schubert
Friedrich-Schiller-Universität Jena
Institute für Organische Chemie and Makromolekulare
Chemie, Humbolattstr. 10, 07743 Jena, Germany

Daniel Steppich
University of Augsburg, Experimental Physics I, Universitätsstr.
1, 86159 Augsburg, Germany
e-mail: daniel.steppich@physik.uni-augsburg.de

Stefan Thalhammer
GSF-Institut für Strahlenschutz, Neuherberg, Germany
e-mail: stefan.thalhammer@gsf.de

D. Voigt
Evolutionary Biomaterials Group, Max-Planck-Institut für
Metallforschung, Heisenbergstrasse 3, 70569 Stuttgart,
Germany
e-mail: voigt@mf.mpg.de

Ke Wang
Department of Mechanical Engineering, Texas A& M University,
College Station, TX 77843, USA
e-mail: ke.phwk@gmail.com

Achim Wixforth
University of Augsburg, Experimental Physics I, Universitätsstr.
1, 86159 Augsburg, Germany
e-mail: achim.wixforth@physik.uni-augsburg.de

1 Oscillation Control in Dynamic SPM with Quartz Sensors

Johann Jersch · Harald Fuchs

Key words: Dynamic SPM, Quartz resonator, Oscillator, Sensor, Analytical signal, Signal processing, Modulation, Demodulation, Feedback loop, Plasmon excitation

Abbreviations

SPM	scanning probe microscopy
AFM	atomic force microscopy
SNOM	scanning nearfield optical microscopy
QCR	quartz crystal resonator
I/Q in phase	quadrature phase
TF	tuning fork
MEMS	micro electromechanical system
AGC	automatic gain control
PLL	phase locked loop
FM	frequency modulation
RF	radio frequency

1.1
Introduction

Control and harmonic analysis of oscillations are key objects in the majority of physical and technical applications. Change of oscillation parameters – amplitude, frequency and phase – reflects the physical interactions in a system under investigation. Typically, vibration in a physical system is transformed into variation of amplitude, frequency or phase of an electrical signal. More or less complex signal processing in an analog or digital processor follows thereafter.

With the introduction of dynamic mode in the atomic force microscopy (AFM) systems, the control and analysis of cantilever oscillations became here also a fundamental issue. Subsequently, dynamic mode was introduced in all scanning probe microscopy (SPM) methods: Scanning Nearfield Optical Microscopy (SNOM), Magnetic Force Microscopy (MFM), Scanning Tunnelling Microscopy (STM), Acoustic Scanning Microscopy (ASM) etc.

The electrical signal nature in dynamic mode SPM is the same as in radio communication and in a number of other techniques: microbalance technique, micro electromechanical system (MEMS), sonar, optics communication etc. In general a carrier frequency (e.g., cantilever vibration) is modulated by some physical interaction and demodulated by a baseband signal processing method. For instance, interaction forces in AFM modulate the cantilever vibration, which can be considered as a carrier, in amplitude, phase and frequency. In a simplified form, the similarity between signal flow in a radio transmitter/receiver and an SPM with a quartz crystal resonator (QCR) sensor. The carrier frequency is in either technique modulated by applying a pressure (microphone in radio) or force (in SPM), and is then demodulated for baseband processing. Shown in Fig. 1.1 is the simple synchronous I/Q (in-phase/quadrature phase) signal processing schema. A more expensive scheme with carrier frequency converting can be advantageous in some special situations (e.g. for very fast detection circuits).

The carrier frequency is typically much higher than frequencies in modulation spectra; consequently, we have a narrow-band signal. Because of the non steady state character of the signals, the fast tracking of instantaneous parameter changes (frequency, phase, amplitude) should be implemented. The extensive development in radio narrow-band signal processing method [1] science during the last century provides a good basis also for the dynamic SPM signal processing. SPM techniques have profited so far with large delay from the development in communication techniques.

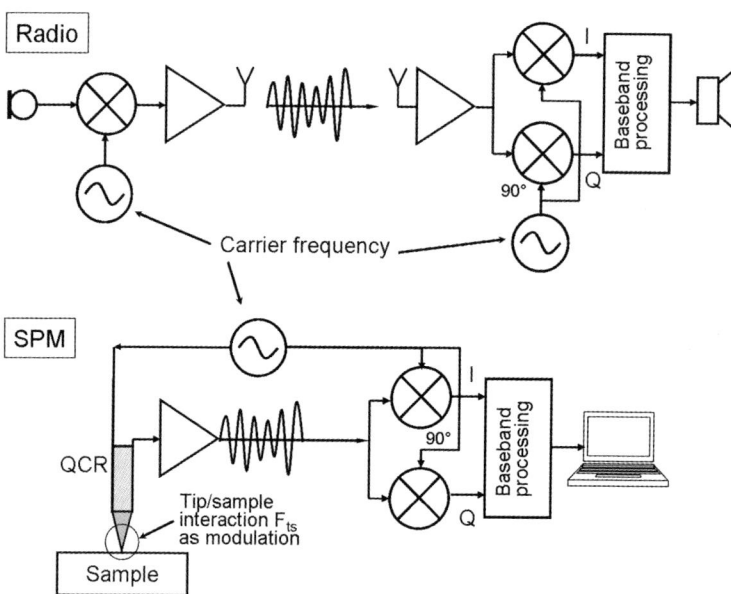

Fig. 1.1. In a radio device, a microphone signal modulates the carrier, on the receiver side I/Q demodulation of the modulated carrier and baseband processing occurs. In SPM, the tip–sample interaction modulates the oscillations of the QCR, the demodulation can occur in the same way as in a radio receiver

Soon after the invention of the AFM, Dürig et al. [2] introduced the dynamic technique to measure forces acting during tunnelling in an STM. The simplest way to detect the interaction, namely the modulation/demodulation of the oscillation amplitude (amplitude modulation technique – AM) was used. The simplest variant of the frequency modulation/demodulation (FM) was introduced [3] only in 1991, the phase modulation technique – in 2006 [4]. Accordingly, the instrumentation also has a variety of developments. From the first simple amplitude peak detector to precise analog–digital oscillation controllers today, allowing arbitrary signal processing algorithms (e.g. oscillation controller OC-4 from Nanonis, Switzerland; and the MFP-3D^{TM} Controller from Asylum Research, US; Easy PLL from Nanosurf, Switzerland).

The considerations above are of particular interest for SPM based on QCR sensors. QCR belong to the most frequently used sensors in engineering because of their low internal energy dissipation, high temperature as well as mechanical and chemical stability, low temperature compatibility and relatively high piezo-electrical constants. Therefore, despite of a wide choice of industrial cantilever-based force sensors in SPM, complementary approaches such as symmetrical QCR, for example tuning fork (TF), trident tuning fork and needle quartz as a force-sensing platform are of great interest. Recently, various QCR were introduced as a platform for an attached sensor in different SPM techniques such as acoustic [5], force [6], SNOM, magnetic force microscopy [7], scanning tip enhanced Raman spectroscopy (TERS) [8] etc. (see also review articles [9, 10] and literature therein). A high quality factor, even in air, and high spring constants are preconditions for resolving short-range forces and stable operation in force microscopy. Very low energy dissipation at low excitation amplitudes, the possibility to maintain a sub-nanometre tip–sample gap without contact (snap-in) and low dielectric losses even at high frequencies are crucial factors in specific applications, for example in scanning microwave or capacitance microscopy.

It should be noted that application and analyses of oscillator schematics in quartz crystal microbalance techniques is similar and was introduced long before tuning fork sensors in SPM techniques were applied. The main difference is that in the quartz crystal microbalance technique tiny changes in the effective oscillating mass of the sensor are measured, and not forces as in SPM. A good introduction and overview of quartz crystal microbalance techniques is provided in the textbook edited by A. Arnau [11].

1.2
Definition and Measurement of Signal Parameters

For signals that are not periodic, the notion of frequency, amplitude and phase are ambiguous and often altogether unclear; this has led to a multitude of frequency-measurement techniques and discrepancy of results. In the case of narrow-band signals, the situation is somewhat better. In this case, for example, dynamical SPM, MEMS, radar, Doppler vibration and velocity measurements etc., the most important parameters are the appropriately defined instantaneous frequency and amplitude signals.

Advanced signal processing theory makes use of the concept of analytical signal and the related *I/Q* (*I* for in-phase and *Q* for quadrature phase signals) modulation/demodulation (modem) technique. Let's consider the model of the formation of an instantaneous frequency and amplitude of a real signal. The commonly used form of a narrow-band real signal is written as:

$$x(t) = u(t) \cos (2\pi f_0 t + \theta), \tag{1.1}$$

where $u(t)$ is the low frequency envelope (amplitude of the oscillation). This signal can be represented as a projection of the vector in a complex plane onto the abscissa axis. The vector length is $u(t)$, and the rotation velocity is $2\pi f_0 t$.

The notation of an analytical signal corresponding to the real narrow-band signal $x(t)$ is widely used in signal processing. An analytical signal is a complex signal that forms if the real signal $x(t)$ is added to its Hilbert transform as an imaginary part. Particularly for narrow-band signals that is the quadrature, which is actually a projection onto the ordinate axis [12]:

$$\psi(t) = x(t) + jy(t) \tag{1.2}$$

$$y(t) = u(t) \sin (2\pi f_0 t + \theta). \tag{1.3}$$

The *analytical signal* concept prevents the occurrence of signals with physical senseless negative frequencies in Fourier transforms and ensures for causality principle. Consider a simple case with an envelope $u(t) = 1 + A \sin \Omega t$, where Ω is a low frequency modulation (e.g. periodic force in AFM) and A the degree of modulation. The real signal $x(t) = \{1 + A \sin (\Omega t)\} \cos (2\pi f_0 t + \theta)$ after direct demodulation exhibits in Fourier spectra a negative frequency $-2\pi\Omega$ and by filtering processes it can formally lead to causality violation. In contrast, demodulation of *I* and *Q* components and subsequent subtracting in baseband processing, results in right positive spectra signals.

Furthermore, an analytical signal provides for a reasonable definition of the instantaneous frequency $f(t) = d\varphi/dt$ as a time derivative of the instantaneous phase $\varphi = (2\pi f_0 + \theta) = arctg(y/x)$, and the amplitude $u(t)$ as a vector length in a complex plane is now straightforward:

$$f(t) = \frac{1}{(x^2 + y^2)} \left[\frac{dy(t)}{dt} x(t) - \frac{dx(t)}{dt} y(t) \right] \tag{1.4}$$

The denominator of the fraction is proportional to the square of the signal $\psi(t)$ envelope (amplitude) and is independent of the choice of the frequency f_0

$$|\psi(t)|^2 = x^2(t) + y^2(t) = u^2(t). \tag{1.5}$$

Technically; the concept of an analytical signal is realized as follows: The Hilbert transform of the real signal, formed by using frequency mixing in an *I/Q* modulator, and the real signal itself are sent in separate "orthogonal" channels. After appropriate frequency shifting, amplification and filtering, they are demodulated and processed in a baseband processor in order to readout the amplitude, phase and frequency information.

In dynamic SPM and above-mentioned techniques, the main tasks of the oscillation controller are: (a) to maintain the oscillation amplitude and frequency in a distinct range, (b) to generate the readout signals that tracks instantaneous frequency and amplitude.

The amplitude of oscillations, however, is typically held constant with a fast automatic gain control (AGC) and instead the AGC signal is used as the output signal (that is reversely proportional to the amplitude). Accordingly, signal processing typically includes two feedback loops: (a) the instantaneous frequency tracker loop that holds oscillation in resonance and (b) the amplitude tracker – AGC or amplitude detector. It should be noted that there is another reason for introduction of an AGC loop: it is obvious that amplitude modulation generally causes parasitic frequency modulation; therefore stabilizing of the amplitude improves the precision of frequency measurements.

Instantaneous frequency demodulation. Two modern elaborate techniques mainly used for frequency demodulation. The first is based on a straightforward realization of the frequency demodulation algorithm implemented in Eq. (1.4). Figure 1.2 shows the block diagram of a device that forms and determines the instantaneous frequency of the analytical signal. For this purpose, in-phase $\sin(2\pi f_0 t)$ and quadrature $\cos(2\pi f_0 t)$ external signals are sent to multipliers with filters *LP* that translate the signal spectrum to zero frequency, forming the in-phase I and quadrature Q components of Eq. (1.2). The differentiating modules form signals of the derivatives, which are multiplied by the Q and I signals and the resulting signals are subtracted according to Eq. (1.4) by a subtracting module *SA*.

This type of frequency demodulation is widely implemented in Doppler shift measurement techniques [13] (radar, laser Doppler velocimetry). This technique was first used as late as 2004 as an FM demodulator in dynamic SPM and named symmetrical quadrature demodulation (SQL) [14], but to date it has not been realized in SPM as a frequency tracking device (in Doppler systems it is a part of a feedback loop of the frequency tracker, rather than making use of a simple FM demodulator).

The second FM demodulation technique is the most widely used technique in frequency processing – the well-known phase-locked loop (PLL) [15] technique. The importance of the PLL concept in SPM was recognized by U. Dürig et al. [16] soon after introduction of dynamic mode in SPM. Furthermore, an instantaneous frequency tracker based on PLL in the same paper was introduced and analyzed.

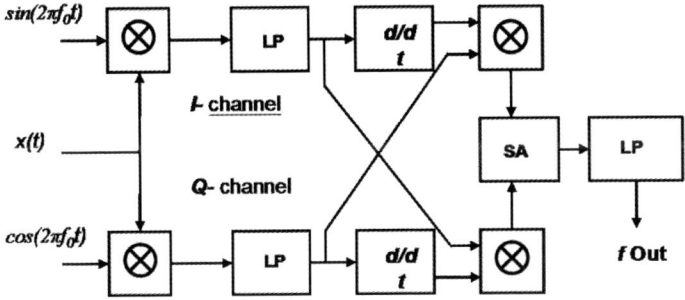

Fig. 1.2. Block diagram of the frequency demodulator based on generation and processing of analytical signal. LP – low-pass filter, *d/dt* differentiator, ⊗ signal multiplication

Generally, because of high performance and relatively simple structure, PLL-based demodulators and tracking systems are widely used in all related techniques. Note that for an amplitude-stabilized signal the instantaneous frequency demodulated with a PLL is coincident with results by using Eq. (1.4). The simplified principle of the two loop oscillation controller – namely an amplitude stabilization loop and PLL with *I/Q* demodulation is shown in the block schematic in Fig. 1.3. This principle of separation in the independent phase/frequency loop and amplitude loop (arrows in Fig. 1.3) is generally popular in signal processing and is very appropriate for signal processing in SPM. The reason is the physics of interaction in SPM: any frequency change in signal, due to tip–sample interaction, is entirely determined by conservative interaction force and reflected only in the *Q* channel. The amplitude change depends on both dissipative and conservative forces (typically more on dissipation) and is reflected in the *I* channel. This means that both *I* and *Q* components should be analyzed to investigate dissipative processes.

In applications, the first of these two FM demodulation techniques (corresponding to Fig. 1.2) shows better tracking performance at high frequency deviations or fast changes of the instant frequency, but introduces more noise because of differentiation. The PLL operates better with very narrow-band signals and very high precision requirements to frequency measurements, that is the case in SPM. Therefore, the state of the art SPM electronics is based on analog PLL circuitry or digital PLL algorithm realization. Resonance *LC*-based circuits as frequency demodulators, as sometimes used in SPM publications are less appropriate and will not be discussed here.

Both these FM techniques and the amplitude processing can be realized in analog or digital electronic schemes. The technical trend is towards more digitalizing in devices that are more flexible and accurate. Today the full or dominant analog design is mostly used either in very high frequency applications (hundreds of MHz and higher) or for simple and cheap applications and in rapid prototyping of experimental

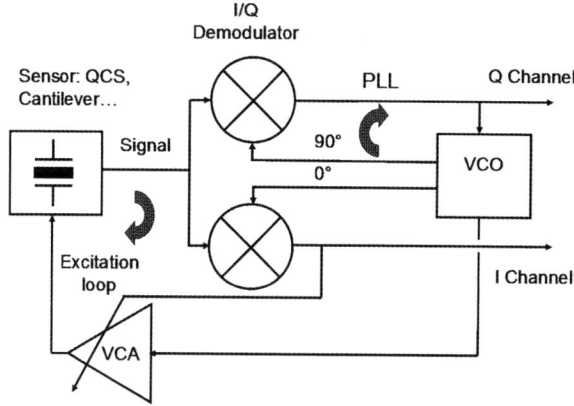

Fig. 1.3. Principle of two feedback loop control with *I/Q* demodulation: *arrows* denote the AGC loop on the base of the VCA (voltage controlled amplifier) and the PLL with a VCO (voltage controlled oscillator). Frequency changes are detected in the *Q* channel, amplitude changes – in the *I* channel

devices. The first electronics with digital PLL for application in dynamic SPM was introduced in 1998 by Ch. Loppacher [17] et al. Since then, the digital technology has developed rapidly, a number of other specific processing algorithms have been introduced and now nearly all commercial SPM systems implement (at least partly) digital signal processing. Typically analog front and I/Q block and digital baseband interfaces are used.

1.3
Connection Between Oscillation Parameters and Tip–Sample Interactions

Let us consider an oscillating dynamic SPM system that continuously tracks the resonant frequency with high precision, using an AGC providing a constant oscillating amplitude (i.e. the AGC control signal tracks the changes in oscillator dissipation). This operation mode is now used in modern dynamic SPM and related techniques. The description can be extended to modifications of this operation mode, such as constant excitation mode, phase modulation mode, Q-control mode etc., however it becomes then more complicated. In addition, the performance of the operation may be biased by the interaction between the feedback loops. For instance, if the amplitude of the oscillations fluctuates (bad or no AGC) the spectra of these fluctuations is mixed with the oscillator excitation (carrier) frequency and distorts the frequency estimation.

If the real signal x represents the deflection in a mechanical oscillator with effective mass m, spring constant k, and initial damping b (far from the surface), a second-order differential equation governing the motion is:

$$m\ddot{x} + b\dot{x} + kx = F_{ts}(x, \dot{x}) + F_{exc} \tag{1.6}$$

where $F_{ts}(x, \dot{x})$ is the interaction force experienced by the tip, $F_{exc} = g\dot{x}$ is the driving force that excites the oscillator with 90° phase shift, g is the AGC signal. At low oscillation amplitudes, typical for quartz sensors and stiff cantilevers, Eq. (1.6) can be rewritten as:

$$m\ddot{x} + (b + \Gamma - g)\dot{x} + (k - k_{cons})x = 0 \tag{1.7}$$

Here we introduced "dissipative" and "conservative" interaction forces $F_{diss} + F_{cons} = F_{ts}$; $F_{diss} = -\Gamma\dot{x}$; $\Gamma = \Gamma(x)$ is the damping caused by the dissipative force. $F_{cons} = F_{cons}(x)$ is the coordinate dependent conservative force, $\frac{d}{dx}F_{cons} = k_{cons}$. The decomposition of tip–sample forces into dissipative and conservative forces (or into "odd" and "even" forces) used here corresponds to commonly used notation in theoretical descriptions of the dynamic SPM, for example see [18, 19] and in particular the discussion in [20].

If both (amplitude and frequency) tracking feedback loops are turned on, Eq. (1.7) becomes a very simple form:

$$m\ddot{x} + (k - k_{cons})x = 0 \text{ and } g = b + \Gamma, \tag{1.8}$$

that means the output signal of the frequency tracker is $\omega = \sqrt{\frac{k-k_{\text{cons}}}{m}}$, for example the *instantaneous frequency*, and the AGC output signal $g = b + \Gamma(x)$ monitors the total *damping*, i.e. the internal sensor damping b and interaction caused damping $\Gamma(x)$.

The reconstruction of the tip–sample forces from measured frequency shift, excitation force and initial damping reduces to the solving of the following integral equations (see e.g. [18, 19]):

$$\frac{\omega - \omega_0}{\omega} = -\frac{1}{akT}\int_0^T F_{\text{cons}}(x)\cos(\omega t)dt \tag{1.9}$$

$$F_{\text{exc}} + ab\omega = \frac{2}{T}\int_0^T F_{\text{diss}}(x)\sin(\omega t)dt \tag{1.10}$$

where ω_0 is the resonance frequency of the oscillating system without interaction, a is the oscillation amplitude and T the oscillation period.

The QCR sensors are generally analyzed by an electrical equivalent circuit, which consists of a parallel series *LRC* and a stray capacitance C_0 (the Butterworth–Van-Dyke equivalent circuit). By use of a stray capacitance compensation circuit (described in detail in [21]), it is possible to make the effect of the stray capacitance negligible and the form of the equation for the piezo-charge oscillations is then similar to that for the mechanical oscillator:

$$\ddot{q} + \frac{R}{L}\dot{q} + \omega^2 q = \frac{U_{ts} + U_{exc}}{L} \tag{1.11}$$

A modified form where the QCR is excited through a feedback loop ($U_{exc} = GR_f\dot{q}$) reads:

$$\ddot{q} + \left(\frac{R - GR_f}{L}\right)\dot{q} + \omega^2 q = \frac{U_{ts}}{L} \tag{1.11a}$$

In Eqs. (1.11, 1.11a) q is the charge on QCR electrodes, R_f – the transimpedance resistance of the amplifier, G – the feedback gain and $U_{ts} = \frac{1}{\alpha}F_{ts}$ – the voltage produced by forces F_{ts}. The electromechanical coupling coefficient α permits us to establish the electromechanical correspondence and transfer from the electrical Eq. (1.11) to a mechanical motion equation by substitutions: $q = \alpha x$, $L = \frac{m}{\alpha^2}$, $C = \frac{\alpha^2}{k}$, $R = \frac{b}{\alpha^2}$.

1.4
Oscillation Control for QCR, Technical Realization

There are two excitation modes in QCR-based techniques: the mechanical excitation (i.e. the QCR is typically attached to a dither-piezo and the mechanically induced piezoelectricity is measured), and the electrical excitation mode (both electromechanical and mechanoelectrical piezo effects used for excitation and detection).

The mechanical excitation mode is more popular because of the simplicity of the principle. However, one of the main advantages of the symmetrical quadruple configuration of the tuning forks – the immunity to environmental mechanical and electrical distortions, higher Q-factor – in the mechanical excitation mode is sacrificed (the fundamental asymmetric mode of a TF cannot be excited by shaking the TF as whole, that's why TF are used in watches). The so-called qPlus sensor [22] can be seen as a limiting case: here one prong of the TF is fixed to a large substrate and the free prong with a tip attached is now a mechanical dipole and can be excited by mechanical shaking. Introduction of a qPlus sensor is argued with assumption of a negative influence on the Q-factor of the ringing mode and symmetry breaking through tip–sample interaction in symmetrical TF. Our experience with an electrical QCR self-excitation showed no ringing problem in appropriately designed circuits; the change of Q-factor through tip–sample interaction by operating in environmental conditions is relatively low and may be taken into account. Nevertheless, even the first experiments with a qPlus sensor were successful, yielding AFM images of silicon with atomic resolution [23].

This example emphasizes the potential of QCR. It seems plausible that in electrical excitation mode the quadruple symmetry of the tuning fork or needle quartz positively manifests on the ultimate sensitivity, stability against acoustic and electromagnetic interferences and time stability. Consequently, by use of a self-excitation QCR sensor an SPM system with less expensive acoustical and EM isolation can be developed, although the fundamental sensitivity limitations in both modes are nearly the same. Very stiff QCR's such as needle quartz or trident tuning fork driven at sub-nanometre amplitudes are particularly appealing for use in ambient conditions with air flows always present. Note that in the self-excitation mode in order to achieve high-quality figures [24] the balancing of the QCR prongs is important.

The typical realization of an oscillation control schematic in SPM based on QCR uses a self-made low-noise preamplifier, an industrial lock-in amplifier and a frequency synthesizer in combination with an industrial XYZ-control platform. For example, in this way atomic resolution in ambient conditions with a trident [25] and needle [26] QCR is achieved. The use of comprehensive and rather expensive commercial devices however, does not release us from problems with low-noise signal pre-amplification, parasitic capacitance compensation etc. These problems may be treated by carefully designing the oscillation control electronics together with (or near) the QCR. A system on a chip is the ideal vision. Therefore, we introduced [27] a wideband bridge oscillator controller in miniature design with grounded QCR placed in immediate vicinity and an auxiliary PLL frequency demodulator.

A further development [28] of the oscillation controller with implementation of the above-mentioned features, a bridge scheme with parasitic capacity compensation, PLL, I/Q demodulation and AGC is next described. The development was initiated to supply our specific SPMs [RF (radio frequency) scanning noise microscope [29] and SNOM] with precise distance control. The resulting simple but versatile QCR oscillator controller realization with two feedback loops (PLL and AGC loop) and instantaneous frequency shift ΔF, amplitude and resonant frequency signal outputs is shown in Fig. 1.4.

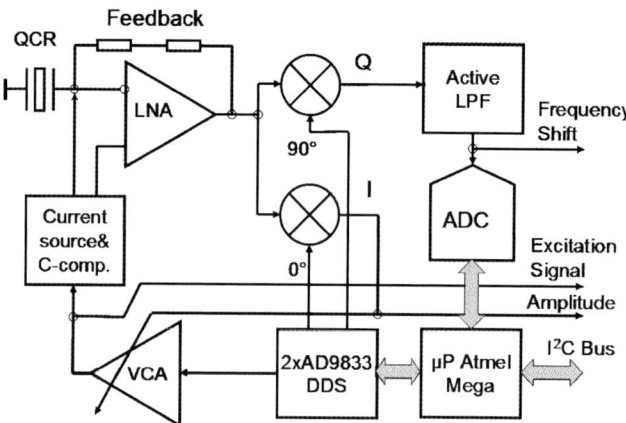

Fig. 1.4. Diagram of the QCR oscillation controller. Abbreviations: LNA – low-noise amplifier, Q_{ch} quadrature channel, I_{ch}-in phase channel, LF loop filter with time constant τ_{PLL}, CE – constant excitation mode, CA – constant amplitude mode, PI + LPF – Proportional-Integral regulator with low-pass filter und time constant τ_{VCA}, VCA – voltage controlled amplifier, RMS – root-mean-square with time constant τ_{RMS}, S1 und S2 mode switches, C_{comp} – compensation capacity

The grounded QCR, low-noise operational amplifier (LNA), feedback resistors (two low-noise metal-film resistors in serial) and excitation string with voltage-controlled amplifier (VCA), R-R divider and compensation of the parasitic OCR capacity C_{comp} form the bridge circuit. The compensation of the parasitic OCR capacity occurs if C_{comp} is trimmed equal to the parasitic capacity; the compensation current is then identical to the parasitic current but in opposite phase. The comparator, Q_{ch} multiplier, loop filter (LF) with a time constant τ_{PLL}, and voltage-controlled oscillator (VCO) forms a phase locked loop (PLL). The PLL maintains the oscillation of QCR in resonance (a precondition is the excitation phase matching). The control signal of the VCO represents the frequency shift ΔF. If amplifiers with very high bandwidth are used (here as LNA an OPA657 with 1,500 MHz bandwidth and as VCA an AD8337 with 280 MHz bandwidth) the phase is matched automatically with appropriate precision.

The amplitude control loop consists of the I_{ch} multiplier (synchronous amplitude demodulator) followed by low-pass filter (LPF), proportional-integral amplitude set regulator (PI) and VCA. The switches S1 and S2 allow operation in constant amplitude (CA) and constant excitation mode, well known in the AFM technique. The RMS measure for amplitude with a time constant τ_{RMS} is formed with use of a true power detector AD8361.

The described scheme is very flexible: wideband electronics allow use of QCR with resonance frequencies from some kHz up to tens of MHz, all typical SPM measurements modes are implemented, grounded QCR platform is essential in a variety of SPM techniques etc.

An analog-digital variant with direct digital synthesizer (DDS) as frequency generator and an ATMEGA88 μP (Atmel) is shown in Fig. 1.5. One of the features

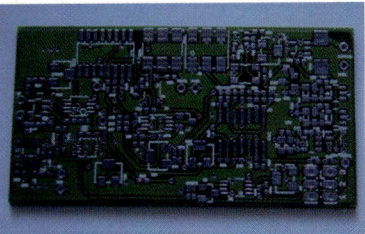

Fig. 1.5. Scheme of the oscillation controller developed on the μP ATMEGA base. On the *right image* the electronic designed on a 4-layer printed board, with dimensions $70 \times 30\,mm^2$, is shown

in this development is a current source drive discussed in detail in [30], the other is the possibility of phase modulation mode [4]. On the input we use wideband, low-noise JFET operational amplifiers OPA657 or AD8067, providing high performance with any QCR up to 10-MHz resonance frequency. Other main IC's: the ADC is an AD7680, VCA – AD 8337. The developed device is shown in Fig. 1.6.

Programming, control and communication with the oscillation controller are realized under the LabView (National Instruments) graphical programming system.

If optimal imaging and circuit parameters are used, the noise limit at a given tip–sample interaction is defined by the parameters of the QCR (dissipative resistance,

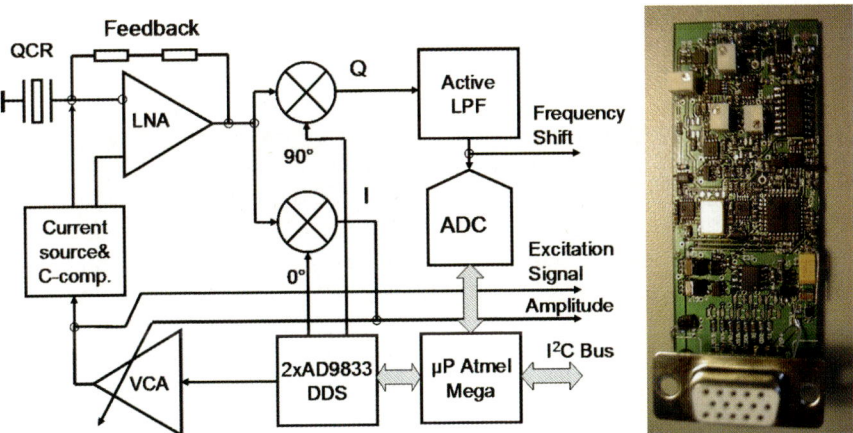

Fig. 1.6. The oscillation controller is mounted on a 30×55-mm^2 multilayer printed board and is located together with the high-frequency detecting part in a shielding box on a NMTD Solver P47 base (*right*). The head shown is the main part of the RF scanning noise microscope

material constants and dimensions) and not by the electronic elements. Consequently, the analysis of the noise limit is similar to that in the AFM self-excitation mode [31–33].

The developed oscillation controllers are applicable with nearly all industrial SPM's with optional signal inputs. For example, we used external analog inputs Ex1 and Ex2 for realization of dissipative and conservative measurement channels in NTMDT (Moscow, Russia) Solver SPM as well as in a self-made SNOM based on a PI (Waldbronn, Germany) XYZ piezo system [34]. In the SNOM a complex self-excitation circuit with dither piezo was successfully replaced through the described circuit.

1.5
Applications of the Oscillation Controllers

A tapping-mode force image made on a clean cleaved mica surface under ambient conditions by use of a 1-MHz needle quartz with a standard cantilever tip attached is shown in Fig. 1.7a. On a 500-nm^2 area, steps of an atomic monolayer are well resolved in amplitude modulation (dissipative) mode. In Fig. 1.7b an image at atomic resolution on a mica surface under ambient conditions is shown. The image is made with the same QCR and tip in frequency demodulation mode on an area of 4 nm^2.

In order to test the capability of measurements on soft samples we used monomolecular DPPC Langmuir–Blodgett (LB) strips on a plasma-etched Si surface. Figure 1.8 shows an image made with use of a 1-MHz needle quartz with a standard cantilever tip attached. The approaching (red) and retracting (blue) curves in Fig. 1.8 (right) corresponds to the frequency shift signal versus distance to the surface. The region where the tip is experiencing attractive or repulsive forces lies near 1 nm. The hysteresis is due to the thin water film on surfaces at ambient conditions.

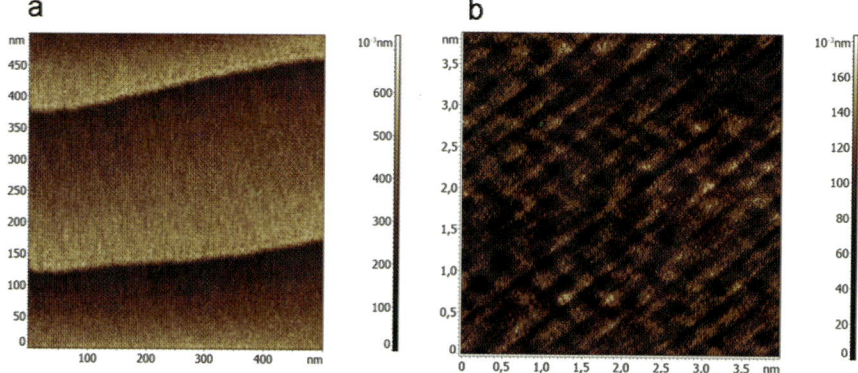

Fig. 1.7. The images made on a clean cleaved mica surface under ambient conditions: (**a**) atomic steps with height approximately 0.3 nm; (**b**) at atomic resolution

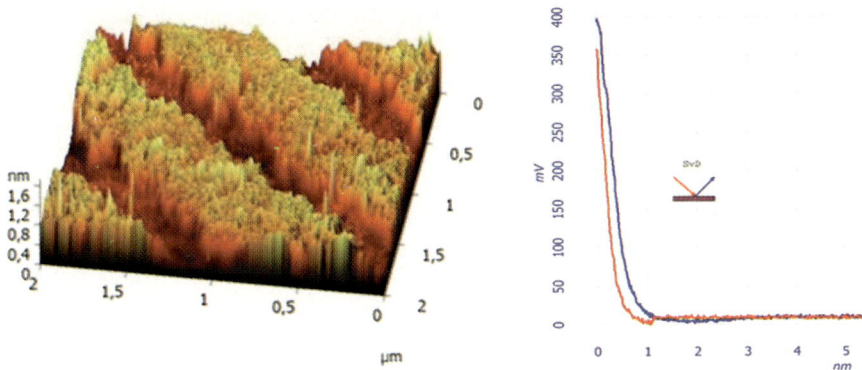

Fig. 1.8. Image of LB strips on a plasma-etched Si surface under ambient conditions and a signal versus distance curve

Further, in Fig. 1.9 we present images obtained by use of developed circuits in the SNOM distance control. As a force sensor, the relatively big tetrahedral glass tip [35] attached on a 32-kHz tuning fork was used. The top line of the image is the topography (full vertical scale 2 nm) of the metalized SNOM pattern, in the middle – optically detected plasmon waves and on the bottom – the SNOM tip attached to a 32-kHz TF.

1.6
Summary

The rigorous definition and mathematical basis of signal processing and measurement of instantaneous frequency and amplitude for narrow-band signals are briefly discussed herein. Oscillation control, signal processing and principles of QCR-based dynamic SPM techniques in comparison with related techniques are also discussed. A realization of oscillation control electronics, designed on a small and cheap board located near the QCR is presented. The possibility to create very simple, cheap and nearly optimal devices for the SPM on the basis of circuitry and ideas developed in radio-communications have been demonstrated. Some applications of our development demonstrated performance comparable with the best commercial SPM. With a view to commercialization, it is clear that market demand relates to the complete integration of the different aspects of an SPM: sensor, detection, interface electronic, protocol analysis and data analysis. The development toward a "lab on a chip" device is an integral part of future efforts also in SPM techniques. Our microprocessor-aided development can be seen as a step towards this goal.

Finally, it should be noted that quartz has a relatively low electromechanical constant and industrial QCR's are very stiff. It would be desirable to test new materials with a much higher electromechanical constant such as $GaPO_4$ and $Ca_3Ga_2Ge_4O_{14}$ (CGG)-type structures (langasite, langanite, langatate) as to their applicability as SPM sensors.

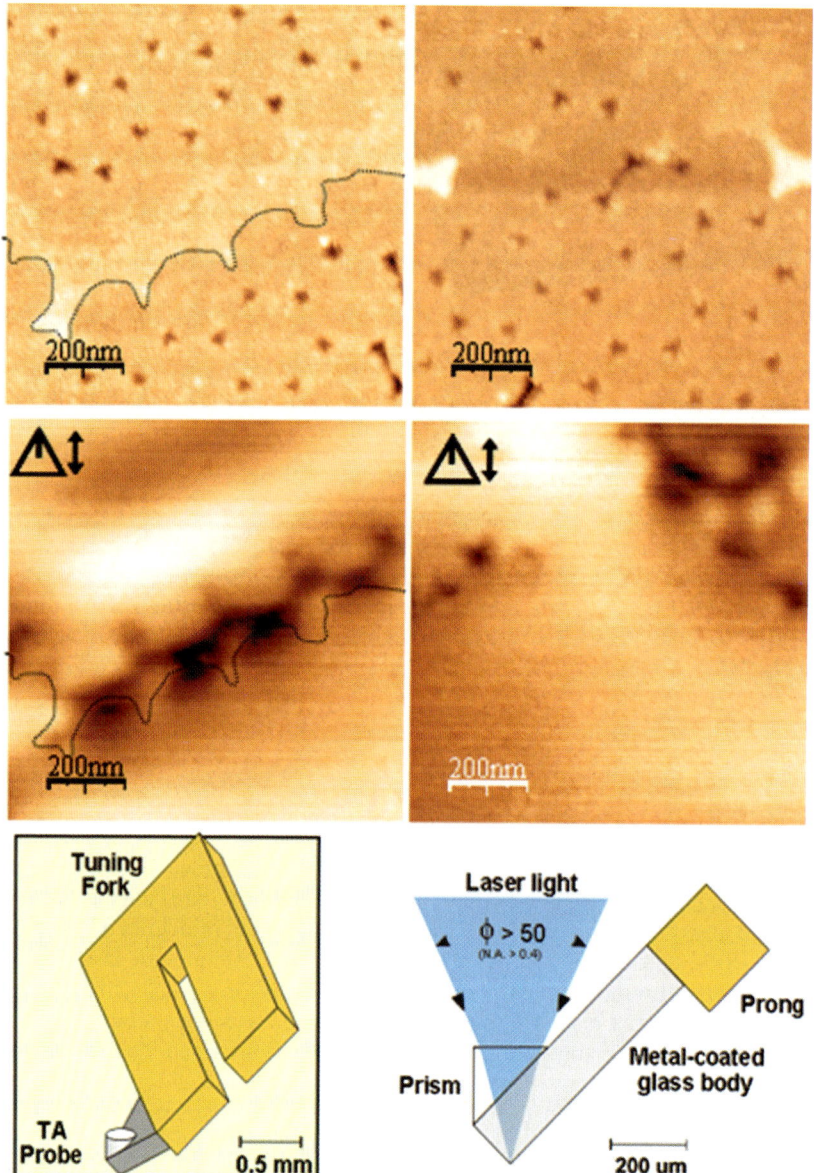

Fig. 1.9. Topography and optical images on special structures for plasmonic excitation (in cooperation with T. Maletzky and U. Ch. Fischer [36])

References

1. Chapman MJ (ed) (1996) Signal processing in electronic communications. Horwood Publishing, Chichester
2. Dürig U, Gimzewski JK, Pohl DW (1986) Phys Rev Lett 57:2403–2406

3. Albrecht TR, Grütter P, Horne D, Rugar D (1991) Frequency modulation detection using high Q cantilevers for enhanced force microscope sensitivity. J Appl Phys 69:668
4. Kobayashi N et al (2006) High-sensitivity force detection by phase-modulation atomic force microscopy. Jpn J Appl Phys 45:L793–L795
5. Gunther P, Fischer UCh, Dransfeld K (1989) Appl Phys B: Photophys Laser Chem 48:89
6. Karrai K, Grober RD (1995) Appl Phys Lett 66:1842
7. Giessibl FJ (2003) Rev Mod Phys 75:949–983
8. Kawata S, Shalaev VM (eds) (2007) Tip Enhancement (Advances in nano-optics and nanophotonics). Elsevier, Amsterdam
9. Seo Y, Hong S (2005) Quartz crystal resonators based scanning probe microscopy. Mod Phys Lett B19:1303–1322
10. Giessibl FJ (2003) Advances in atomic force microscopy. Rev Mod Phys 75:949–983
11. Arnau A (ed) (2004) Piezoelectric transducer and applications. Springer-Verlag, Berlin Heidelberg New York
12. Franks L (1969) Signals theory. Prentice-Hall, Englewood Cliffs, NJ, USA
13. Wilmshurst TH, Rizzo JE (1974) An autodyne frequency tracker for laser Doppler anemometry. J Phys E, Sci Instrum 7:924
14. Kobayashi D, Kawai S, Kawakatsu H (2004) Jpn J Appl Phys 43:4566
15. Stephens DR (2002) Phase-locked loop for wireless communication, 2 edn. Kluwer, Dordrecht
16. Dürig U, Züger O, Stadler A (1992) J Appl Phys 72:1778
17. Loppacher Ch, Bammerlin M, Battiston F, Guggisberg M, Müller D, Hidber HR, Lüthi R, Meyer E, Güntherodt HJ (1998) Appl Phys A 66:S215
18. Hölscher H (2006) Appl Phys Lett 89:123109
19. Sader JE, Jarvis SP (2006) Phys Rev B 74:195424
20. Sader JE, Ushuhashi T, Higgins MJ, Farrell A, Nakayama Y, Jarvis SP (2005) Nanotechnology 16:S94
21. Arnau A, Sogorb T, Jimenez Y (2000)A continuous motional series resonant frequency monitoring circuit and a new method of determining Butterworth–Van Dyke parameters of a quartz crystal microbalance in fluid media. Rev Sci Instrum 71(6):2563–2571
22. Giessibl FJ, Hembacher S, Bielefeldt H, Mannhart J (2000) Science 289:422
23. Giessibl FJ, Hembacher S, Herz M, Schiller Ch, Mannhart J (2004) Nanotechnology 15:S79
24. Smit RHM, Grande R, Lasanta B, Riquelme JJ, Rubio-Bollinger G, Agraït N (2007) Rev Sci Instrum 78:113705
25. Seo Y, Choe H, Jhe W (2003) Appl Phys Lett 83:1860
26. Nishi R, Houda I, Aramata T, Sugawara Y, Morita S (2000) Appl Surface Sci 157:332
27. Jersch J, Maletzky T, Fuchs H (2006) Interface circuits for quartz crystal sensors in scanning probe microscopy applications. Rev Sci Instrum 77:083701
28. Jersch J, Fuchs H (2007) Quartz crystal resonator as sensor in scanning probe microscopy. In: IEEE Proceedings of TimeNav'07
29. ASPRINT (STRP 001601 project under the Sixth Framework Programme of the European Community 2002–2006), see the WWU Muenster contribution
30. Cherkun AP, Serebryakov DV, Sekatskii SK, Morozov IV, Letokhov VS (2006) Double-resonance probe for near-field scanning optical microscopy. Rev Sci Instrum 77:033703
31. Cherkun AP, Serebryakov DV, Sekatskii SK, Morozov IV, Letokhov VS (2006) Double-resonance probe for near-field scanning optical microscopy. Rev Sci Instrum 77:033703
32. Hölscher H, Gotsmann B, Allers W, Schwarz UD, Fuchs H, Wiesendanger R (2001) Phys Rev B 64:075402
33. Hölscher H, Gotsmann B, Schirmeisen A (2003) Phys Rev B 68:153401

34. Höppener C, Molenda D, Fuchs H, Naber A (2003) J Microscopy 210(Pt 3):228
35. Fischer UC (1998) In: Wiesendanger R (ed) Scanning probe microscopy: analytical methods. Springer, Berlin Heidelberg New York
36. Maletzky T, Jersch J, Fuchs H, and Fischer UCh (2006) Revealing the interaction of metal-nanostructures with dye-molecules by near-field-microscopy. EOS Topical Meeting on Molecular Plasmonic Devices, 27–29 April 2006, Engelberg, Switzerland

2 Atomic Force Microscope Cantilevers Used as Sensors for Monitoring Microdrop Evaporation

Elmar Bonaccurso · Dmytro S. Golovko · Paolo Bonanno · Roberto Raiteri ·
Thomas Haschke · Wolfgang Wiechert · Hans-Jürgen Butt

Abstract. For studying the evaporation of millimetre-sized drops of liquids techniques such as video-microscope imaging and ultra-precision weighing with electronic microbalances or with quartz crystal microbalances have been employed in the past decades. Similar techniques are, however, hardly applicable to microscopic drops. Moreover, they do not provide a measure of the interfacial stresses arising at the contact area between liquid and solid. Here we demonstrate the use of atomic force microscope (AFM) cantilevers as sensitive stress, mass, and temperature sensors for monitoring the evaporation of microdrops of water from solid surfaces. Starting from considerations on drops in equilibrium, we will further discuss evaporating drops and details of the experimental technique. We will show how the evaporation of water microdrops on a hydrophobic surface differs from the evaporation on a hydrophilic surface, and how this difference becomes more pronounced towards the end of evaporation. We further show that one-side metal-coated cantilevers, acting as bimetals, allow measuring the average temperature of an evaporating microdrop. Finally, we will discuss two further applications of microdrops evaporating on cantilevers, namely testing the local cleanliness of cantilevers' surfaces and calibrating cantilevers' spring constants.

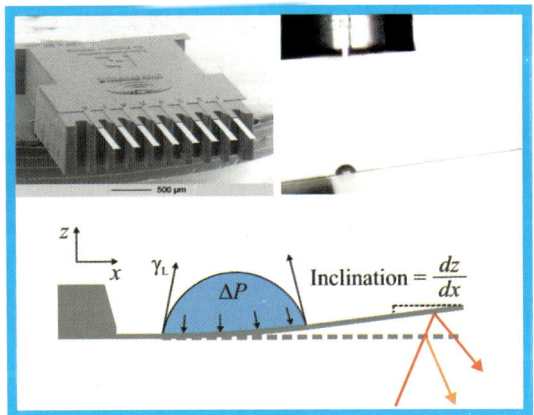

Key words: Microdrop evaporation, Evaporation law, Atomic force microscopy, Micromechanical cantilevers, Surface tension, Young's equation, Vaporization heat, Spring constant calibration, Contamination control

Abbreviations and Symbols

NPT	Normal pressure and temperature
RH	Relative humidity
dpi	Dots per inch
AFM	Atomic force microscope
QCM	Quartz crystal microbalance
TPCL	Three-phase contact line
CCR	Constant contact radius
CCA	Constant contact angle
std. dev.	Standard deviation
SEM	Scanning electron microscope

$\Delta\sigma$	Surface stress [N m^{-1}]
Θ	Drop contact angle [°]
γ	Liquid surface tension [N m^{-1}]
τ	Evaporation time [s]
η	Viscosity [mPa s]
ρ	Density [g cm^3]
α	Thermal expansion coefficient [K^{-1}]
ν	Poisson's ratio
a	Drop contact radius [m]
m	Drop mass [kg]
g	Gravitational acceleration [$g = 9.8\,\mathrm{m\,s^{-2}}$]
f	Frequency [Hz]
k	Cantilever spring constant [N/m]
d	Cantilever thickness [m]
w	Cantilever width [m]
l_0	Cantilever length [m]
V	Drop volume [m^3]
K	Capillary length [m]
ΔP	Laplace-, or capillary pressure [Pa]
P_0	Vapor pressure [Pa]
D	Diffusion coefficient [cm^2 s^{-1}]
T	Temperature [K]
E	Young's or elasticity modulus [Pa]
R	Drop radius of curvature [m]

2.1
Introduction

Understanding the kinetics of evaporation or drying of microscopic, sessile drops
from solid surfaces is a key factor in a variety of technological processes, such as:
(1) printing [1–3] and painting [4]; (2) heat-transfer applications, for example in
the electronic industry to cool integrated circuits (ICs) and electronic components

[5–8], or for fire fighting [9]; (3) micro lithography, for example on polymer [10–13] or on biomaterial [14] surfaces. Such microscopic drops are primarily generated by spray nozzles and atomizers [15], or by inkjet devices and drop-on-demand generators [16]. Spray nozzles and atomizers are capable of simultaneously producing a large number of drops by one nozzle, but with a large size distribution and little control over the size (from below $1\,\mu m$ to above $100\,\mu m$). The second type of apparatus is only capable of generating single consecutive drops by one nozzle, however monodisperse and with a good control over the size. In fact, the resolution of inkjet printers is steadily increasing as the size of the drops decreases. A commercial standard inkjet printer has nowadays a resolution of around 1,200 dpi, which means that the drops have a diameter of around $20\,\mu m$ and a volume of around 4 pL. If such a drop would be pure water, instead of a mixture of water and dye, and be deposited on a flat surface, it would evaporate in less than 150 ms at normal pressure and temperature (NPT) and at a relative humidity (RH) of 50%. In comparison to inkjetted drops, rain drops have diameters between 1 and 2 mm, while drops in a fog have diameters below $10\,\mu m$ [17].

Drop evaporation has been classically monitored by means of video-microscope imaging [18, 19], by ultra-precision weighing with electronic microbalances [20, 21] and with quartz crystal microbalances (QCM) [22]. Recently also atomic force microscope (AFM) cantilevers have been successfully employed for this purpose [23–25]. Using the first two, since long established techniques, a wealth of information was gained and the evaporation of macroscopic drops of simple liquids from inert surfaces is now well understood. These techniques are, however, not sensitive enough to characterize microscopic drops. Furthermore, they can not directly measure the interfacial stresses arising at the contact area between liquid and solid, which are known to play a key role in the evaporation kinetics of small drops [18, 20, 21, 26–28], nor are they capable of sensing the heat absorbed by the liquid during evaporation.

In this chapter evaporation studies of microscopic water drops on solid surfaces performed with a nonstandard technique are presented. It will be shown how it is possible to simultaneously measure surface forces, the mass, and the vaporization heat of a microdrop.

2.2
Background, Materials and Methods

2.2.1
Drop in Equilibrium

For the sake of simplicity, let us first consider a drop deposited onto a nondeformable and nonsoluble substrate (Fig. 2.1). In equilibrium, i.e. when the drop is not evaporating, Young's equation must hold. It establishes the relation among the three surface tensions acting at the rim of the drop (three-phase contact line, TPCL).

$$\gamma_S - \gamma_{SL} = \gamma_L \cos\Theta \qquad (2.1)$$

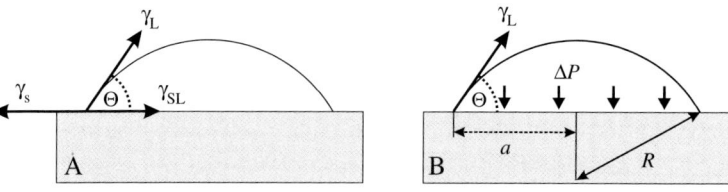

Fig. 2.1. (**A**) Sessile drop in equilibrium on a solid surface, with contact angle Θ and surface tensions γ_L, γ_S, and γ_{SL}. (**B**) Action of liquid surface tension γ_L and Laplace pressure ΔP

γ_L is the surface tension at the interface liquid/gas, γ_S is the surface tension at the interface solid/gas, and γ_{SL} is the surface tension at the interface solid/liquid. Θ is called contact angle, or wetting angle, of the liquid on the solid. Equation (2.1) is strictly valid only if the drop is not evaporating and if gravity can be neglected. We can neglect gravity, and thus the effect of the hydrostatic pressure that would flatten the drop, when the drop is smaller than the capillary length

$$K = \sqrt{\frac{\gamma_L}{\rho g}} \qquad (2.2)$$

where ρ is the density of the liquid and g the gravitational acceleration. For water, $\gamma_L = 0.072\,\text{N/m}$, $\rho = 1\,\text{g/cm}^3$, and $g = 9.8\,\text{m/s}^2$. The shape of the drop is thus not influenced by gravity if the radius of curvature is well below 2 mm. This requirement is fulfilled for all results presented in the following, where drops smaller than $100\,\mu\text{m}$ were always used. For such sizes the shape of the drop is determined solely by surface forces and it has the form of a spherical cap.

In addition to the surface tensions acting at the TPCL, another force plays a major role. Because of its curvature, the pressure inside the drop is higher than outside. The difference between in and out is called Laplace or capillary pressure

$$\Delta P = \frac{2\gamma_L}{R} = \frac{2\gamma_L \sin \Theta}{a} \qquad (2.3)$$

where R is the radius of curvature of the drop, which is related to the contact radius a and the contact angle Θ (Fig. 2.1).

As an example, for a drop of water forming a contact angle of $60°$ with a surface, the pressure difference is $\Delta P = 1.2\,\text{mbar}$ when $a = 1\,\text{mm}$, and $\Delta P = 1,200\,\text{mbar}$ when $a = 1\,\mu\text{m}$.

Summarizing, one can say that when a small, nonevaporating microdrop is sitting on a surface and forms a finite contact angle with it, the general picture is:

(1) The drop has a spherical shape.
(2) Young's equation accounts for the in plane (horizontal) balance of forces at the TPCL.
(3) The vertical component of the surface tension $\gamma_L \sin \Theta$ is pulling upwards at the TPCL and is counterbalanced by the Laplace pressure ΔP, which is pushing uniformly downwards over the whole contact area πa^2.

2.2.2
Evaporating Drop

Why do drops evaporate at all? A liquid (condensed phase) with a planar surface evaporates only when its vapor pressure P_0 is higher than the pressure of its vapor (gas phase) in its surroundings. As a consequence, if the surroundings are saturated with its vapor the liquid does not evaporate. It is in equilibrium, because at any time the number of molecules evaporating from and condensing to the surface is similar. However, drops have a slightly higher vapor pressure in comparison to a planar surface due to their curvature. For this reason, they evaporate also in a saturated atmosphere. This is quantified by the Kelvin Equation

$$P_V = P_0 e^{\lambda/R} \tag{2.4}$$

where the vapor pressure of the liquid in the drop is P_V, and the parameter λ is a function of the temperature and the nature of the liquid. Thus, the vapor pressure increases with decreasing drop size. As an example, a planar water surface has a vapor pressure $P_0 = 31.69$ mbar at NPT. If the surface is curved and the radius of curvature is $R = 1\,\mu$m, the vapor pressure is $P_V = 31.72$ mbar, and if $R = 100$ nm, $P_V = 32.02$ mbar. The difference between the planar and the curved surfaces is small, but high enough for the drop to evaporate.

The evaporation law for microscopic drops was derived for two cases for drops of pure liquids, assuming $\Theta = \text{const}$ and neglecting the cooling resulting from the vaporization [29].

1. Drop in its saturated vapor

$$V_L = V_{L0} - \alpha D \cdot P_0 \cdot t \tag{2.5}$$

where V_L is the drop volume, V_{L0} is the initial drop volume, D is the diffusion coefficient of the molecules in the vapor, P_0 is the vapor pressure, and α is a known parameter that depends on the drop properties, on the contact angle, and on the temperature (for details see [29]). The equation contains no free parameters, and states that the volume of the drop decreases linearly with time.

2. Drop in nonsaturated vapor

$$V_L^{2/3} = V_{L0}^{2/3} - \beta D \cdot \varphi P_0 \cdot t \tag{2.6}$$

where φP_0, with $\varphi < 1$, represents the reduced vapor pressure, and β is a known parameter that depends on the drop properties, on the contact angle, and on the temperature. The equation contains no free parameters, and states that the volume of the drop to the power of 2/3 decreases linearly with time.

Both evaporation processes, in saturated and in nonsaturated atmosphere, are "diffusion limited", i.e. the evaporation is limited by the diffusion of the liquid molecules through a saturated vapor layer around the drop.

As an application of the above evaporation laws, the evaporation of microdrops of water with different initial volumina on a silicon surface coated with a 30-nm thick fluoropolymer film (perfluoro-1,3-dimethylcyclohexane) was observed. The initial contact angle was $\Theta = 90°$ and remained constant for more than half of the evaporation time. During the experiments, the temperature ($T = 25\,°C$) and the relative

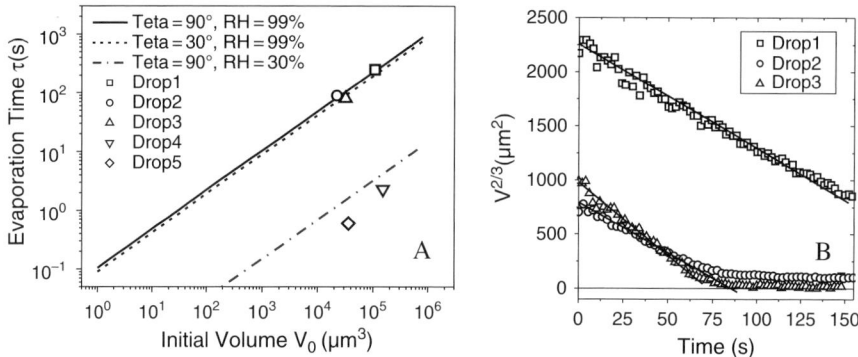

Fig. 2.2. (**A**) Calculated evaporation times τ versus initial drop volume V_0 for three cases: Θ = 90° and RH = 99% (*solid line*), Θ = 30° and RH = 99% (*dashed line*), and Θ = 90° and RH = 30% (*dashed-dotted line*). Corresponding experimental evaporation times (*open symbols*). (**B**) $V^{2/3}$ versus time for three drops; *solid lines* serve only as guides for the eye. (From [29])

humidity (RH ~ 99%) were constant. The dimension of the evaporating drop was monitored from the side with a video microscope [23, 29].

Figure 2.2A shows the calculated evaporation times τ versus the initial drop volume V_0 (in double logarithmic scale), as calculated by Eq. (2.5) for RH = 99%. The similarity between the two upper lines, calculated for contact angles of Θ = 90° and 30°, emphasizes that τ depends more strongly on V_0 than on Θ. The slope of the curves is exactly 2/3. The first three hollow symbols represent evaporation times of microdrops with different initial volumina, all other parameters are unchanged. The agreement with the calculated times is very satisfying, especially since no free parameters were used. The model is also applicable for smaller RHs, as shown for two microdrops evaporating at RH = 30%: the calculations yield respectively τ = 4.3 s and 1.6 s, the measurements yield τ = 2.2 s and 0.6 s. The evaporation time is strongly dependent on the vapor saturation: at RH = 30% a microdrop with a mass of 150 ng evaporates in τ = 4.3 s, at RH = 99% in τ = 290 s, and at RH = 100% in τ = 7000 s. It must be noted, that it is experimentally very difficult to set a constant RH = 100% for a prolonged time.

By representing $V^{2/3}$ versus time (Fig. 2.2B), the experiments reveal that at the beginning of the process, when the three evaporating drops are still large, the dependence is linear with time (solid lines). At the end of the process there are deviations and the evaporation appears to slow down. This can be due, for example to the presence of solid impurities in the water, which get enriched as the drop evaporates so that the vapor pressure decreases. Another explanation might be that the resolution limit of the optical technique is reached. In order to track the evaporation until the end an alternative technique will be introduced in the following paragraph. It will allow us to test the evaporation law also for extremely small drops.

2.2.3
Experimental Setup

When a drop is sitting on a surface, its surface tension pulls upwards at the TPCL, while the Laplace pressure pushes uniformly downwards on the entire contact area.

If the substrate is thin enough the surface forces can cause its bending: the thinner the plate, the stronger the bending. This can be used as a sensor principle. Silicon cantilevers, which look like microscopic diving boards, were employed as a suitable thin plate and a technique was developed for measuring the degree of cantilever bending. We employed silicon cantilevers like those imaged in Fig. 2.3A, where eight identical cantilevers are supported by a common silicon chip (Micromotive, Mainz, Germany). Nominal cantilever dimensions are: Length $l_0 = 750\,\mu m$, width $w = 90\,\mu m$, and thickness $d = 1\,\mu m$. Using an inkjet capillary, water microdrops were deposited onto the upper side of a cantilever, close to its base (Fig. 2.3B). The working principle is as follows (Fig. 2.3C): (1) before drop deposition no forces act on the cantilever, which is thus straight (the effect of gravity induces a neglectable bending); (2) upon drop deposition the cantilever bends upwards. This bending can be measured with the so-called light lever technique, as usually done in AFM, where a laser beam is pointed at the free end of the cantilever; (3) the measured signal is not the "actual bending," but the "inclination" dz/dx at the free end of the cantilever.

Simultaneously with cantilever inclination, a video microscope from the side records a movie of the evaporating drop from which the contact angle Θ and the contact radius a are calculated versus time. The width w of the cantilever needs to be at least two times larger than the contact radius a of the drop, i.e. if the drop touches the edges of the cantilever, border effects distort the shape of the drop, flawing the measurement of the contact angle. The experimental set-up used to deposit the microdrops onto the cantilever surface, monitor its inclination and resonance frequency and record the video of the evaporating drop is described in [23–26].

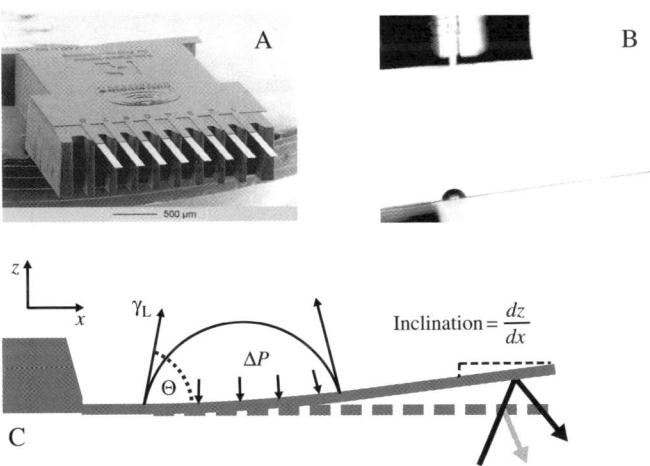

Fig. 2.3. (**A**) SEM image of a silicon chip with eight attached cantilevers. (**B**) Side view of a cantilever with drop deposited at its base and inkjet capillary used for drop generation. (**C**) Configuration of the equilibrium of forces between a drop and a cantilever: Θ is the equilibrium contact angle, γ_L is the liquid surface tension, ΔP is the Laplace pressure

2.3
Evaporation Results on Microdrops

2.3.1
Evaporation Curve

A typical evaporation curve of a water microdrop on a silicon cantilever, acquired at NPT and RH $\sim$ 30%, is a plot of the inclination of the cantilever versus time (Fig. 2.4A). At the same time, the contact angle Θ and the contact radius a can be recorded by video microscopy and plotted versus time (Fig. 2.4B, C). The water microdrop is deposited onto the cantilever at $t = 0$ with the inkjet device, and immediately starts to evaporate. The evaporation is accomplished after $\sim$0.6 s, as the cantilever's inclination returns to its initial value. In the contact angle and contact radius curves, the black lines are simply guides for the eye. They show that two evaporation regimes take place: at the beginning, the drop evaporates in the Constant-Contact-Radius (CCR) mode, and after $\sim$0.3 s both, Θ and a, decrease linearly with time. Plots of V and of $V^{2/3}$ versus time demonstrate the agreement with the evaporation law derived in Eq. (2.6) for a drop evaporating in nonsaturated vapor.

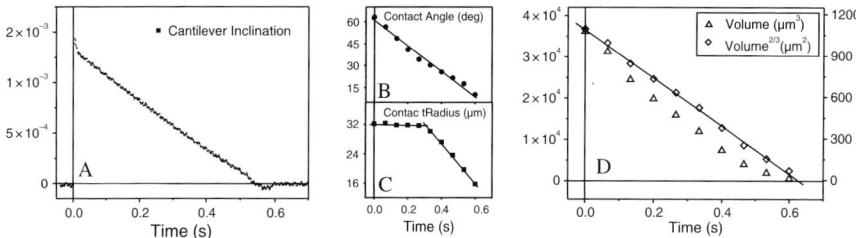

Fig. 2.4. T $\sim$ 25 °C, RH $\sim$ 30%; drop data: water, $a = 32\,\mu$m, $\Theta = 63°$, $\gamma_L = 0.072\,$N/m, $m_0 = 36\,$ng. Cantilever inclination (**A**), contact angle (**B**), contact radius (**C**), and volume and volume$^{2/3}$ versus time (**D**). *Solid lines* are guides for the eye

At this point a model is needed to analyze the acquired inclination data and relate it to surface forces, drop shape, and cantilever properties.

2.3.2
Force Model

First, for a simplified treatment, some assumptions on the drop and on the cantilever have to be made:

(1) The drop is in thermodynamic equilibrium, i.e. it does not evaporate, during the acquisition of a single data point. This is verified, since the acquisition time is typically < 1 ms.
(2) Beam theory is used for modeling the cantilever. It must hold $l_0 >> w$ and d, which means that the bending of the cantilever is considered to be one dimensional ($z = f(x)$) and that the transversal cross sections are flat. The Poisson's

ratio is zero ($v = 0$) in beam theory, which means that during its deformation the volume of the cantilever is not conserved.

The inclination at the end of the cantilever, which is given by the overall balance of forces acting on it, then is [23]:

$$\frac{dz}{dx} = \frac{3\pi a^3}{Ewd^3}\left[\gamma_L \sin\Theta + \frac{2d}{a}(\gamma_L \cos\Theta - \gamma_S + \gamma_{SL})\right] \tag{2.7}$$

where E is Young's modulus of the cantilever material, while all other parameters have been introduced before. The first term contains the vertical contribution of the surface tension and the Laplace pressure, while the second term is basically Young's equation. It is significant to note here that all parameters in this equation are known, i.e. there are no fitting parameters.

Figure 2.5 shows both the observed evaporation curve (Fig. 2.4A) and the curve calculated according to Eq. (2.7). The agreement between the two curves is fair, especially when considering that no fitting parameters were used and that the cantilever model is a simplification (Fig. 2.5).

It appears, however, that all curves acquired in similar experimental conditions lie systematically below the calculated curves. Thus, the model needs a refinement. According to the simplifications made, mainly three issues can cause this deviation:

(1) Beam theory was employed to model the cantilever and Poisson's ratio was zero ($v = 0$).
(2) The cantilever also has a lateral extension. Therefore a drop causes also a transversal bending of the cantilever, which increases its effective stiffness.
(3) The cantilever is clamped at the base. That edge can thus not be deformed, which increases its effective stiffness.

A more sophisticated 3D finite element (FE) model of the cantilever can be therefore implemented, which takes into account the above three issues [24]. In order to check the model the mechanical equilibrium deformation of a cantilever caused by a microdrop was analyzed. A nonevaporating ionic liquid (1-butyl-3-methylimidazolium hexafluorophosphate from Merck KGaA, Darmstadt,

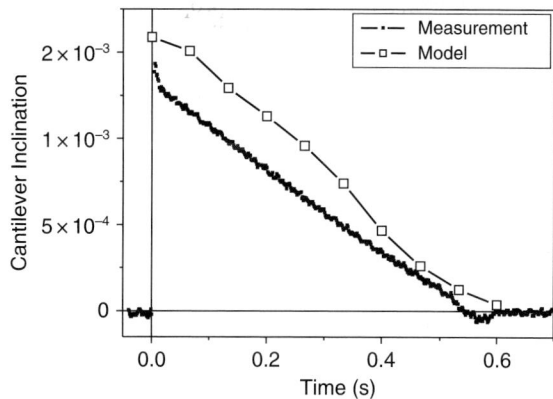

Fig. 2.5. $T \sim 25°$ C, RH $\sim 30\%$; drop data: water, $a = 32\,\mu\text{m}$, $\Theta = 63°$, $\gamma_L = 0.072\,\text{N/m}$, $m_0 = 36\,\text{ng}$; cantilever data: silicon, $E = 180\,\text{GPa}$, $l_0 = 500\,\mu\text{m}$, $w = 100\,\text{mm}$, $d = 0.9\,\mu\text{m}$. Experimental and calculated curves in comparison. (From [23])

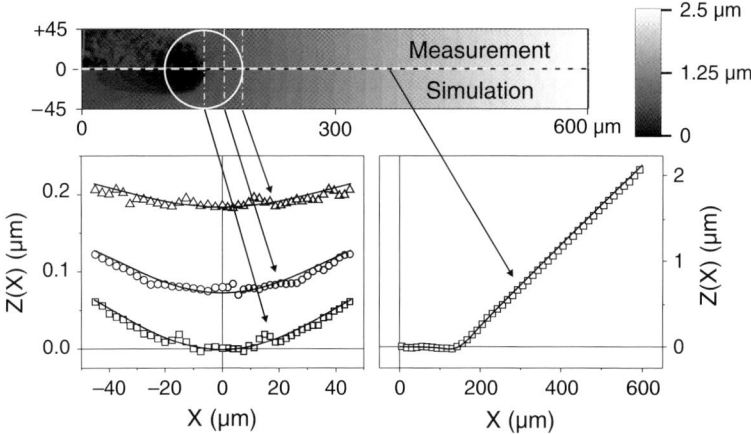

Fig. 2.6. Ionic liquid drop: $a = 45\,\mu m$, $\Theta = 64°$, $\gamma_L = 0.045\,N/m$; cantilever: $E = 180\,GPa$, $v = 0.26$, $l_0 = 600\,\mu m$, $w = 90\,mm$, $d = 0.7\,\mu m$. *Solid lines* represent simulations, *open symbols* experimental data points. (From [24])

Germany) was deposited onto a silicon cantilever. The face of the cantilever opposite to the one where the drop was sitting was imaged with a confocal profilometer (μ Surf from NanoFocus AG, Oberhausen, Germany) and a 3D image of the bent cantilever was obtained. It was then compared with the simulation results (Fig. 2.6)

A Young's modulus $E = 180\,GPa$ and a Poisson's ratio $v = 0.26$, which are standard values for silicon, were used in the simulation. The upper half of Fig. 2.6 represents the measurement, the lower half the simulation, the white circle the position of the drop. The agreement between simulation and measurements is very satisfactory. In the lower left graph three transversal profiles of the cantilever, taken at the three places indicated by the arrows, are shown. Modeling this bending is beyond the capabilities of the analytic model. In the lower right graph the longitudinal profiles, experimental and simulated, are shown. The overall relative error is below 6% in this case, and below 10% on the average. This is an extremely good agreement, especially considering that all parameters of the simulation are fixed: they are determined from independent measurements (cantilever and drops dimensions) or are known from literature (silicon and ionic liquid properties).

From further parameter studies it resulted that the transversal bending and the clamping at the base cause negligible effects, so that the consideration of Poisson's ratio appears to be the dominating factor that causes the discrepancy between the analytic model and the experimental results [24].

2.3.3
Negative Inclination

Evaporation curves recorded with water drops deposited on very clean hydrophilic cantilevers systematically differ from curves recorded with drops on very clean hydrophobic cantilevers. Inclination and evaporation time have been normalized in

Fig. 2.7. Evaporation curves of two water microdrops on a cantilever with a hydrophilic (*blue circles*) and a hydrophobic (*red triangles*) surface.
T $\sim$ 25 °C, RH $\sim$ 30%.
Hydrophilic case:
$a = 37\,\mu$m, $\Theta = 70°$,
$m_0 = 66$ ng. Hydrophobic case: $a = 34\,\mu$m, $\Theta = 116°$, $m_0 = 187$ ng

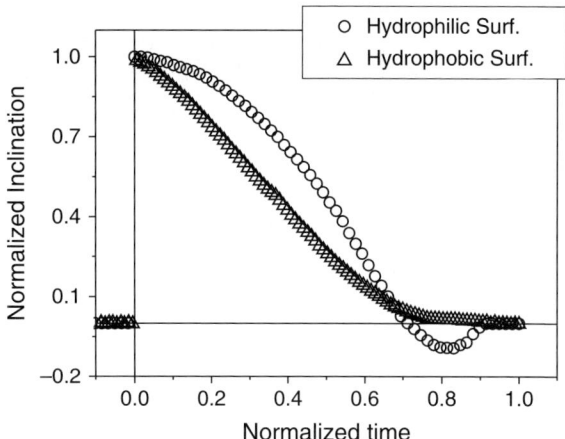

order to compare two measurements (Fig. 2.7). The difference is that the inclination becomes negative at the end of the "hydrophilic curve." This is not an artifact and is reproducible. The initial drop volumina were different, as well as the evaporation modi of the two drops. On the hydrophilic cantilever, the TPCL of the drop remained pinned for almost the entire evaporation, as far as could be concluded from video microscope images, while the contact angle of the drop gradually decreased. This corresponds to the CCR evaporation mode. Conversely, on the hydrophobic cantilever, both, contact radius and contact angle, changed during evaporation.

The negative inclination can not be explained by the model proposed before, because Eq. (2.7) contains only positive terms:

(1) The first is the contribution from the Laplace pressure ΔP and from the vertical component of the surface tension $\gamma_L \sin \Theta$.
(2) The second is basically Young's equation, which should be zero under the assumption that the drop is in thermodynamic quasi-equilibrium.

It is thus necessary to introduce an additional term, which must be negative and which describes the mechanical stress applied to the cantilever towards the end of evaporation of the drop. It can be called $\Delta\sigma$, for example, and one should keep in mind that it is not constant

$$\frac{dz}{dx} = \frac{3\pi a^3}{Ewd^3}\left[\gamma_L \sin\Theta + \frac{2d}{a}(\gamma_L \cos\Theta - \gamma_{SL} + \gamma_{SL}) + \Delta\sigma\right] \tag{2.8}$$

This additional stress is for now just a free parameter, which helps to describe the experimental curves. A tentative explanation of its physical origin could be as follows:

(1) When the contact angle is "large enough," the drop pulls at the TPCL and the cantilever bends upwards (Fig. 2.8A). The first term of the equation, $\gamma_L \sin\Theta$, dominates over the second term, $\gamma_L \cos\Theta - \gamma_{SL} + \gamma_{SL}$, which is zero under the

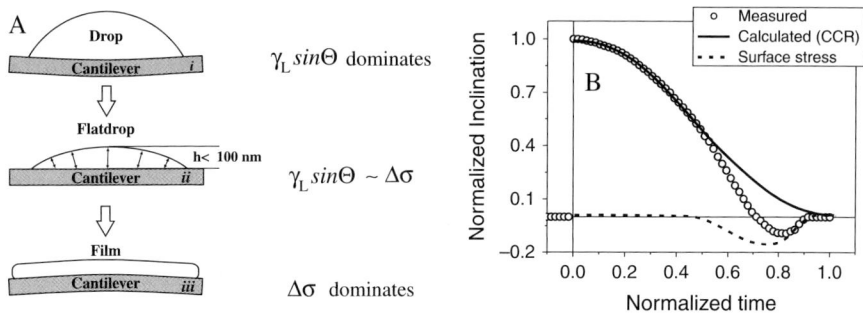

Fig. 2.8. (**A**) Schematic of the model to explain the negative cantilever inclination, and range of action of the forces involved. (**B**) Experimental evaporation curve (*circles*), calculated curve assuming CCR evaporation (*solid line*), and calculated surface stress (*dashed line*)

assumption of Young's equation being valid. The third term, $\Delta\sigma$, must also be negligible.

(2) When the drop becomes thinner than some 100 nm, the surface forces inside the flat drop start to play a role, favoring the formation of a thin film and acting to stabilize it. At this stage, ΔP, γ_L, and $\Delta\sigma$ have a similar magnitude and their effects on the bending of the cantilever cancel out each other. The cantilever crosses the "zero inclination" axis for the first time.

(3) Then, what is left of the drop wets the surface and forms a thin film, which could span a larger area than the original drop contact area. Now $\Delta\sigma$ dominates over the other two terms, which become vanishingly small. The interfacial tension between the water film and the cantilever is smaller than between air and the cantilever, so that the latter bends away from the drop. The measured signal is negative.

It may also be possible to determine $\Delta\sigma$ experimentally from the inclination curve (Fig. 2.8B). In fact, from video images it is known that the drop evaporates in the CCR mode. Assuming that the CCR mode holds until the end (last 100 ms), from the contact radius and contact angle data one could calculate the inclination as if only ΔP and γ_L were acting on the cantilever (solid line). Subtraction of these two curves yields the curve of $\Delta\sigma$ (dashed line). So, although measurable, still a quantitative description for $\Delta\sigma$ has to be found.

In summary, using cantilevers as sensors an effect arising with microscopic, pinned drops was observed which, to the best of our knowledge, was never observed using other methods. The tentative explanation is that a thin liquid film wets the surface, reduces the surface tension on the top side, and causes the cantilever to bend towards the bottom side.

2.3.4
Mass and Inclination

Surface forces exerted by the drop on the cantilever cause its bending. With the light lever technique the resulting inclination at the end of the cantilever is measured.

This signal can be called "static," since it changes continuously, but slowly, during the evaporation of the drop which takes usually one second. Additionally to this use, as surface stress sensors, cantilevers can also be employed as microbalances. Cantilevers are harmonic oscillators, whose resonance frequency depends, among other parameters, on their own mass and on their load. The change of the resonance frequency caused by the mass change of the evaporating drop can be recorded. This signal can be called "dynamic" with respect to the "static" signal described before, because a cantilever oscillation takes less than a millisecond.

Assuming that Young's equation is valid at all times, that the additional surface stress $\Delta\sigma$ is negligible, and that the evaporation takes place in the CCA mode (these issues are practically met using a hydrophobized cantilever), the "static" equation simplifies to

$$\frac{dz}{dx} \approx \frac{3\pi a^3}{Ewd^3} \gamma \sin\Theta \tag{2.9}$$

while the "dynamic" equation states that the mass added to the cantilever (load) is inversely proportional to the resonance frequency squared [25, 30].

$$m \propto \frac{1}{(2\pi f)^2} \tag{2.10}$$

The two signals, inclination and resonance frequency, are acquired simultaneously, but are independent of each other. They yield the stress and the mass (Fig. 2.9). As an example of a simultaneous static and dynamic signal analysis, a water drop was deposited onto a silicon cantilever hydrophobized with a monolayer of hexamethyldisilazane (HMDS). The initial contact angle was $\sim 80°$. It decreased nearly linearly during evaporation, and was $\sim 70°$ at the end. The initial contact radius was $\sim 33\,\mu m$, and decreased nearly linearly during evaporation. At the end it was below $10\,\mu m$. At present, we can record the inclination curve with a temporal resolution of $\sim 0.1\,ms$ between data points, and the frequency curve with $\sim 5\,ms$. The mass calculated from the resonance frequency of the cantilever and from video

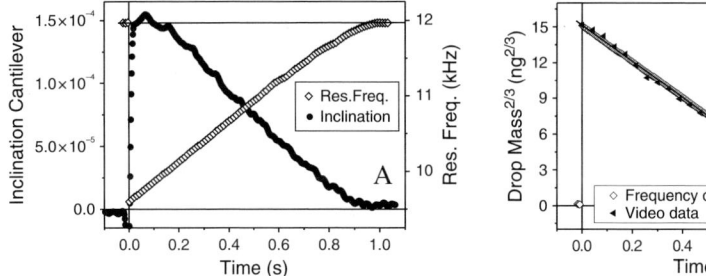

Fig. 2.9. (**A**) Simultaneously acquired inclination (*blue full circles*) and resonance frequency (*red hollow diamonds*) of a cantilever versus time upon evaporation of a water microdrop on a hydrophobized silicon cantilever. T $\sim 25\,°C$, RH $\sim 30\%$. Drop data: $a = 33\,\mu m$, $\Theta = 80°$, $\gamma_L = 0.072\,N/m$, $m_0 = 60\,ng$. Cantilever data: $l_0 = 500\,\mu m$, $w = 90\,\mu m$, $d = 2.1\,\mu m$. (**B**) Drop mass$^{2/3}$ versus time from frequency (*hollow diamonds*) and video (*black triangles*) data, and linear fit (*solid line*)

microscope images is similar (Fig. 2.9B), although the time resolution (~ 5 ms) and the sensitivity (~ 50 pg) is much higher for the frequency-derived mass. This offers the possibility of studying drop evaporation closer to its end (see inset). If the mass is plotted as "$m^{2/3}$ versus time" it results that the evaporation law "$V^{2/3}$ linear vs. time," which is valid for macroscopic drops, can be extended to microscopic drops and is valid until the evaporation ends.

In summary, using cantilevers as drop evaporation sensors we can simultaneously record the mass of the drop by tracking the resonance frequency of the cantilever, and the surface forces of the drop by tracking the inclination of the cantilever.

2.3.5
Vaporization Heat

Upon evaporation, a drop absorbs heat from its surroundings (air and cantilever). For example, a water drop cools down by 1–2 °C [31], while benzene cools down by 15–20 °C [32]. The processes involved in heat dissipation in this case are mainly conduction and convection. If the cantilever is made of pure, crystalline silicon, the cooling does not affect its bending, as shown in Fig. 2.10A [33]. The cantilever inclination measured upon deposition of a microdrop of water, which is the result of the combined action of surface forces and thermal effect, and the inclination simulated taking into account the surface forces only, from Eq. (2.7), are in good agreement. The difference between the two curves is smaller than 10% over all the evaporation.

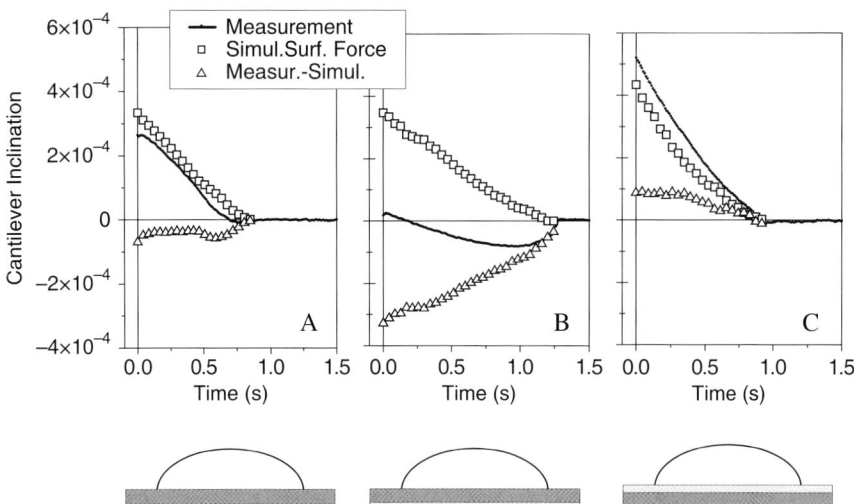

Fig. 2.10. Experimental (*solid line*), simulated (*hollow squares*), and difference (*hollow triangles*) inclination of silicon cantilevers versus time upon evaporation of water microdrops. T $\sim$ 25 °C, RH $\sim$ 30%. Drop data: various initial volumes, contact radii, and contact angles; $\gamma_L = 0.072$ N/m. Cantilever data: $l_0 = 750\,\mu$m, w = 90 μm, $d_A = 1.8\,\mu$m, $d_B = 1.5\,\mu$m, and $d_C = 1.7\,\mu$m; Gold layer thickness = 30 nm; Young's Moduli: $E_{Si} = 180$ GPa, $E_{Au} = 78$ GPa; Poisson's ratios: $\nu_{Si} = 0.26$, $\nu_{Au} = 0.44$

However, if the cantilever is, for example, gold coated on one of its sides, it behaves like a bimetal, since silicon and gold have different linear thermal expansion coefficients of $\alpha_{Si} = 2.6 \times 10^{-6}\,\mathrm{K}^{-1}$ and $\alpha_{Au} = 14.2 \times 10^{-6}\,\mathrm{K}^{-1}$, respectively. The direction of the bending depends on which side the gold layer is with respect to the drop. The thermal effect alone (experimental inclination minus inclination simulated for surface forces) would cause the cantilever to bend downwards (negative inclination) if the gold layer is on the bottom (Fig. 2.10B), or upwards (positive inclination) if the gold layer is on the top (Fig. 2.10C). The three water drops used here had different initial volumina, and the three cantilevers had different thicknesses. The evaporation times and the inclinations are therefore not directly comparable. However, contact angle and radius of each drop were recorded during evaporation, and the cantilever properties are known. With these data FE simulations were performed.

In summary, with the proper choice of the cantilever coating it is possible to sense the temperature of the evaporating drop, along with its mass and the effect of its surface tension.

2.4
Further Applications of Drops on Cantilevers

2.4.1
Spring Constant Calibration

The calibration of the spring constant of cantilevers is a fundamental issue in all AFM applications that involve force measurements [34, 35]. If cantilevers are employed to determine surface forces, a quantitative statement is possible only when the spring or force constant of the cantilever is known. The spring constant of commercially available cantilevers lies between $k \approx 0.01$ and $k \approx 50\,\mathrm{N/m}$. "Soft" cantilevers with k between 0.01 and 1 N/m are usually employed for the measurements of surface forces like adhesion, van der Waals interactions, and electrostatic forces. Because of their small spring constant, they allow for a high force resolution. "Stiff" cantilevers are commonly used for indentation measurements and nanolithography [34, 35] on soft samples, and for imaging in tapping mode. In the last 15 years a variety of methods have been introduced for the experimental determination of spring constants. Out of them all, around four methods have been established as standards [36, 37]. The first method makes use of the dimensions and the material properties of the cantilever, as well as its experimental resonance frequency and the quality factor of the resonance spectrum [38–45]. This approach is suited for rectangular cantilevers with known dimensions, for which it provides accurate results (std. dev. < 10%). The second method is based on the acquisition of the thermal noise spectrum of the cantilever [46, 47]. It is referred to as the "thermal noise method." It can be applied to cantilevers of any shape and does not require precise measurements of the cantilever dimensions, but suffers from a disadvantage: at least one force curve has to be acquired with the cantilever on a hard substrate in order to calibrate the spectrum. This can contaminate or damage the tip. The results obtained by this method are also accurate and reproducible (std. dev. $< 10\%$). The third method is a so-called "direct" method. A known force is applied to the cantilever and the resulting bending is measured with the light lever technique. This force can be of hydrodynamic origin and act along [48–50] or at the free end [51] of the cantilever. The force can also be exerted by a second cantilever with a known spring constant [52, 53], by electrostatics [54], or by excitation with microdrops of known mass and velocity [23]. For all these techniques, special instruments need to be developed. Further, contamination of the tip or of the surface of the cantilevers is possible. The results obtained by this method are usually a little less accurate (std. dev. $< 15\%$). The fourth method, and also the most used/cited according to ISI – Web of Science[SM], was presented by Cleveland et al. [30]. It provides extremely precise values (std. dev. $\sim 5\%$). A cantilever is loaded at known positions with small weights (in the nanogram range) and the resulting shift of the resonance frequency is measured [see Eq. (2.10)]. This method, also referred to as the "added mass method," can be applied to all types of cantilevers and has no major restrictions. However, it is somehow time consuming:

(1) weights or particles of different masses must be placed at the very free end of the cantilever without damaging it
(2) the thermal noise spectrum must be acquired for each particle
(3) the particles must be removed without damaging the cantilever, and be positioned on a sample holder for having later their dimensions characterized by a SEM to calculate their mass.

This last technique can be refined and simplified [25]. Instead of weights, microdrops have been used. The method has two advantages: first, contamination is avoided by working "contactless." Water drops are generated and deposited on the cantilever surface by a XYZ-controlled inkjet nozzle; second, since the drops evaporate after some time, the weight does not need to be removed manually.

The technique has been validated in two ways:

(1) Microdrops with different masses were deposited at the free end of a cantilever (similar to the "added mass method") and the resonance frequency of the cantilever was measured. It is not necessary any more to acquire a thermal noise spectrum, because the cantilever is excited by the drop impact and oscillates with an amplitude. The result was compared to results from beam theory, and the spring constant was calculated according to Eq. (2.10) (Fig. 2.11A). This can be done because the cantilever with the drop at its end is a harmonic oscillator, and for the dependence of its resonance from the drop mass a closed solution does exist.
(2) Microdrops with similar mass were deposited at different positions along a cantilever, the resonance frequency was measured for each drop, the result was compared to predictions from FE simulations, and the spring constants were calculated (Fig. 2.11B). FE simulations are necessary because the system with the drop along the cantilever is still a harmonic oscillator, but the dependence of its frequency from the drop mass does not have an analytic solution.

The obtained values are in good agreement with calibration results of the same cantilevers obtained by the "thermal noise method" (TN) (Table 2.1).

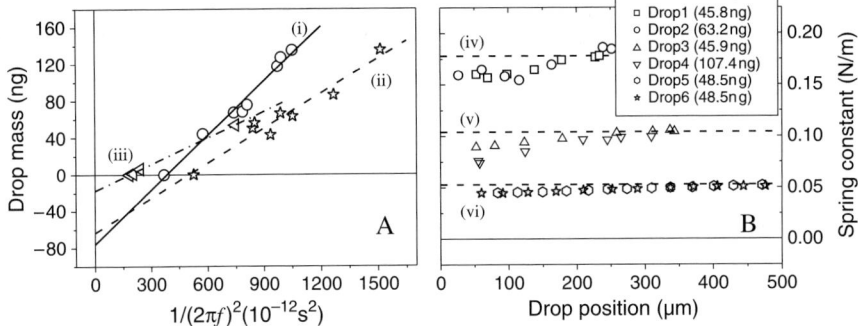

Fig. 2.11. (A) Mass of different water drops versus $1/(2\pi f)^2$ for three cantilevers. (i) Rectangular cantilever: $l_0 = 500\,\mu m$, $w = 90\,\mu m$. The linear regression of the data yields $k \approx 0.198\,N/m$. (ii) Rectangular cantilever: $l_0 = 460\,\mu m$, $w = 50\,\mu m$. $k \approx 0.126\,N/m$. (iii) Triangular cantilever: $l_0 = 200\,\mu m$, $w = 22\,\mu m$. $k \approx 0.096\,N/m$. **(B)** Spring constants versus drop position for the rectangular cantilevers (iv), (v), and (vi). The *dashed horizontal line* represents the results from the "thermal noise method". (From [25])

Table 2.1. Values of the spring constant, calibrated by the thermal noise method (TN) and by the microdrops (EXP), for six different commercial cantilevers, with different dimensions and by different manufacturers

Cantilever	l_0 [μm]	f_0 [Hz]	K_{TN} [N/m]	Nr. of drops	K_{EXP} [N/m]	E_{rel}[%] $(K_{TN} - K_{EXP})/K_{TN}$
(i) Micromotive, rectangular	500	10,404	0.19 ± 0.01	7	0.198 ± 0.012	4.2
(ii) Nanosensors, rectangular	460	11,782	0.13 ± 0.01	7	0.126 ± 0.012	3.1
(iii) Veeco, triangular	200	19,253	0.10 ± 0.01	3	0.096 ± 0.002	4.0
(iv) Micromotive, rectangular	297	16,900	0.178 ± 0.01	2 × 8	0.170 ± 0.010	4.5
(v) Micromotive, rectangular	372	11,215	0.104 ± 0.01	2 × 7	0.098 ± 0.006	5.8
(vi) Micromotive, rectangular	503	6,400	0.053 ± 0.01	2 × 12	0.048 ± 0.002	9.4

2.4.2
Contamination Control of Cantilevers

The resolution of AFM imaging strongly depends on the shape and chemical composition of the cantilever tip. Scanning a surface with a double-tip results in an image with a so-called "ghost." Scanning a surface with a contaminated tip results in a large adhesion between tip and sample, which increases the contact area and decreases the resolution [55]. AFM is not only used for imaging, but also for friction measurements [56–59], surface force measurements in air or liquids [60, 61], force spectroscopy between biomolecules [62–66], and adhesion [67,68], just to cite a few. Reliable and reproducible results can be obtained only with a clean, or at least known, AFM tip chemistry. Cleanliness is also important for the quality of further surface modifications of the AFM tip. Thin layers of contaminants may change the reactivity or adsorptivity of its surface. Also cantilevers without tips are used as force transducers and micromechanical stress sensors [69–72], above all for biochemical applications. In this case the whole cantilever must be chemically modified and/or functionalized. For this, a noncontaminated surface is required. Because the AFM tip is so small, characterization of its surface chemistry is difficult. One indirect approach is to characterize a reference surface that is made of the same material as the AFM tip and treated with the same process as the tip, by means of macroscopic techniques. Contact angle measurements [73] or X-ray photoelectron spectroscopy (XPS) [74] are often employed. However, unless the reference surface was subjected to the same storage conditions, it might not be representative of the cantilever and tip surface chemistry. Surface sensitive methods, like scanning Auger microscopy (SAM [75]) and time-of-flight secondary-ion mass spectroscopy (TOF-SIMS [74]), have been employed to perform surface microarea analyses (areas of hundreds of square micrometers) on the cantilevers. These methods allow for a precise characterization of the contaminants, however they do not allow for the characterization of the sole tip area, and are indeed time consuming and expensive. A simple and straightforward method for checking tip contamination was proposed by Thundat et al. [55]: by monitoring the change of the adhesion between tip and sample during imaging one can infer the degree of contamination.

Several procedures for cleaning the AFM tip-cantilever assembly have been discussed in the literature. Such procedures include ultraviolet ozone treatment [75,76], aggressive acid-based baths [63, 74] and plasma etching [66, 77]. The majority of AFM users directly use as-received cantilevers, or they simply clean them by rinsing with organic solvents. This demonstrates that despite the fact that the AFM community has grown in the last years, apparently contamination is still considered a minor issue. Earlier attempts to draw attention to it reported that contamination undoubtedly affects AFM experiments [55], and that the main source of organic contamination on new cantilevers comes from their packaging [74]. To simplify the characterization avoiding the use of such expensive apparatus cited above, a straightforward, low-cost, and effective way of characterizing the wetting behavior of cantilevers and tips was proposed [78]. A small water drop is deposited directly on the cantilever by an inkjet nozzle. On deposition, the drop spreads depending on the hydrophobicity (contact angle) of the cantilever surface, and finally attains a contact radius a and a contact angle Θ. They are determined by video microscopy (Fig. 2.12).

One of the most used rectangular contact mode cantilevers, the Pointprobe (Nanosensors, Neuchatel, Switzerland), with a length $l_0 = 470\,\mu m$, a width $w = 50\,\mu m$, and a thickness $d = 2\,\mu m$ was examined. The average contact angle of a water microdrop on an as-received cantilever was $\Theta \approx 88°$ ($\pm 1°$) (Fig. 2.12A). After cleaning in a plasma reactor (PDC-002, Harrick Scientific Corp., NY) under an argon atmosphere at medium power over 20 s, the contact angle decreased to $\Theta \approx 11°$ ($\pm 2°$). Since the drop strongly wets the surface, the drop shape is not

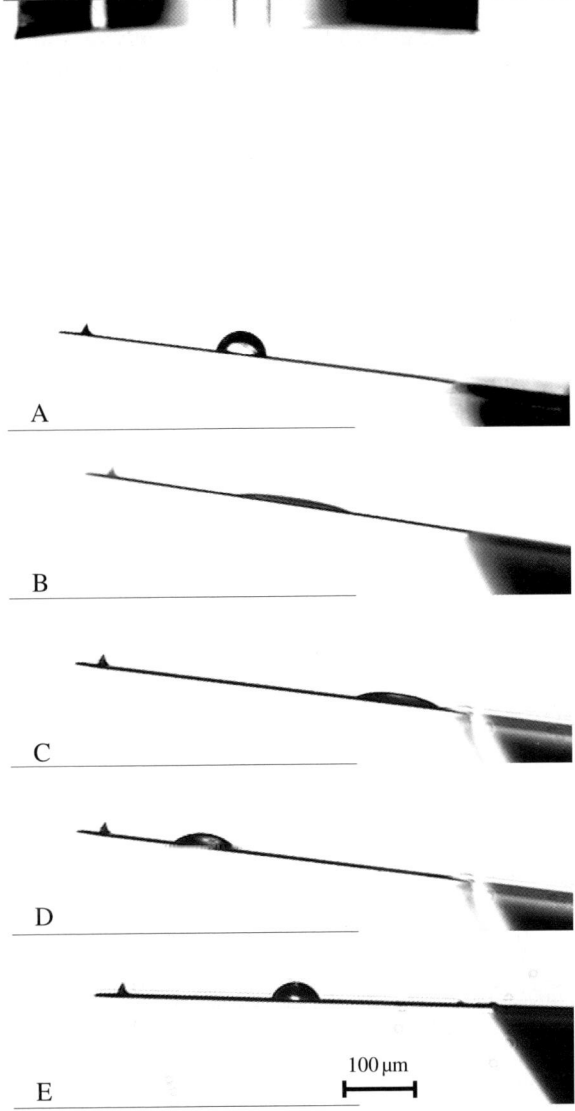

Fig. 2.12. Sequence of five images of water microdrops on a rectangular silicon cantilever. Before plasma cleaning (**A**) the drop forms a contact angle $\Theta = 88°$ ($\pm 1°$); after plasma cleaning (**B**) the contact angle is $\Theta = 11°$ ($\pm 2°$); after storage of the cantilever in a Gel-Box$^{®}$ for 1 (**C**), 3 (**D**) and 24 (**E**) hours, the drop forms contact angles of $\Theta = 26°$ ($\pm 2°$), $45°$ ($\pm 3°$), and $72°$ ($\pm 3°$). (From [78])

spherical anymore (Fig. 2.12B). The recontamination was controlled by storing the cantilever in a Gel-Box® (Gel-Pak Inc., Hayward, CA) for different long periods of time. Micro contact angle measurements were performed after 1, 3, and 24 hours (Fig. 2.12C–E). The contact angle increased stepwise, and after 24 hours was $\Theta \approx 72°$ ($\pm 3°$). On the basis of these results can be concluded that plasma cleaning removes satisfactorily the silicone oil contamination. However, cleaned cantilevers are fully recontaminated if they are stored in plastic boxes for 24 hours.

2.5
Conclusions

Drop evaporation has been commonly observed by means of video-microscope imaging and ultra-precision weighing with electronic microbalances or with quartz crystal microbalances. Abundant information was gained over the years with these techniques, so that the evaporation of macroscopic drops of simple liquids from inert surfaces is nowadays well understood. The same techniques are, however, not applicable to microscopic drops. Furthermore, they do not directly provide a measure of the interfacial stresses arising at the contact area between liquid and solid, which are known to play a key role in the evaporation kinetics of small drops. It was shown that atomic force microscope (AFM) cantilevers can be used as sensitive stress, mass, and temperature sensors, and how they can be employed to monitor the evaporation of microdrops of water from solid surfaces. The technique has some advantages with respect to state-of-the-art techniques cited above, since it allows one to measure more drop parameters simultaneously and for smaller drop sizes. The technique further allows detection of differences between water microdrops evaporating from clean hydrophilic and hydrophobic surfaces. The difference is especially manifest close to the end of evaporation. The evidence arises that on the hydrophilic surface a thin water film forms, while this is not the case for the hydrophobic surface. Metal-coated cantilevers can be used as thermometers, and allow one to precisely measure the temperature of an evaporating microdrop. This can be relevant for further applications of cantilevers as micro-sized calorimetric sensors for chemical reactions taking place in drops on their surface.

Further applications of the inkjet technology combined with cantilever sensors allows one to test the local cleanliness of cantilever surfaces with micro contact angle measurements, or to calibrate the spring constant of cantilevers in a contactless and noncontaminating way.

Acknowledgments. E.B. acknowledges the Max pLanck Society (MPG) for financial support. This work was partially funded by the German Research Foundation (DFG) (Forschergruppe FOR 516, WI 1705/7).

References

1. Heilmann J, Lindqvist U (2000) J Imag Sci Technol 44:491
2. Socol Y, Berenstein L, Melamed O, Zaban A, Nitzan B (2004) J Imag Sci Technol 48:15

3. Kim EK, Ekerdt JG, Willson CG (2005) J Vac Sci Technol B 23:1515
4. Tullo AH (2002) Chem Eng News 80:27
5. Chou FC, Gong SC, Chung CR, Wang MW, Chang CY (2004) Jap J Appl Phys 1 43:5609
6. Fabbri M, Jiang SJ, Dhir VK (2005) J Heat Mass Transf 127:38
7. Amon CH, Yao SC, Wu CF, Hsieh CC (2005) J Heat Transf Trans ASME 127:66
8. Shedd TA (2007) Heat Transf Eng 28:87
9. "Fogtech International". www.fogtec-international.com
10. Kawase T, Sirringhaus H, Friend RH, Shimoda T (2001) Adv Mater 13:1601
11. Bonaccurso E, Butt HJ, Hankeln B, Niesenhaus B, Graf K (2005) Appl Phys Lett 86:124101
12. de Gans BJ, Hoeppener S, Schubert US (2006) Adv Mater 18:910
13. Karabasheva S, Baluschev S, Graf K (2006) Appl Phys Lett 89:031110
14. Ionescu RE, Marks RS, Gheber LA (2003) Nano Letters 3:1639
15. Lefebvre AH (1989) Atomization and Sprays. Taylor & Francis
16. Lee ER (2003) Microdrop Generation. CRC Press, Taylor and Francis
17. Edwards BF, Wilder JW, Scime EE (2001) Eur J Phys 22:113
18. Bourges-Monnier C, Shanahan MER (1995) Langmuir 11:2820
19. Rowan SM, Newton MI, McHale G (1995) J Phys Chem 99:13268
20. Birdi KS, Vu DT, Winter A (1989) J Phys Chem 93:3702
21. Picknett RG, Bexon R (1977) J Colloid Interface Sci 61:336
22. Pham NT, McHale G, Newton MI, Carroll BJ, Rowan SM (2004) Langmuir 20:841
23. Bonaccurso E, Butt HJ (2005) J Phys Chem B 109:253
24. Haschke T, Bonaccurso E, Butt H-J, Lautenschlager D, Schönfeld F, Wiechert W (2006) J Micromech Microeng 16:2273
25. Golovko DS, Haschke T, Wiechert W, Bonaccurso E (2007) Rev Sci Instrum 78:043705
26. Obrien RN, Saville P (1987) Langmuir 3:41
27. Cordeiro RM, Pakula T (2005) J Phys Chem B 109:4152
28. Soolaman DM, Yu HZ (2005) J Phys Chem B 109:17967
29. Butt HJ, Golovko DS, Bonaccurso E (2007) J Phys Chem B 111:5277
30. Cleveland JP, Manne S, Bocek D, Hansma PK (1993) Rev Sci Instrum 64:403
31. David S, Sefiane K, Tadrist L (2007) Colloid Surf A 298:108
32. Blinov VI, Dobrynina VV (1971) J Eng Phys Thermophys 21:229
33. Golovko DS, Bonanno P, Lorenzoni S, Raiteri R, Bonaccurso E (2008) J Micromech Microeng 18:095026
34. Butt HJ, Cappella B, Kappl M (2005) Surface Science Reports 59:1–152
35. Cappella B, Dietler G (1999) Surf Sci Rep 34:1
36. Burnham NA, Chen X, Hodges CS, Matei GA, Thoreson EJ, Roberts CJ, Davies MC, Tendler SJB (2003) Nanotechnology 14:1
37. Ralston J, Larson I, Rutland MW, Feiler AA, Kleijn M (2005) Pure Appl Chem 77:2149
38. Green CP, Lioe H, Cleveland JP, Proksch R, Mulvaney P, Sader JE (2004) Rev Sci Instrum 75:1988
39. Green CP, Sader JE (2002) J Appl Phys 92:6262
40. Green CP, Sader JE (2005) J Appl Phys 98:114913
41. Green CP, Sader JE (2005) Phys Fluids 17:073102
42. Higgins MJ, Proksch R, Sader JE, Polcik M, Mc Endoo S, Cleveland JP, Jarvis SP (2006) Rev Sci Instrum 77:013701
43. Sader JE (1998) J Appl Phys 84:64
44. Sader JE, Chon JWM, Mulvaney P (1999) Rev Sci Instrum 70:3967
45. Sader JE, Larson I, Mulvaney P, White LR (1995) Rev Sci Instrum 66:3789
46. Butt H-J, Jaschke M (1995) Nanotechnology 6:1
47. Hutter JL, Bechhoefer J (1993) Rev Sci Instrum 64:1868
48. Craig VSJ, Neto C (2001) Langmuir 17:6018

49. Notley SM, Biggs S, Craig VSJ (2003) Rev Sci Instrum 74:4026
50. Senden TJ, Ducker WA (1994) Langmuir 10:1003
51. Maeda N, Senden TJ (2000) Langmuir 16:9282
52. Torii A, Sasaki M, Hane K, Okuma S (1996) Meas Sci Technol 7:179
53. Tortonese M, Kirk M (1997) SPIE Proc 3009:53
54. Bonaccurso E, Schonfeld F, Butt HJ (2006) Phys Rev B 74:085413
55. Thundat T, Zheng X-Y, Chen GY, Sharp SL, Warmack RJ (1993) Appl Phys Lett 63:2150
56. Mate CM, McClelland GM, Erlandsson R, Chiang S (1987) Phys Rev Lett 59:1942
57. Marti O, Colchero J, Mlynek J (1990) Nanotechnology 1:141
58. Meyer E, Lüthi R, Howald L, MBammerlin M, Guggisberg M, Güntherodt H-J (1996) J Vac Sci Technol B 14:1285
59. Tsukruk VV, Bliznyuk VN (1998) Langmuir 14:446
60. Butt H-J (1991) Biophys J 60:1438
61. Ducker WA, Senden TJ, Pashley RM (1991) Nature 353:239
62. Florin E-L, Moy VT, Gaub HE (1994) Science 264:415
63. Hinterdorfer P, Baumgartner W, Gruber HJ, Schilcher K, Schindler H (1996) Proc Nat Acad Sci USA 93:3477
64. Radmacher M (1999) Physics World 12:33
65. Heinz WF, Hoh J (1999) Nanotechnology 17:143
66. Lee GU, Kidwell DA, Colton RJ (1994) Langmuir 10:354
67. Hoh J, Cleveland JP, Prater CB, Revel J-P, Hansma PK (1992) J Am Chem Soc 114:4917
68. Heim L-O, Blum J, Preuss M, Butt H-J (1999) Phys Rev Lett 83:3328
69. Berger R, Gerber C, Lang HP, Gimzewski JK (1997) Microel Eng 35:373
70. Fritz J, Baller MK, Lang HP, Rothuizen H, Vettiger P, Meyer E, Güntherodt H-J, Gerber C, Gimzewski JK (2000) Science 288:316
71. Raiteri R, Grattarola M, Butt H-J, Skladal P (2001) Sens Actuators B 79:115
72. Lang HP, Hegner M, Meyer E, Gerber C (2002) Nanotechnology 13:R29
73. Knapp HF, Stemmer A (1999) Surf Interface Anal 27:324
74. Lo Y, Huefner ND, Chan WS, Dryden P, Hagenhoff B, Beebe TP (1999) Langmuir 15:6522
75. Arai T, Tomitori M (1998) Appl Phys A 66:S319
76. Fujihira M, Okabe Y, Tani Y, Furugori M, Akiba U (2000) Ultramicroscopy 82:181
77. Senden TJ, Drummond CJ (1995) Coll Surf A: Physicochem Eng Asp 94:29
78. Bonaccurso E, Gillies G (2004) Langmuir 20:11824

3 Mechanical Diode-Based Ultrasonic Atomic Force Microscopies

M. Teresa Cuberes

Abstract. Recent advances in mechanical diode-based ultrasonic force microscopy techniques are reviewed. The potential of Ultrasonic Force Microscopy (UFM) for the study of material elastic properties is explained in detail. Advantages of the application of UFM in nanofabrication are discussed. Mechanical-Diode Ultrasonic Friction Force Microscopy (MD-UFFM) is introduced, and compared with Lateral Acoustic Force Microscopy (LAFM) and Torsional Resonance (TR) – Atomic Force Microscopy (AFM). MD-UFFM provides a new method for the study of shear elasticity, viscoelasticity, and tribological properties on the nanoscale. The excitation of beats at nanocontacts and the implementation of Heterodyne Force Microscopy (HFM) are described. HFM introduces a very interesting procedure to take advantage of the time resolution inherent in high-frequency actuation.

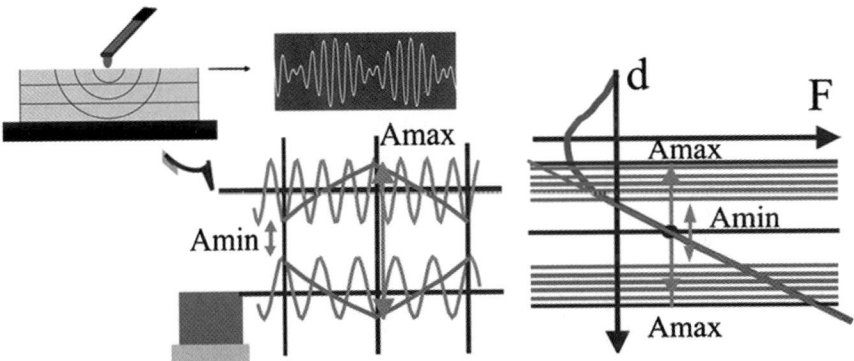

Key words: Ultrasonic Force Microscopy, Ultrasonic Friction Force Microscopy, Heterodyne Force Microscopy, Nanomechanics, Nanofriction, Mechanical-diode effect, Nanoscale ultrasonics, Beat effect

3.1 Introduction: Acoustic Microscopy in the Near Field

3.1.1 Acoustic Microscopy: Possibilities and Limitations

Acoustic microscopy uses acoustic waves for observation in a similar way as optical microscopy uses light waves. In acoustic microscopy, a sample is imaged by ultrasound, and the contrast is related to the spatial distribution of the mechanical properties. The procedure can be implemented in transmission as well as in reflection. The first Scanning Acoustic Microscope (SAM) was introduced in 1974 [1], and was mechanically driven and operated in the transmission mode. Today most commercial acoustic microscopes work in the reflection mode; by using pulsed acoustic systems the reflections of the acoustic beam from the specimen may be separated from spurious reflections. Applications of Acoustic Microscopy [1–10] include mapping of inhomogeneities in density and stiffness in materials, measurement of coating thicknesses, detection of delaminations in electronic integrated circuit chips, detection of microcracks and microporosity in ceramics, identification of the grain structure and anisotropy in metals and composites, and evaluation of elastic properties of living cells, etc.

A schema of a SAM operating in reflection mode is shown in Fig. 3.1. The acoustic microscope works on the principle of propagation and reflection of acoustic waves at interfaces where there is a change of acoustic impedance ($Z = density \times velocity$). Acoustic waves initiated at the piezoelectric transducer refract at the lens/coupling medium interface and focus to a diffraction limited spot. The sound wave is propagated to the sample through a couplant, usually water. When a sudden change in

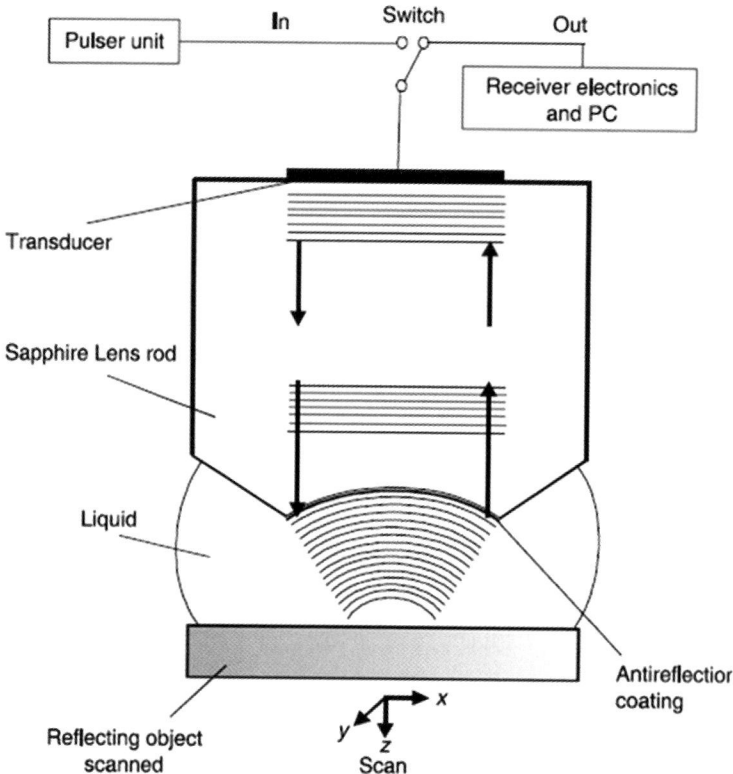

Fig. 3.1. SAM operating in reflection mode (from Ref. [7])

acoustic impedance is encountered, like at a material boundary, a portion of the sound energy is reflected, and the remainder propagates through the boundary. The transducer detects and converts the reflected acoustic waves into an electrical signal, which is digitized and stored at appropriate points during scanning of the lens. The amplitude and phase of the reflected acoustic signal determine the contrast in the acoustic images.

SAM also permits the implementation of time-of-flight (TOF) measurements. In time-resolved acoustic microscopy a short sound pulse is sent towards a sample. The time-of-flight method monitors the time required for the pulse sent into the sample to return back to the acoustic lens. TOF images provide a means to determine relative depth variations in the location of inhomogeneous or defective sites within a sample.

The spatial resolution and depth of penetration in SAM are inter-related, and dependent on the operating frequency. When using acoustic waves of frequencies around 1–2 GHz, SAM images with conventional optical resolution of the order of microns can be obtained. Nevertheless, when using low frequency ultrasound, in the 2–10 MHz range, the spatial resolution is typically limited to the millimeter range. On the contrary, the depth of penetration decreases as the frequency increases. In technical materials, the attenuation of pressure waves is given by the microstructure, and increases at least with the square of frequency. In fine-grained or fine-structured materials, the absorption, which increases linearly with frequency, limits the penetration. For GHz frequencies, the penetration depth may be of the order of microns. When the frequencies are in the MHz range the penetration may be in the millimeter range. The pressure waves, regardless of frequency, are more heavily attenuated in air than in liquids, so water is usually used as a convenient couplant between the transmitter/receiver of acoustic waves and the specimen.

Many advanced techniques based on SAM have emerged. In phase-sensitive acoustic microscopy (PSAM) [7–9], phase and amplitude SAM images are simultaneously obtained. The phase detection mode can be implemented in transmission, and it permits the observation of propagating waves emitted from an acoustic lens in a holographic manner, as shown in Fig. 3.2. Hybrid SAM-based technologies, such as photoacoustic microscopy (PAM) have proved to be of extreme value (see for instance [10]).

In spite of the advantages of the aforementioned acoustic techniques, the resolution achievable when using acoustic waves for observation at best is still poor for applications in nanotechnology. Acoustic imaging with nanometer-scale resolution can be realized using Ultrasonic Atomic Force Microscopy techniques, as described in this chapter.

3.1.2
Ultrasonic Atomic Force Microscopies

The main motivation for the initial development of Ultrasonic Atomic Force Microscopies was to implement a near-field approach that provided information such as that obtained with the Acoustic Microscope, but with a lateral resolution on the nanometer scale. The area of mechanical contact between the tip of an atomic

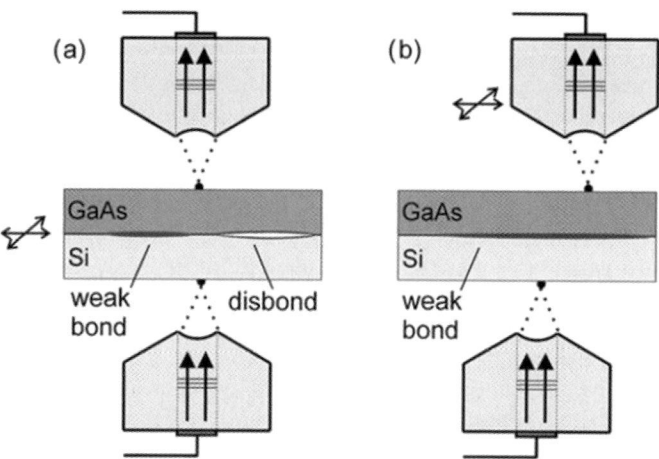

Fig. 3.2. Set-up for (**a**) Scanning Transmission Acoustic Microscopy and (**b**) Scanning Acoustic Holography (from Ref. [9])

force microscope (AFM) cantilever and a sample surface is typically of the order of nanometers in diameter. An AFM cantilever with the tip in contact with a sample surface follows a small-amplitude out-of-plane surface vibration linearly, provided its frequency is below the cantilever resonance frequency. One might expect that the cantilever would react in the same way to the pressure exerted at the tip–sample contact by an acoustic wave of millimeter wavelength, realizing ultrasound detection in the near-field. However, due to the inertia of the cantilever, the linear behavior is not evident in the limit of high-frequency signals. As a matter of fact, if the cantilever is regarded as a simple point mass, the amplitude of vibration at the driving frequency vanishes in the limit of very high frequencies.

Basically, we may distinguish two different procedures for the detection of high-frequency surface mechanical vibrations with the tip of an AFM cantilever. The first is based on the fact that actually the cantilever is not a point mass, but a tiny elastic beam that can support high-frequency resonant modes [11, 12]. When a cantilever tip is in contact with the sample surface and high-frequency surface vibration is excited at the tip–sample contact, the so-called *contact resonances* of the cantilever are excited at certain characteristic frequencies. Those depend on both the cantilever and the sample elastic properties [11–14]. Techniques such as Acoustic Atomic Force Microscopy (AFAM) [11] and Ultrasonic Friction Force Microscopy (UFFM) [15–17] are based on the study of cantilever contact resonances. Scanning Microdeformation Microscopy (SMM) [18,19] and Scanning Local Acceleration Microscopy (SLAM) [20] also monitor the vibration of an AFM cantilever with the tip in contact with the sample surface at the ultrasonic excitation frequency. A second approach is based on the so-called *mechanical-diode* effect [21, 22], which will be explained in more detail in Sect. 3.2.1. In this case, the operating ultrasonic frequency is extremely high, or such that the cantilever does not follow the high-frequency surface vibration amplitude due to its inertia. Nevertheless, if the surface ultrasonic vibration amplitude is sufficiently high that the tip–sample distance is varied over the nonlinear regime of the tip–sample interaction force, the cantilever experiences an additional force, or *ultrasonic force*. This can be understood as the averaged force acting upon the tip in each ultrasonic cycle. As a result of the ultrasonic force, the tip experiences a deflection – ultrasonic deflection, or mechanical-diode response – that can be monitored, and which carries information about the elastic properties of the sample, and the adhesive properties of the tip–sample contact [22]. Techniques such as Scanning Acoustic Force Microscopy (SAFM) [23, 24], Ultrasonic Force Microscopy (UFM) [25], Mechanical-Diode Ultrasonic Friction Force Microscopy (MD-UFFM) [26], and Heterodyne Force Microscopy (HFM) [27] utilize mechanical-diode type responses.

Up-to-now, most ultrasonic-AFM studies have been performed with the tip in contact with the sample surface, although in principle noncontact ultrasonic-AFM techniques can be implemented using either the high-order cantilever resonance frequencies, or the mechanical diode effect, as long as the distance between the tip of an inertial cantilever and a sample surface is swept over a nonlinear interaction regime. Recently, the possibility to detect high-frequency vibration using dynamic force microscopy has been demonstrated [28]; the detection of acoustic vibration in this case is apparently also facilitated because of the activation of the mechanical diode effect (see Sect. 3.2.1 for further discussions).

This chapter is mostly devoted to reviewing the fundamentals and recent advances in mechanical diode-based ultrasonic force microscopy techniques implemented in contact mode. In Sect. 3.2, Ultrasonic Force Microscopy will be explained in detail. The use of UFM in nanofabrication provides unique advantages [29]. In Sect. 3.3, Mechanical-Diode Ultrasonic Friction Force Microscopy [26] will be introduced. MD-UFFM is based on the detection of shear ultrasonic vibration at a sample surface via the lateral mechanical-diode effect. This is a new method for the study of shear elasticity, viscoelasticity, and tribological properties on the nanoscale. Section 3.4 discusses the technique of Heterodyne Force Microscopy [28]. HFM provides a novel and very interesting procedure to take advantage of the time resolution inherent in high-frequency actuation. In HFM, mechanical vibration in the form of beats is induced at the tip–sample contact by simultaneously launching ultrasonic waves towards the tip–sample contact region from the cantilever base and from the back of the sample, at slightly different frequencies. If the launched cantilever and sample vibration amplitudes are such that the tip–sample distance is varied over the nonlinear tip–sample force regime, the cantilever vibrates additionally at the beat frequency due to the mechanical-diode effect (beat effect). HFM monitors the cantilever vibration at the beat frequency in amplitude and phase. As has been demonstrated, Phase-HFM provides information about dynamic relaxation processes related to adhesion hysteresis at nanoscale contacts with high time sensitivity [28]. Recently, Scanning Near-Field Ultrasound Holography (SNFUH) [30] has been introduced. The principle of operation is very similar to that of HFM. The experimental data reported by SNFUH demonstrate its capability to provide elastic information of buried features with great sensitivity. Also, the technique of Resonant Difference-Frequency Atomic Force Ultrasonic Microscopy (RDF-AFUM) [31] has been proposed, based on the beat effect. Discussions of the beat effect and HFM will be included in Sect. 3.4.

3.2
Ultrasonic Force Microscopy: The Mechanical Diode Effect

3.2.1
The Mechanical Diode Effect

The first observation that the tip of an AFM cantilever can be used to detect out-of-plane high frequency vibration of a sample surface was reported in [21]. In these experiments, Surface Acoustic Waves (SAWs) were excited at (slightly) different frequencies by means of interdigital transducers (IDTs) and the frequency of surface vibration was detected using a cantilever tip in contact with the sample surface. The technique of Scanning Acoustic Force Microscopy has been demonstrated for the characterization of SAW field amplitudes [24] and phase velocities [32]. Acoustic fields in bulk acoustic-wave thin-film resonators have also been imaged with this method [33].

The physical mechanism that allows a cantilever to detect out-of-plane surface ultrasonic vibration excited at the tip–sample contact is based on the nonlinearity of the tip–sample interaction force [22]. Even though it is expected that inertia prevents a cantilever tip in contact with a sample surface to move fast enough to keep up with surface atomic vibrations at ultrasonic frequencies, the displacement of the surface leads to modification of the tip–sample force F_{t-s} provided the ultrasonic vibration amplitude is sufficiently high and the tip–sample distance d is varied over the nonlinear tip–sample force regime. In Fig. 3.3, it is assumed that the tip is in contact with a sample surface, in the repulsive force regime. When out-of-plane surface ultrasonic vibration is switched on the tip–sample distance d is varied at ultrasonic frequencies between some minimum and maximum values, corresponding to the amplitude of ultrasound excitation. If the ultrasonic amplitude is small, the tip–sample distance sweeps a linear part of the tip–sample interaction force curve. In this case, the net averaged force that acts upon the cantilever during an ultrasonic time period is equal to the initial set-point force, and hence the deflection of the cantilever remains the

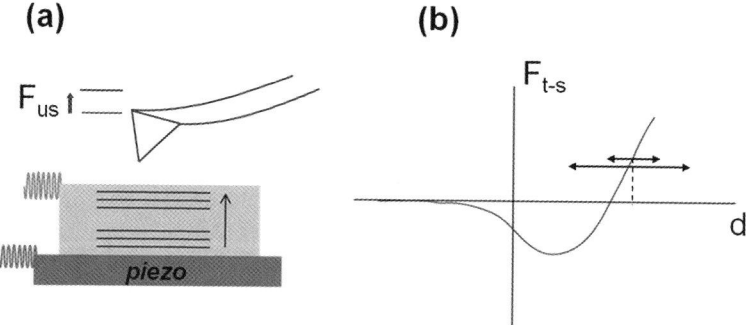

Fig. 3.3. Detection of surface out-of-plane ultrasonic vibration with the tip of an AFM cantilever via the mechanical-diode effect (**a**) When the surface vibration amplitude is sufficiently high, the tip experiences an *ultrasonic force* F_{us}. (**b**) Tip–sample force F_{t-s} versus tip–sample distance d curve

same as in the absence of ultrasound. However, if the amplitude of ultrasonic vibration is increased, the tip–sample distance sweeps over the nonlinear part of the force curve, and the averaged force includes an additional force F_{ult} given by

$$F_{ult}(d) = \frac{1}{T_{ult}} \int_0^{T_{ult}} F_{t-s}\left(d - a \cdot \cos\left(\frac{2\pi}{T_{ult}}t\right)\right) dt \tag{3.1}$$

where d the tip–sample distance, a is the amplitude of ultrasonic vibration, and T_{ult} the ultrasonic period.

Because of this additional force, named hereafter the *ultrasonic force*, the cantilever experiences an additional deflection which can be easily detected by means of the optical lever technique, and is the physical parameter which is monitored in Ultrasonic Force Microscopy [25]. The *UFM deflection* is a quasi-static cantilever deflection that occurs as long as out-of-plane ultrasonic vibration of sufficiently high amplitude is present at the tip–sample contact. The quasi-static equilibrium deflection is given by:

$$k_c z_{eq} = F_{ult}(d_{eq}, a) \tag{3.2}$$

where k_c is the cantilever stiffness, and z_{eq} and d_{eq} are the new cantilever deflection and tip indentation depth, respectively. As the surface ultrasonic vibration amplitude is further increased, F_{ult} increases due to the nonlinearity of the tip–sample force curve, and hence the cantilever deflection increases too until a new equilibrium position is reached. In this sense, the cantilever behaves as a "mechanical diode" [20], and deflects when the tip–sample contact vibrates at ultrasonic frequencies of sufficiently high amplitude.

To perform UFM, the ultrasonic excitation signal is typically modulated in amplitude with a triangular or trapezoidal shape (see Fig. 3.4). In UFM, the ultrasonic amplitude modulation frequency is chosen to be much lower than the first cantilever resonance, but higher than the AFM feedback response frequency to avoid that the feedback compensates for the ultrasonic deflection of the cantilever. Hence, contact-mode AFM can be performed to obtain a surface topographic image, in spite of the presence of ultrasound. To record a UFM image, the ultrasonic deflection of the cantilever is tracked at the amplitude modulation frequency using a lock-in amplifier.

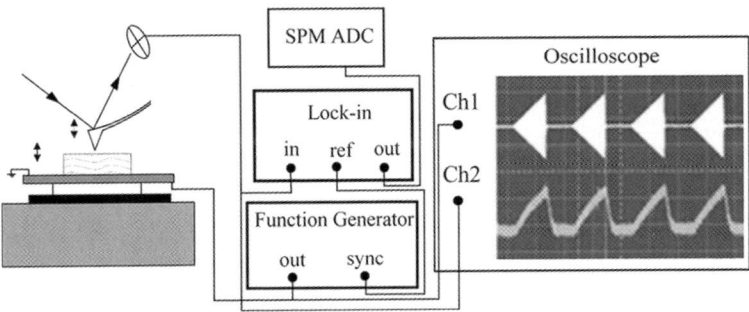

Fig. 3.4. Set-up for UFM measurements

To detect surface ultrasonic vibration with dynamic force microscopy in ref. [28], a resonator is used as a sample, and its acoustic vibration (at about 1.5 GHz) is modulated in amplitude with a sinusoidal shape. The ultrasonic amplitude modulation frequency is chosen to be coincident with the second eigenmode of the cantilever. Typical Dynamic Force Microscopy is performed using the first eigenmode of the AFM cantilever (at about 72 KHz) in order to obtain a surface topographic image, which can be properly done in spite of the presence of acoustic vibration. To obtain acoustic information, the cantilever vibration in the second eigenmode (at about 478 KHz) is monitored with a lock-in amplifier. The surface acoustic vibration occurs in the GHz range, and the cantilever tip oscillates in the 10^2 KHz range. Hence, it can be considered that at each point of the cantilever tip vibration cycle at the ultrasonic amplitude modulation frequency in the 10^2 KHz range, the tip–sample distance varies many times, due to the sample vibration in the GHz range. The cantilever cannot vibrate at the resonator GHz frequencies due to its inertia. This results in a periodic ultrasonic force acting upon the cantilever, with a period corresponding to the ultrasonic amplitude modulation frequency, i.e. the second cantilever eigenmode.

3.2.2
Experimental Implementation of UFM

The experimental set-up for UFM can be implemented by appropriately modifying a commercial AFM [25, 34]. A schema of a UFM apparatus is shown in Fig. 3.4.

An ultrasonic piezoelement is located on the sample stage and the sample is directly bonded to the piezo using a thin layer of crystalline salol, or just honey, to ensure good acoustic transmission. In this way, longitudinal acoustic waves may be launched from the back of the sample to the sample surface. A function generator is needed to excite the piezo and generate the acoustic signal (Fig. 3.4). The ultrasonic deflection of the cantilever is monitored using the standard four-segment photodiode. As mentioned in Sect. 3.2.1, the ultrasonic signal is modulated in amplitude with a triangular or trapezoidal shape, with a modulation frequency above the AFM feedback response frequency. The UFM response (ultrasonic deflection) can be monitored with a lock-in amplifier using the synchronous signal provided by the function generator at the ultrasonic modulation frequency. In this way, contact-mode AFM topographic images and UFM images can be simultaneously recorded over the same surface region.

In addition to the described configuration, it is also possible to perform UFM by exciting the ultrasonic vibration at the tip–sample contact using a piezotransducer located at the cantilever base [35, 36]. This latter procedure has been named Waveguide-UFM. In this case, the ultrasonic vibration is propagated through the cantilever to the sample surface. Here, the cantilever tip should vibrate at the ultrasonic excitation frequency. However, if the ultrasonic frequency is sufficiently high, the amplitude of high-frequency cantilever vibration can be very small, and it has been experimentally demonstrated that a mechanical-diode response (i.e. an ultrasonic deflection of the cantilever) is activated well under these conditions [35, 36].

UFM responses are also detected in liquid environments [37]. To perform UFM in liquid, the ultrasonic piezoelectric transducer is simply attached with honey to the back of the sample-holder stage of the AFM liquid cell [37].

3.2.3
Information from UFM Data

3.2.3.1
UFM Curves

In order to study the UFM response, UFM data are typically collected in the form of *ultrasonic curves*, obtained at each surface point by monitoring the cantilever ultrasonic deflection or mechanical-diode response as a function of the ultrasonic excitation amplitude. In the following, a description of the current understanding of those curves is provided.

As discussed in Sect. 3.2.1, the UFM signal stems from the time-averaged force exerted upon a cantilever tip in contact with a sample surface when ultrasonic vibration of sufficiently high amplitude is excited at the tip–sample contact, in such a way that the tip–sample distance is varied over the nonlinear tip–sample force regime at each ultrasonic period. The forces acting at a cantilever tip in contact with a sample surface are often described in the context of continuum mechanics. In particular, a tip–sample force–indentation curve with shape as depicted in Fig. 3.5a can be derived from the Johnson–Kendall–Roberts (JKR) model that describes a sphere pressed against a flat surface. The pull-off distance is defined as the tip–sample distance at which the tip–sample contact breaks when the tip is withdrawn from the sample surface. If the tip–sample indentation is varied over the linear tip–sample force regime, as is the case for the amplitude a_o in Fig. 3.5a, the average force is F_i. If the vibration amplitude is a_1, the pull-off point is reached and the tip–sample contact is broken for a part of the ultrasonic cycle; the ultrasonic curve – cantilever deflection versus ultrasonic amplitude – shows a discontinuity at this amplitude value. Figure 3.5b schematically shows the ultrasonic deflection (UFM signal) that will be received as the surface ultrasonic vibration amplitude is linearly increased. The mechanical diode response or ultrasonic cantilever deflection experiences a discontinuity attributed to an *ultrasonic force jump* when the vibration amplitude reaches the so-called *threshold amplitude* a_1.

Figure 3.6 displays experimental UFM curves recorded on highly oriented pyrolitic graphite (HOPG) when the surface ultrasonic vibration amplitude is varied with a trapezoidal shape as indicated (lowest curve in the figure) for various initial loads. Notice that, as expected, when the initial tip–sample force is increased the ultrasonic thresholds occur at higher vibration amplitudes. The ultrasonic deflection of the cantilever is dependent on both the initial set-point force and the ultrasonic excitation amplitude. For a given set-point force, the threshold amplitude is needed to reach the pull-off point and induce the jump of the ultrasonic cantilever deflection. Consistently, the threshold amplitude increases as the set-point force is increased. From the analysis of the ultrasonic curves, information about the tip–sample interaction force can be obtained, and the elastic and adhesive properties of the tip–sample contact can be derived.

Fig. 3.5. (a) Schematic plot
of a force-indentation curve.
(b) Schematic ultrasonic
cantilever deflection
(mechanical-diode signal)
induced by out-of-plane
sample vibration of
increasing amplitude (from
Ref. [38])

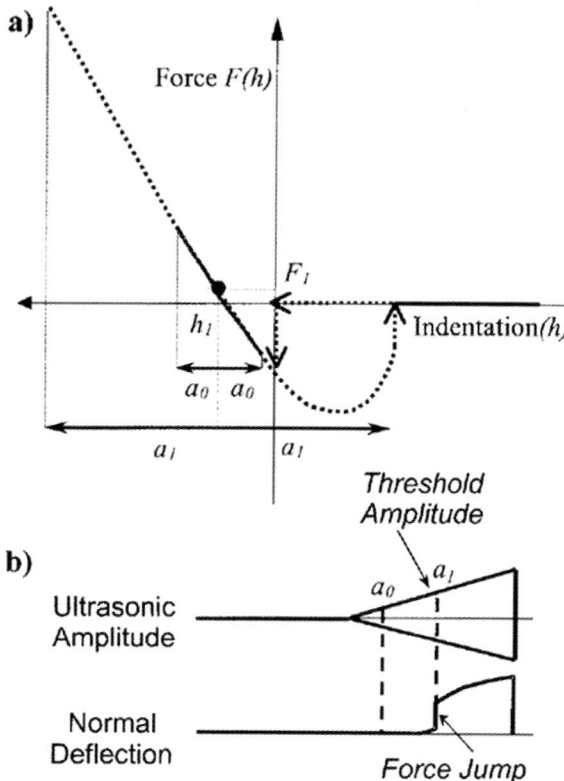

The procedure of *differential UFM* has been proposed to extract quantitative information about the sample stiffness with nanoscale resolution [38], based on the measurement of the threshold amplitudes a_i of the ultrasonic curves for two different initial tip–sample normal forces F_i. If the normal forces do not differ much, the effective contact stiffness S_{eff} can be obtained as follows

$$S_{eff}(F_{av}) = \frac{F_2 - F_1}{a_2 - a_1} \tag{3.3}$$

$$F_{av} = \frac{F_2 + F_1}{2} \tag{3.4}$$

This method has the advantage that for the derivation of the contact stiffness it is not necessary to consider the details of a specific contact-mechanics model for the tip–sample interaction.

Simulations of the UFM curves have been done introducing the concept of *modified tip–sample force curves* [38] (see Fig. 3.7). When the tip–sample distance is varied because of the excitation of ultrasound, the tip–sample interaction forces are modified because of the mechanical-diode effect. In order to simulate the UFM

Fig. 3.6. Experimental UFM curves. Cantilever deflection z_c recorded for different tip–sample forces F_n on highly oriented pyrolitic graphite (HOPG)

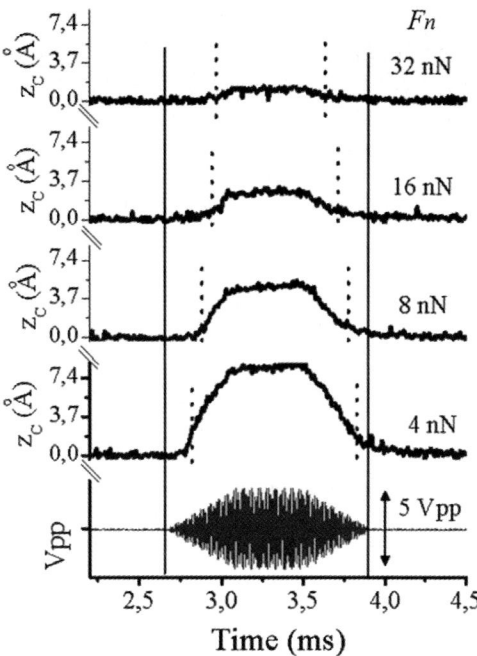

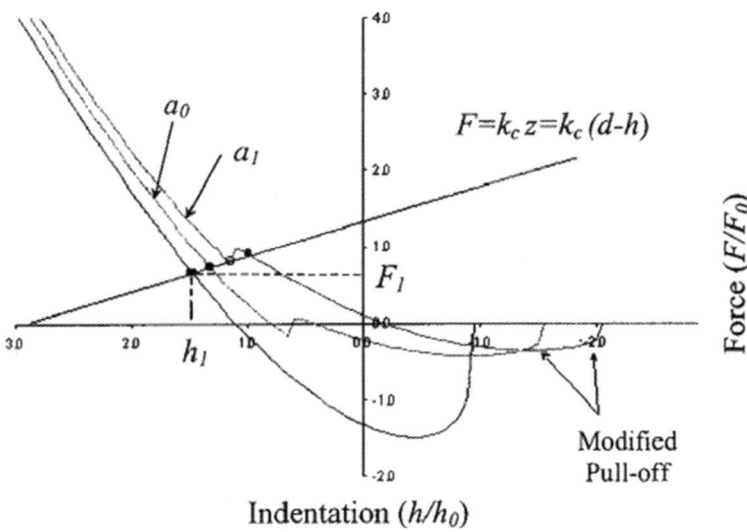

Fig. 3.7. Modified force-indentation curves (retraction branches) for a JKR solid–solid (tip–sample) contact in the presence of normal ultrasonic vibration of different amplitudes a_i. The *black line* is the force exerted by the cantilever considered as a point mass (from Ref. [38])

curves, a series of modified tip–sample force curves are generated, each of them corresponding to a specific value of the ultrasonic amplitude. In Fig. 3.7, force curves obtained for ultrasonic vibration amplitudes a_o and a_1 ($a_o < a_1$) have been plotted together with the original force curve in the absence of ultrasonic vibration, derived from the JKR model. The straight line in Fig. 3.7 represents Hooke's law, which relates the force acting on the tip to the cantilever normal deflection. Here, the cantilever tip is modeled as a point mass on a spring. The equilibrium positions of the tip in contact with the surface are obtained from the intersection of the line with the corresponding force curve, which varies depending on the ultrasonic excitation amplitude. It can be noticed that the pull-off forces and the indentation values are modified at the new modified force curves; for a given indentation value, the new force value is generally higher than the one obtained for zero ultrasonic amplitude. It may also be noted that for the amplitude a_1, which corresponds to the threshold amplitude, there are two solutions for the new equilibrium position of the tip, which accounts for the discontinuity in the cantilever displacement or force jump at this amplitude value.

The stiffness values chosen to generate the original JKR curve can be derived from the application of differential UFM to the simulated ultrasonic curves, giving confidence in the reliability of this method [38]. From the analysis of the dependence of the simulated UFM curves on the sample Young's modulus and adhesion, it can be concluded that (1) the threshold amplitude increases when the normal force is increased, the Young's modulus is low or the work of adhesion is high, and (2) the force jump increases when the Young's modulus is low and the work of adhesion is high [34]. Analysis of the ultrasonic curves with other contact models for the tip–sample interaction yield similar conclusions [39].

Figure 3.6 shows that when the ultrasonic amplitude at the tip–sample contact is linearly decreased, the cantilever returns to its original equilibrium position, experiencing a sudden jump-in in the force. The ultrasonic amplitude at which the jump-off in the UFM response is observed when increasing the excitation amplitudes, i.e. the threshold amplitude, is different from that at which the jump-in occurs when the amplitude is decreased. A method has been proposed [40] to determine both the sample elastic modulus and the work of adhesion from such force jumps in the ultrasonic curves. In [41, 42] the area between experimental ultrasonic curves obtained increasing and decreasing the ultrasonic amplitude – due to the different jump-off and jump-in threshold amplitudes – is defined as the *UFM hysteresis area* (UH), and related to the local adhesion hysteresis. Correlations between the adhesion hysteresis and the local friction were theoretically and experimentally investigated [43, 44]. Using the ability of UFM to provide information about local adhesion hysteresis, the protein–water binding capacity was investigated in protein films at different relative humidities, with the proteins in hydrated and dehydrated states [45].

The transfer of ultrasound to an AFM cantilever in contact with a sample surface has also been evaluated by numerically solving the equation of motion, taking into account the full nonlinear force curve and considering that the cantilever is a rectangular beam that supports flexural vibrations. By this procedure, the change in the mean cantilever position that results from the nonlinear tip–sample interactions is also demonstrated [46].

3.2.3.2
UFM Images

The ability of UFM to map material properties simultaneously with the acquisition of contact-mode AFM topographic images over the same surface area has been extensively demonstrated. Given the set-point force and the maximum ultrasonic amplitude, if it is assumed that the tip–sample adhesion is invariant, the UFM signal corresponding to a locally stiff region is large in magnitude, and gives rise to a bright contrast in the UFM image. Since the UFM signal depends on both adhesion and elasticity, the contrast in the UFM images must be carefully analyzed. The UFM brings advantages for the study of both soft and hard materials. In the presence of surface out-of-plane ultrasonic vibration of sufficiently high amplitude, nanoscale friction reduces or vanishes [47, 48], which facilitates the inspection of soft samples without damage. The elastic properties of hard materials can also be investigated by UFM. Because of the inertia of the cantilever, in the presence of surface ultrasonic vibration a cantilever tip effectively indents hard samples [20, 25].

Important applications of UFM rely on its capability to provide subsurface information. The subsurface sensitivity of the UFM has been experimentally demonstrated [20, 25]. Subsurface dislocations in HOPG have been observed and manipulated (see Fig. 3.8) using the ultrasonic AFM [25, 49, 50].

The penetration depth in AFM with ultrasound excitation is determined by the contact-stress field, which increases when the set-point force and the ultrasonic

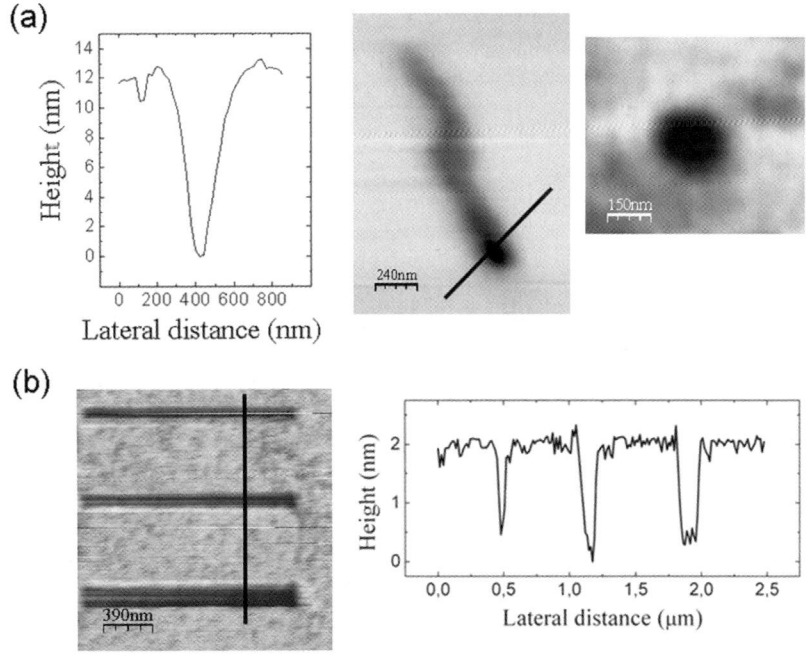

Fig. 3.8. Machining of nanotrenches and holes on silicon using a UFM (from Ref. [59])

amplitude are increased. The penetration depth and the minimum detectable over-layer thickness in Atomic Force Acoustic Microscopy are defined in [51], on the basis of the detectable minimum contact stiffness change. In [52] it is concluded from the change in contact stiffness of SiO_2/Cu that buried void defects (≈ 500 nm) at nm distance from the dielectric surface can be detected using the UFM. In [20], a GaAs grating buried under a polymeric layer is clearly imaged using Scanning Local Acceleration Microscopy. Changes in contact stiffness of cavities in Si within about 200 nm from the Si(100) surfaces have been detected using UFM [53]. In [30], Au particles with a diameter of 15–20 nm buried under a 500-nm polymeric film have been observed using Scanning Near-Field Ultrasound Holography. UFM has been applied to characterize defects such as debonding, delaminations, and material inhomogeneities [54–56]. Subsurface information is also apparent in Resonant Difference-Frequency Atomic Force Ultrasonic Microscopy (RDF-AFUM) [31]. The UFM has been used to map stiffness variations within individual nanostructures such as quantum dots [57] or nanoparticles (NP) [58].

3.2.4
Applications of UFM in Nanofabrication

Ultrasonic AFM techniques provide a means to monitor ultrasonic vibration at the nanoscale, and open up novel opportunities to improve nanofabrication technologies [49, 59]. As discussed above, in the presence of ultrasonic vibration, the tip of a soft cantilever can dynamically indent hard samples due to its inertia. In addition, ultrasound reduces or even eliminates nanoscale friction [47, 48]. Typical top-down approaches that rely on the AFM are based on the use of a cantilever tip that acts as a plow or as an engraving tool. The ability of the AFM tip to respond inertially to ultrasonic vibration excited perpendicular to the sample surface and to indent hard samples may facilitate nanoscale machining of semiconductors or engineering ceramics in a reduced time.

Figure 3.8 demonstrates the *machining of nanotrenches and holes* on a silicon sample in the presence of ultrasonic vibration. Interestingly, no debris is found in the proximity of lithographed areas. Figure 3.8a refers to results performed using a cantilever with nominal stiffness in the 28–91 N m^{-1} range and a diamond-coated tip. Figure 3.9b refers to results achieved using a cantilever with nominal stiffness 0.11 Nm^{-1} and a Si_3N_4 tip; in the absence of ultrasound, it was not possible to scratch the Si surface using such a soft cantilever. In the machining of soft materials, as for instance plastic coatings, the ultrasonic-induced reduction of nanoscale friction may permit eventual finer features and improved surface quality in quasi-static approaches.

In bottom-up approaches, ultrasound may assist in *self-assembly or AFM manipulation of nanostructures* [49]. Effects such as sonolubrication and acoustic levitation have been studied at the microscale. These phenomena may facilitate a tip-induced motion of nanoobjects. In the manipulation of nanoparticles on ultrasonically excited surfaces with the tip of an AFM cantilever, both the tip–particle and particle–surface frictional properties change [59–61]. Moreover, the excitation of NP high-frequency

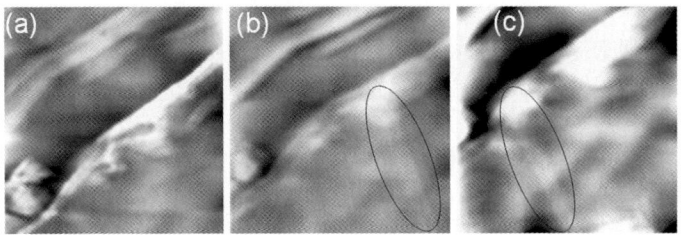

Fig. 3.9. (**a**) Topography on the HOPG surface (700 × 700) nm. (**b, c**) Ultrasonic-AFM images recorded in sequence over nearly the same surface region as in (**a**). A subsurface dislocation not noticeable in the topographic image is enclosed by the ellipse in (**b**) and (**c**). In (**c**) the dislocation is laterally displaced (from Ref. [49])

internal vibration modes may also modify the NP dynamic response, and introduce novel mechanisms of particle motion. Using the UFM mode for manipulation allows us to monitor the mechanical diode response of the cantilever while individual nanoparticles are being laterally displaced over a surface by tip actuation, and receive information about the lateral forces exerted by the tip.

Eventually, it should be pointed out that the sensitivity of ultrasonic-AFM to subsurface features makes it feasible to monitor *subsurface modifications* [49]. We have recently demonstrated that actuation with an AFM tip, in the presence of ultrasonic vibration can produce stacking changes of extended grapheme layers, and induce permanent displacements of buried dislocations in Highly Oriented Pyrolitic Graphite. This effect is illustrated in Fig. 3.10. In the presence of normal surface ultrasonic vibration, both AFM and lateral force microscopy (LFM) images reveal subsurface features [49,59]. Subsurface modification was brought about in this case by scanning in contact mode, with high set-point forces, and high surface ultrasonic excitation amplitudes [49].

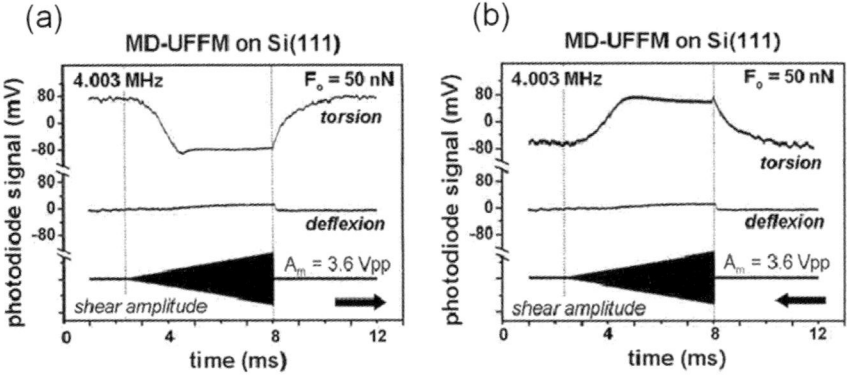

Fig. 3.10. Experimental evidence of the lateral MD effect (see text) (from Ref. [26])

3.3
Mechanical Diode Ultrasonic Friction Force Microscopy

3.3.1
The Lateral Mechanical Diode Effect

A lateral MD effect has also been experimentally observed. Similar to the UFM cantilever deflection that switches on in the presence of out-of-plane surface ultrasonic vibration of sufficiently high amplitude, an additional torsion of the cantilever is activated when the cantilever tip is in contact with a sample surface and scans laterally over the surface at low frequency. This is done in the presence of shear surface ultrasonic vibration of sufficiently high amplitude [24, 26].

The lateral MD-effect is exploited in Lateral Scanning Acoustic Force Microscopy (LFM-SAFM) [24] to obtain information about the amplitude and phase velocity of in-plane polarized SAWs. Recently, the technique of MD-UFFM has been proposed [26] to study the shear contact stiffness and frictional response of materials at the nanoscale. In MD-UFFM, shear ultrasonic vibration is excited at a tip–sample contact using a shear piezoelectric element attached to the back of the sample. Shear acoustic waves originated at the piezo propagate through the sample to reach the tip–surface contact area. An ultrasonic-induced additional torsion of the cantilever or MD-UFFM cantilever torsion is observed while the cantilever tip in contact with the surface is laterally scanning at low frequencies [26]. Experimental evidence of the lateral MD effect is provided in Fig. 3.10.

Figure 3.10a, b show typical MD-UFFM cantilever responses recorded on a Si sample in forward and backward scans respectively, in the presence of shear ultrasonic vibration at the tip–sample contact modulated in amplitude with a triangular shape. In both scanning directions, the ultrasound-induced torsion of the cantilever diminishes initially due to friction. As the shear ultrasonic excitation amplitude is increased, the MD-UFFM cantilever torsion increases in magnitude until a critical shear ultrasonic amplitude is reached, after which it remains invariant or decreases.

The lateral MD effect can be understood by considering the lateral ultrasonic force emerging from the interaction of the tip with the lateral surface sample potential [26].

Figure 3.11 illustrates a physical explanation for the MD effect, in agreement with experimental results [26]. In the absence of ultrasound, when scanning at low velocity on a flat surface, the cantilever is subjected to an initial torsion due to friction. At the typical low AFM scanning velocities, nanoscale friction proceeds by the so-called stick-slip mechanism [62]. At a sticking point, the tip is located at a minimum of the sum of the periodic surface potential and the elastic potential of the cantilever; the lateral displacement of the cantilever support relative to the sample introduces an asymmetry in the total potential that facilitates the jumping of the tip to the next energy minimum site. Most of the time, the tip sticks to a surface point, and then slips to a next sticking point with some energy dissipation. In Fig. 3.11, E corresponds to the total potential acting upon the tip when scanning forward at low velocity. Because of this potential, the tip is subjected to the force given by the derivative curve. When the tip lies in the minimum energy site crossed by the dashed

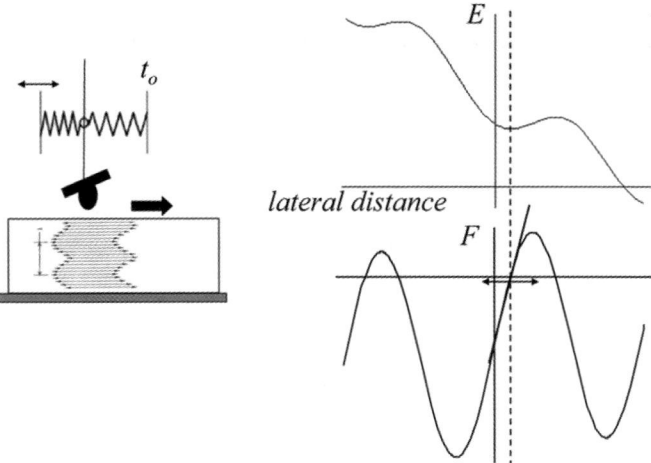

Fig. 3.11. Physical model for MD-UFFM. The surface atoms are laterally displaced due to the shear surface vibration, but due to its inertia, the cantilever cannot follow the surface lateral displacements at ultrasonic frequencies not coincident with a torsional cantilever resonance (from Ref. [27])

line, the corresponding force is zero. Because of the different time-scales, we may consider that the tip–sample potential brought about by scanning at low velocity is frozen during a shear ultrasonic vibration period. The shear ultrasonic wave transmitted through the sample introduces in-plane oscillations at the sample surface, in the direction perpendicular to the long cantilever axis. Atomic species within the tip–sample contact area are subjected to shear ultrasonic vibration, but the inertia of the cantilever hinders its out-of-resonance rotation. The lateral displacement of the surface atoms relative to the tip leads to a time-dependent variation of the total potential acting upon the tip at ultrasonic time scales. We define the lateral ultrasonic force as the average force that acts upon the cantilever during each ultrasonic cycle in the presence of shear ultrasonic vibration,

$$F_{ult}^{l}(x_l, A) = \frac{1}{T_{ult}} \int_0^{T_{ult}} F\left(x_l - A * \cos\left(\frac{2\pi}{T_{ult}}t\right)\right) dt \qquad (3.5)$$

where x_l is the lateral equilibrium location of the tip in the presence of lateral ultrasonic vibration of amplitude A, which defines the new equilibrium torsion of the cantilever, A is the amplitude of shear ultrasonic vibration, and T_{ult} refers to the ultrasonic time period. Once a critical lateral vibration amplitude is reached, sliding sets in, and the MD-UFFM signal does not increase further. A study of the MD-UFFM cantilever torsion may provide information about the sample shear stiffness and frictional response.

In Fig. 3.10, a vertical lift-off of the cantilever or *MD-UFFM cantilever deflection* is observed as a result of the shear surface ultrasonic vibration. Samples such as silicon are known to be covered by a liquid layer under ambient conditions. In such samples, the observed lift-off may originate from an elastohydrodynamic response of an ultrathin viscous layer sheared at the tip–sample contact at ultrasonic velocities [16,63]. The study of the MD-UFFM cantilever deflection may provide information about the elastohydrodynamic properties of thin confined lubricant layers.

3.3.2
Experimental Implementation of MD-UFFM

The experimental set-up for Mechanical Diode Ultrasonic Friction Force Microscopy measurements can be implemented by appropriately modifying a commercial AFM [26]. The set-up required for MD-UFFM is similar to that required for the UFM substituting the longitudinal ultrasonic piezoelectric transducer with a shear-wave type. The shear-wave piezotransducer is mounted below the sample with its polarization perpendicular to the longitudinal axis of the cantilever. The sample should be attached to the sample with an appropriate couplant, as for instance crystalline salol. The changes in the cantilever torsion due to the lateral MD effect can be monitored in both forward and backward scans using the laser deflection method with a standard four-segment photodiode, simultaneously with the acquisition of contact-mode topographic images. MD-UFFM images can be collected by modulating the amplitude of the shear ultrasonic excitation and using a lock-in amplifier to detect the MD-UFFM signal. We distinguish torsion and deflection MD-UFFM modes, depending on whether the shear-ultrasonic-vibration-induced cantilever torsion or deflection response is studied.

In a shear-wave piezoelectric transducer, parasitic out-of-plane vibration may arise due to the existence of boundaries, etc. In the presence of out-of-plane ultrasonic vibration of sufficiently high amplitude, the normal mechanical diode effect described in Sect. 3.2.1 would lead to the excitation of an additional out-of-place cantilever deflection related to the sample elastic properties. In the absence of out-of-plane ultrasonic vibration, but with shear ultrasonic vibration excited on the sample surface, a lift-off or deflection of the cantilever is expected as a result of elastohydro-dynamic lubrication effects of ultrathin viscous layers compressed at the tip–sample contact [15,26]. In order to distinguish between those two effects, the UFM response of the sample under study and the used shear-wave piezotransducer should be very well characterized before establishing definitive conclusions from MD-UFFM measurements.

3.3.3
Comparison of MD-UFFM with UFFM and TRmode AFM

In UFFM, also named Lateral-Acoustic Friction Force Microscopy (L-AFAM) or Resonant Friction Force Microscopy (R-FFM) [15–17,64,65] surface in-plane vibration polarized perpendicular to the long axis of the cantilever is excited with a shear-wave piezotransducer bonded to the back of the sample, as in MD-UFFM. UFFM monitors the torsional vibration of the cantilever at the shear ultrasonic frequency excited, being the cantilever tip in contact with the sample surface. At shear ultrasonic frequencies, the torsional cantilever vibration is only significant near the

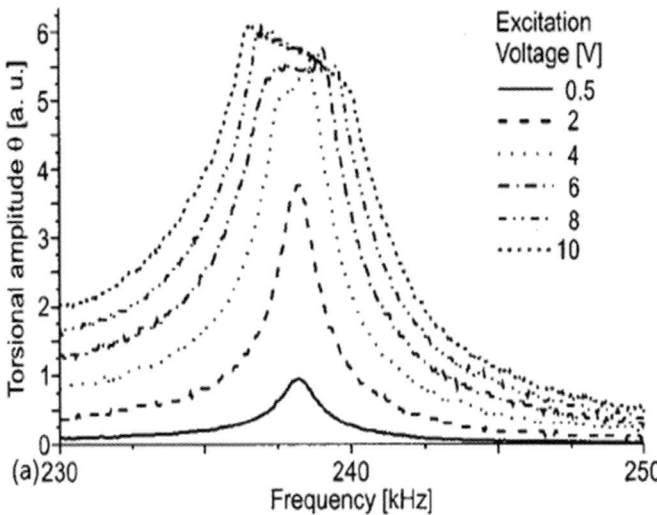

Fig. 3.12. Torsional vibration amplitude of the cantilever as a function of the excitation frequency. Measurements on bare silicon. The different curves correspond to increasing excitation voltages applied to the shear-wave piezotransducer (from Ref. [17])

cantilever torsional contact resonances. Figure 3.12 shows UFFM measurements at a torsional contact resonance, cantilever torsional vibration amplitude versus surface shear ultrasonic excitation frequency, for different shear ultrasonic excitation amplitudes. At low shear excitation voltages, the resonance curve has a Lorentzian shape with a well-defined maximum [17]. The cantilever behaves like a linear oscillator with viscous damping, with the AFM tip stuck to the sample surface and following the surface motion. Above a critical surface shear ultrasonic vibration amplitude, typically 0.2 nm, the amplitude maximum of the resonance curves does not increase further, and the shape of the resonance curves change indicating the onset of sliding friction [17]. The information obtained from the analysis of the resonance curves in Fig. 3.12 supports the interpretation of torsional MD-UFFM curves discussed in Sect. 3.3.1. In the MD-UFFM responses in Fig. 3.10, two different regimes are also distinguished. At low shear excitation voltages, the lateral mechanical diode effect leads to an increasing lateral ultrasonic force due to increasing shear vibration amplitude. Above a critical surface shear ultrasonic vibration amplitude, a maximum ultrasonic force is reached, and sliding begins.

In TR-AFM [66–71] torsional vibrations of the cantilever are excited via two piezoelectric elements mounted beneath the holder of the chip, which vibrate out-of-phase, in such a way that they generate a rotation at the long axis of the cantilever. The TR-mode can be implemented in contact, near-contact, and noncontact modes, and provides information about surface shear elasticity, viscoelasticity, and friction. When operating in contact, torsional cantilever resonance curves such as those in Fig. 3.12 have also been observed [68]. In the TR mode, the torsional resonance amplitude (or phase) can be used to control the feedback loop and maintain the tip–sample relative position through lateral interaction. Frequency modulation procedures have also been implemented for TR-AFM measurements [72].

3.3.4
Information from MD-UFFM Data

3.3.4.1
MD-UFFM Curves

As in UFM, in MD-UFFM the data are typically collected in the form of *ultrasonic curves*, obtained by monitoring the mechanical-diode cantilever responses as a function of the shear ultrasonic excitation amplitude.

As discussed in Sect. 3.3.1, the torsional MD-UFFM response stems from the lateral time-averaged force exerted upon a cantilever tip in contact with a sample surface, and scanning laterally over the surface at low typical AFM velocities, when shear ultrasonic vibration of sufficiently high amplitude is excited at the tip–sample contact. Properties such as shear contact stiffness, shear strength, and friction of surfaces at a nanometer scale are obtained in lateral force microscopy, also named Friction Force Microscopy (FFM) [62, 73]. In MD-UFFM, the excitation of shear ultrasonic vibration at the tip–sample contact leads to relative tip-surface velocities of mm s^{-1} or larger, and the evaluation of these properties in these different experimental conditions may bring additional light to the understanding and control of nanoscale friction. Also, it is expected that MD-UFFM will provide subsurface information related to subsurface inhomogeneities.

In the realm of continuum mechanics, for a sphere-plane geometry, the lateral stiffness of a contact is given by [74]:

$$K_{contact} = 8a_c G^* \tag{3.6}$$

where a_c is the contact radius, and G^* is the reduced shear modulus, defined as:

$$\frac{1}{G^*} = \frac{2 - v_t}{G_t} + \frac{2 - v_s}{G_s} \tag{3.7}$$

with G_t, G_s, v_t, v_s being the shear moduli and the Poisson's ratios of the tip and the sample, respectively. This equation is valid for various continuum elasticity models and does not depend on the interaction forces. For small displacements it is reasonable to assume that there is no change in the contact area.

The elastic response of the tip–sample contact in shear can be described by a series of springs. A lateral displacement of the sample Δz is distributed between three springs:

$$\Delta x = \Delta x_{contact} + \Delta x_{tip} + \Delta x_{cantilever} \tag{3.8}$$

And the lateral force F_{lat} at the contact is given by

$$F_{lat} = k_{eff}\Delta x \tag{3.9}$$

with K_{eff} the effective contact stiffness,

$$\frac{1}{K_{eff}} = \frac{1}{K_{contact}} + \frac{1}{K_{tip}} + \frac{1}{c_L} \tag{3.10}$$

where $K_{contact}$ is the lateral contact stiffness, K_{tip} is the lateral elastic stiffness of the tip, and c_L is the lateral spring constant of the cantilever, considered as a point mass. For most commercial cantilevers, only the torsional spring constant is relevant for the estimation of c_L. In FFM experiments, the lateral stiffness of the tip is comparable or even smaller than the lateral stiffness of the cantilever [75].

For larger displacements at the contact, the threshold force to overcome the static friction is reached, and the tip starts to move. In FFM, K_{eff} can be measured from the so-called friction force loops, lateral force vs. lateral position, in which a sticking part where the tip essentially stays at the same position and a sliding part can be easily distinguished. K_{eff} is given by the slope of the sticking part.

The shear strength can be defined as:

$$F_f = \tau A = \tau \pi a_c^2 \tag{3.11}$$

where F_f is the friction force, and A is the contact area. From Eqs. (3.11), (3.6), and (3.7) we obtain an expresion for the shear strength, independent of the contact diameter a_c.

$$\tau = \frac{64 G^{*2} F_f}{\pi (K_{contact})^2} \tag{3.12}$$

It is well known from FFM studies that at typical low AFM scanning velocities, nanoscale friction proceeds by stick-slip. Once static friction at the tip–sample contact is overcome, the tip "slips" to a next static position and "sticks" there until the surface displacement is again large enough so that a threshold force needed for it to slip is reached again. Stick-slip also occurs at the micro and macro scales and can be observed whatever the chemical nature of the solids in contacts, and the state of their surfaces, provided that the loading system is soft enough. Stick-slip friction with atomic periodicity has been demonstrated in numerous LFM experiments with atomic resolution, in which the lateral force exhibits a periodic, sawtooth like behavior [62]. According to the Tomlinson model, the tip is considered to move in the periodic potential field formed by the substrate lattice while being dragged along the surface by means of spring-type interactions. Atomic-scale stick-slip is usually limited to the low load regime, and sharp tips, although, atomic-scale stick slip at high loads have also been observed. The latter may be restricted to layered materials or to the presence of some lubricating contamination films. In the Prandtl–Tomlinson model, the total potential experienced by the tip is given by:

$$V_{tot}(x,t) = -\frac{E_o}{2} cos \frac{2\pi x}{a} + \frac{1}{2} K_{eff}(x - vt)^2 \tag{3.13}$$

where E_o is the peak-to-peak amplitude of the surface potential, a is the lattice constant of the surface, K_{eff} is the effective lateral spring constant and v is the velocity of the sample.

The model for MD-UFFM described in Sect. 3.3.2 is based on the Prandtl Tomlinson model. This accounts qualitatively quite well for the experimental results (see Fig. 3.11 and related text). In principle, the application of this model allows us to obtain K_{eff}, F_f, and τ from MD-UFFM data, and also learn about the relationship of these magnitudes to the surface lateral potential, its amplitude E_o, and periodicity a, and the mechanisms of friction in the presence of shear ultrasonic vibration at the

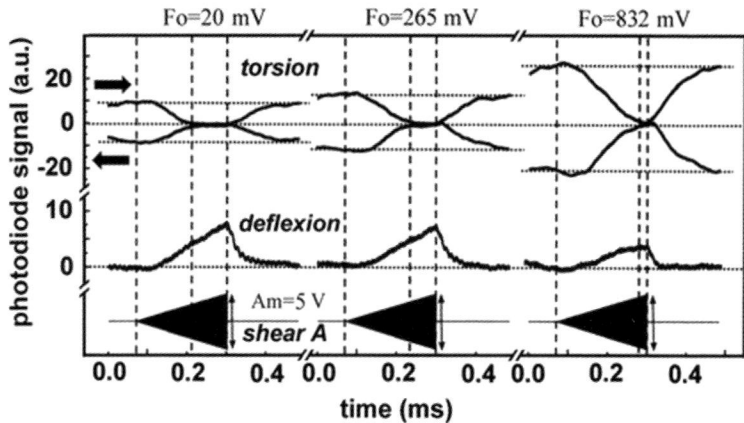

Fig. 3.13. MD-UFFM responses on Si(111) for different normal loads (from Ref. [26])

tip–sample contact. In FFM atomic-scale stick-slip friction experiments performed at low loads, the values obtained for K_{eff} suggest that the area of contact consists of just a few atoms, precluding the application of continuum mechanical models in those cases.

Figure 3.13 shows MD-UFFM responses on Si(111) recorded at different normal set-point forces, including the torsion curves recorded in both forward and backward scans. For higher normal loads, the magnitude of the torsional MD signal increases, and a higher critical shear ultrasonic amplitude is required to reach the flat torsion regime attributed to sliding. These results are also in agreement with the model sketched in Fig. 3.11. For higher loads, the magnitude of the surface interatomic potential is expected to be larger [76].

In Fig. 3.13, the distance between the torsion curves recorded in forward and backward scans is proportional to the magnitude of the friction force. The results indicate that friction reduces as a result of the excitation of shear ultrasonic vibration at the tip–sample contact, and that in this case friction vanishes in the flat MD torsional response regime. Physically, the onset of a lateral ultrasonic force is necessarily related to a reduction of friction (see Fig. 3.11). The effect might be related to the observations. There it was concluded that a cantilever may exhibit apparent stick-slip motion, and hence reveal a nonzero mean friction force, even when the tip–surface contact is completely thermally lubricated by fast activated jumps of the tip apex, back and forth between the surface potential wells. Even though, as mentioned before, in MD-UFFM, the excitation of shear ultrasonic vibration at the tip–sample contact leads there to relative tip–surface velocities of the order of mm s^{-1} or larger within the contact, it is still the displacement of the position of the cantilever center of mass relative to the surface that determines the contact velocity.

The lift-off (deflection) signals that accompany the MD torsional response in Fig. 3.13 have been attributed to the presence of an ultrathin viscous liquid layer at the tip–sample contact that develops hydrodynamic pressure when sheared at ultrasonic velocities [15]. The shape of those lift-off curves is essentially different from the typical UFM MD deflection response that results from the excitation of normal ultrasonic vibration [38]. In the MD-UFFM case, the cantilever deflection increases linearly as the shear ultrasonic vibration amplitude is increased, and no apparent jump-off is noticeable. Slight deviations of the linear shape of the deflection curve when the maximum deviation of the initial cantilever torsion is reached may be related to a coupling of the cantilever lateral and vertical motions at the onset of the sliding regime. The presence of a squeezed liquid layer at the Si surface–Si tip contact has been previously considered to explain a reduction of friction in ambient conditions as a result of the excitation of normal ultrasonic vibration at amplitudes not sufficiently large to break the tip–sample contact during the ultrasonic period [47]. However, such a lift-off has not been observed when performing MD-UFFM experiments on Si in a liquid environment [37]. Figure 3.14a, b shows lateral mechanical diode responses – MD-UFFM signals – measured on silicon, in milliQ water. The torsion MD-UFFM curves in liquid are similar as in air, although in the liquid environment they appear considerably noisier [37]. A lift-off MD-UFFM deflection signal has also not been observed in MD-UFFM experiments performed on highly oriented pyrolitic graphite in air either [78].

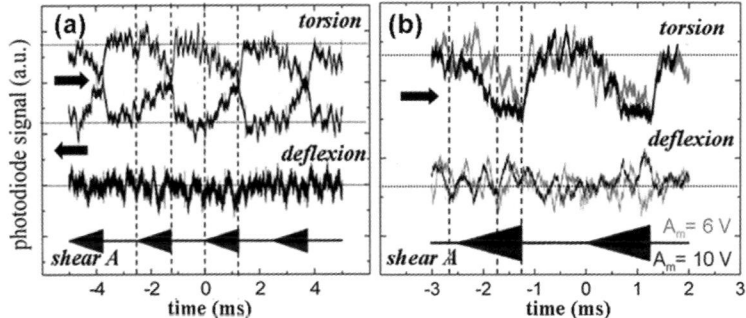

Fig. 3.14. MD-UFFM on Si, in milliQ water (from Ref. [37]). (**a**) Forth and backward torsion and deflection MD-UFFM signals; scanning velocity 2.6 μm s^{-1}; set point force 1.06 V; shear ultrasonic vibration 4.040 MHz; maximum ultrasonic amplitude Am = 6 V. (**b**) Forward torsion and deflection MD-UFFM signals recorded as in (**a**), with Am = 6 V (*grey curves*) and Am = 10 V (*black curves*)

3.3.4.2
MD-UFFM Images

As demonstrated in Fig. 3.15, MD-UFFM can also be implemented in an imaging mode, using a lock-in amplifier to monitor the signal at the amplitude modulation frequency.

Figure 3.15a–d shows FFM images in forward (a) and backward (b) scans, an ultrasonic MD-UFFM torsion image (c) and ultrasonic MD-UFFM curves (d) recorded at different points on the same surface region. As in UFFM [15], MD-UFFM images are independent of the scanning direction, i.e. not influenced by topography-induced lateral forces. Whereas Fig. 3.15 evidences the possibility to map surface properties in MD-UFFM, a precise interpretation of the MD-UFFM

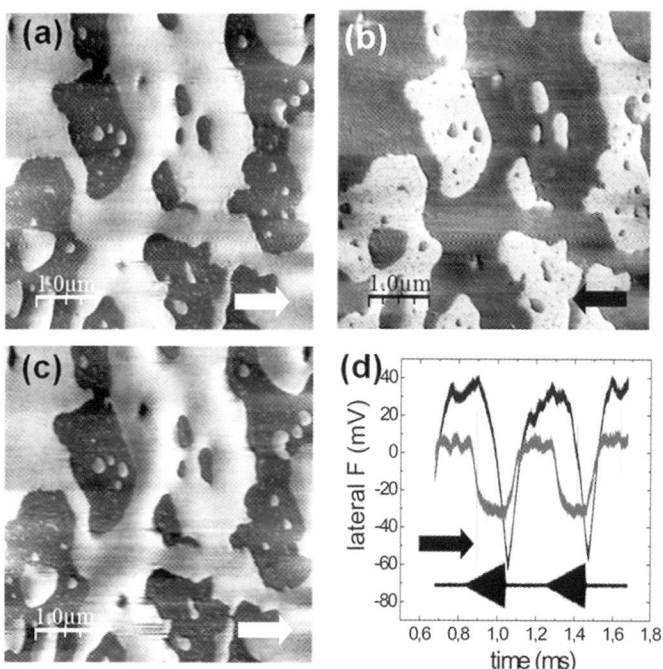

Fig. 3.15. Octadecylamine on mica. (**a, b**) FFM images in the absence of ultrasound. (**c**) MD-UFFM image on the same surface region. (**d**) Torsional MD-UFFM curves on different surface points measured while recording (**c**)

contrast in Fig. 3.15 is nevertheless, not straightforward, and deserves further investigation.

Summarizing, MD-UFFM is an interesting new technique, based on the study of the lateral mechanical diode cantilever response in the presence of shear surface ultrasonic vibration. Although in a very incipient state of development, the technique shows promise for the measurement of shear elasticity, shear strength and friction at the nanometer scale, to probe the surface interatomic potential, for investigation of the atomistic mechanisms involved in nanoscale tribology, the study of elastohydro-dynamic lubrication effects in confined layers at nanogaps, and for the characterization of boundary lubricants, etc.

3.4
Heterodyne Force Microscopy: Beats at Nanocontacts

3.4.1
Beats at Nanocontacts

If at a nanocontact, we excite vibration of frequency ω_1 at one end, and vibration of frequency ω_2 at the other end, the excitation frequencies being different but close to each other ($\omega_1 \neq \omega_2$; $\omega_1 \approx \omega_2$), the separation between both ends d will vary periodically with time, one cycle of this variation including many cycles of the basic vibrations at both ends, and with a frequency equal to the average of the two combining frequencies. The phenomena is actually the description of a beating effect [78] applied to the nanocontact. If y_1 and y_2 are the positions of each nanocontact end,

$$y_1(t) = Asin\omega_1 t \tag{3.14}$$

$$y_2(t) = d_o + Asin\omega_2 t \tag{3.15}$$

d_o its separation in the absence of vibration, and A the vibration amplitude of each end, then, the separation of both ends will vary with time according to:

$$d(t) = y_2(t) - y_1(t) = d_o + 2A \, sin \left(\frac{\omega_2 - \omega_1}{2} t \right) cos \left(\frac{\omega_2 + \omega_1}{2} t \right) \tag{3.16}$$

Eq. 3.16 holds in fact for any values of ω_1 and ω_2, but the description of the beat phenomenon is physically meaningful only if $|\omega_2 - \omega_1| << \omega_2 + \omega_1$. Then, over a substantial number of cycles, the vibration approximates to sinusoidal vibration with constant amplitude and with frequency $(\omega_2 + \omega_1)/2$.

The term $cos \left(\frac{\omega_2 + \omega_1}{2} t \right)$ describes the rapidly oscillating factor in Eq. 3.16, and will always lie between the limits ± 1. The distance between the two ends in the nanocontact will vary between minimum $d_o - 2A$ and maximum $d_o + 2A$ values at a frequency given by $|\omega_2 - \omega_1|$, i.e. at the beat frequency.

If we consider now that the nanocontact is that formed by the tip of an AFM cantilever and a sample surface (see Fig. 3.16), the beat effect implies that the tip–sample distance varies between a minimum value and a maximum value at the beat frequency in the case where we simultaneously excite ultrasonic vibration at the tip and the sample surface at slightly different frequencies. Notice that the beat frequency is in fact much smaller than the actual tip and sample vibration frequencies. Hence, if the tip–sample distance variation in the beats is such that the tip–sample force remains in the linear regime, if we try to detect the force that acts upon the cantilever at the beat frequency, we will find that the tip–sample distance, and hence the force upon the cantilever, is varying from the minimum value to the maximum value many times in the time scale that we will use to track the beat frequency, and we will only be able to detect the averaged value of this force, which will be null in the linear case. However, if the variation of tip–sample distance during a beat cycle extends to the nonlinear tip–sample force regime, when trying to measure the force acting upon the cantilever at the beat frequency, we will detect the average force, which will change with the periodicity of the beats.

In Heterodyne Force Microscopy [27], ultrasound is excited both at the tip (from a transducer at the cantilever base) and at the sample surface (from a transducer at the back of the sample) at adjacent frequencies, and mixed at the tip–sample gap. If d_o is the initial tip–sample indentation, and the vibration amplitude A of the tip and the surface is the same, the tip–sample force will vary according to Eq. (3.16), assuming that for instance ω_1 corresponds to the frequency vibration of the sample, and ω_2 to the frequency vibration of the tip. In HFM, the modulation frequency is usually chosen much lower than the first cantilever resonance frequency. The cantilever will not be able to follow the force exerted at the frequency $(\omega 1 + \omega 2)/2$ due to its inertia. However, provided that the low-frequency varying tip–sample separation is large

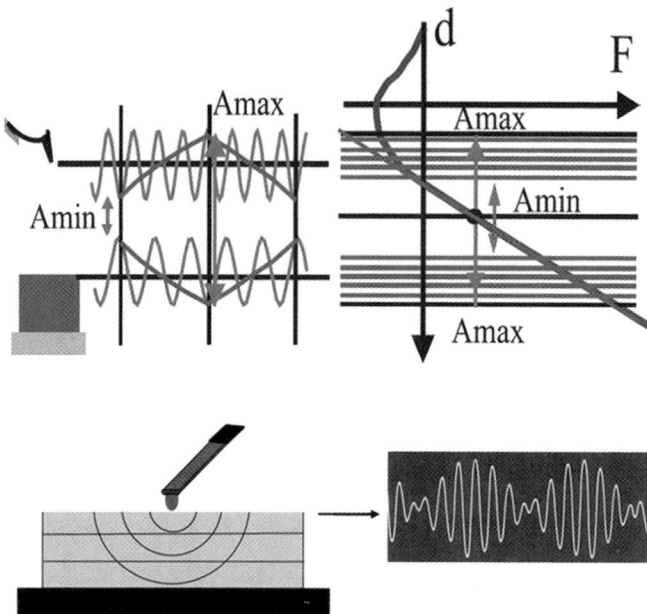

Fig. 3.16. Beats at the tip–sample contact

enough to cover the nonlinear range of the tip–sample interaction force, an ultrasonic force (stronger for larger amplitudes) will act upon the cantilever and displace it from its initial position. Owing to the varying ultrasonic force, the cantilever vibrates at the difference mixed frequency.

In principle, even if the modulation frequency is chosen higher than the first cantilever resonance [30] or coincident with a cantilever contact resonance [31] the beat effect should also lead to the activation of an ultrasonic force at the beat frequency, provided that the tip–sample distance is varied over the nonlinear tip–sample force regime as a result of the tip and sample high frequency vibration. Also, the effect should similarly work if the cantilever is operated in a dynamic AFM mode.

An important feature of the beat effect is that it facilitates the monitoring of phase shifts between tip and sample ultrasonic vibrations with an extremely high temporal sensitivity. In HFM, it has been demonstrated that small differences in the sample dynamic viscoelastic and/or adhesion response to the tip interaction result in a shift in phase of the beat signal that is easily monitored. In this way, HFM makes possible the study of dynamic relaxation processes in nanometer volumes with a time-sensitivity of nanoseconds or even better.

3.4.2
Experimental Implementation of HFM

The experimental set-up for HFM is shown in Fig. 3.17. The technique can be implemented by appropriately modifying commercial AFM equipment [27]. For HFM,

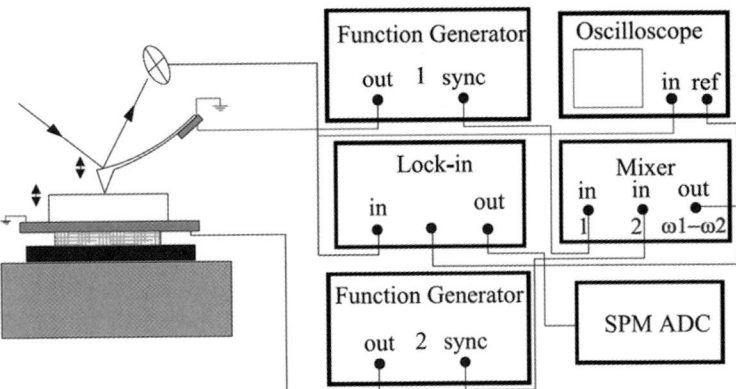

Fig. 3.17. Set-up for HFM (from Ref. [27])

lead zirconium titanate (PZT) ceramic piezos are attached to the sample and the tip holder. Both the sample and the cantilever are bonded to the corresponding piezos using a thin layer of crystalline salol (phenyl salicilate). Two function generators are needed to simultaneously excite sinusoidal vibration of the sample surface and the cantilever tip at two adjacent ultrasonic frequencies. In [27], sample and tip vibrations were excited at frequencies in the MHz range, differing in some KHz. The synchronous signals from both generators at the high-frequency excitation can be electronically mixed using a simple electronic mixer, which provide as an output a reference signal at the difference frequency. By means of the lock-in amplifier, the vibration of the cantilever at the beat frequency, i.e. the HFM signal in this case, can be easily monitored in amplitude and phase.

The recently proposed technique of Scanning Near-Field Ultrasound Holography [30] is implemented in a similar way as HFM, choosing a difference frequency (beat frequency) in the range of hundreds of KHz, above the first cantilever resonance frequency. In Resonant Difference Frequency Atomic Force Ultrasonic Microscopy [31], the difference frequency (beat frequency) is chosen to be coincident with a high-order cantilever contact resonance.

3.4.3
Comparison of HFM with UFM

If in UFM the surface ultrasonic vibration excited from a piezo located at the back of the sample is modulated in amplitude using a sinusoidal shape instead of the customary triangular or trapezoidal modulation shape, the tip–sample distance will vary similarly as it does in the case of HFM (see Fig. 3.16). Actually, in UFM we could also collect an Amplitude-UFM and a Phase-UFM signal using the lock-in amplifier, although up-to-now usually only the Amplitude-UFM response has been considered. The main important difference between UFM and HFM lies in the fact that in UFM the ultrasonic vibration is input into the system only from one end of the tip–sample nanocontact, while in HFM, both nanocontact ends are independently excited.

So far the excitation from one end of the nanocontact will be transmitted through the contact to the other end, i.e. ultrasonic vibration from the sample surface will propagate to the AFM cantilever tip, Amplitude-HFM and UFM signals are expected to be quite similar. In fact, as we mentioned in Sect. 3.2.2, UFM can also be implemented in the so-called Waveguide UFM mode, in which the ultrasonic vibration at the tip–sample contact is excited from a piezo located the cantilever base, and no significant qualitative differences in the UFM response have been encountered when comparing ultrasonic curves received in either case [36]. The comparison of UFM and waveguide UFM studies on the same sample is interesting in order to differentiate surface from subsurface effects. In HFM, this same kind of information may be available by appropriate modification of the sample or tip ultrasonic vibration amplitudes. In any case, for some studies, the use of a triangular or trapezoidal shape for ultrasonic amplitude modulation may be preferred, and UFM may still be the technique of choice.

The great strength of HFM versus UFM relies in the phase measurements. By monitoring the phase of the cantilever vibration at the beat frequency, HFM allows us to detect slight changes in phase of the sample vibration with time resolution of fractions of the sample and cantilever ultrasonic periods. If the excitation frequencies are in the MHz regime, and the difference frequency is of some KHz, phase delays between tip and sample vibrations of the order of nanoseconds are easily detectable [27]. Notice that even though it is possible to perform phase-UFM by monitoring the phase of the cantilever vibration at the ultrasonic modulation frequency because of the mechanical diode response, in the absence of forced ultrasonic excitation of the tip, the phase differences between sample and tip ultrasonic vibrations cannot be straightforwardly measured, and the time-sensitivity to phase-delay-related processes is in the best of cases limited to the ultrasonic period, at least easily three orders of magnitude smaller than in the HFM case.

3.4.4
Information from HFM: Time Resolution

As discussed in Sect. 3.4.3, the big potential of HFM is based on its capability to perform phase-delay measurements with an extremely high sensitivity. Phase delays may originate from different elastic or viscoelastic properties, from different in-depth locations of the same-type of elastic inhomogeneity, and in general from any local dissipative process activated by mechanical vibration. So far ph-HFM provides a means to probe a local response in an extremely short time, the technique may reveal dissipation due to extremely quick transitions, otherwise unresolved from other dissipative effects occurring at larger time scales. Phase-HFM has been applied to PMMA–rubber nanocomposites that consist of an acrylic matrix, a copolymer based upon PMMA, and toughening particles, composed of a core of acrylic enclosed with rubber with a bonded acrylic outer shell to ensure good bonding to the matrix [27] (see Fig. 3.18). Using Phase-HFM, it has been possible to distinguish differences in contrast at identical thin polymer layers with different boundary constraints on the nanometer scale. In the Ph-HFM images a different viscoelastic and/or adhesion hysteresis response time of the PMMA on top of the rubber that is not linked to

Fig. 3.18. HFM on PMMA/rubber nanocomposites (from Ref. [27])

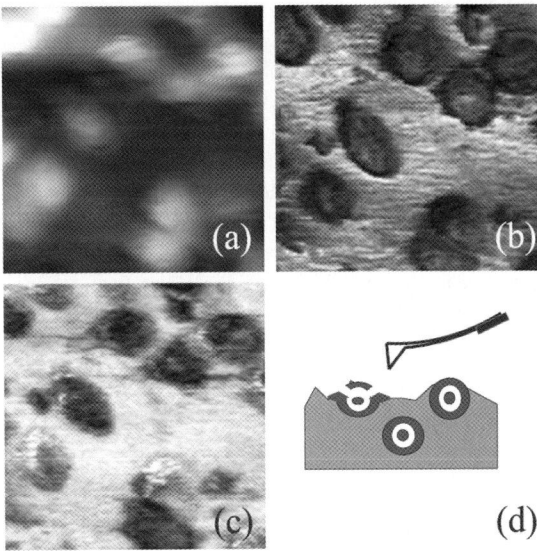

the PMMA rubber matrix is clearly distinguished. Such different PMMA responses cannot, however, be appreciated from the Amplitude-HFM images.

Using the recently proposed SNFUH mode [30], performed similarly as phase-HFM, elastic information of buried features have been obtained from phase measurements with great sensitivity. In the RDF-AFUM procedure, subsurface nanoscale elastic variations have also been observed [31]. In RDF-AFUM the beat effect is used as in HFM, but the beat frequency is chosen to be coincident with a cantilever contact resonance. In Ref. [31], an analytical model is proposed to account for the RDF-AFUM response, considering the interaction of the ultrasonic wave generated at the bottom of the sample with nano-/microstructural features within the sample bulk material, and the nonlinear cantilever tip–sample surface interactions.

Nevertheless, up-to-date, the data reported with beat-effect related AFM techniques is still very limited. The beat effect may facilitate opportunities ranging from the precise evaluation of elastic or viscoelastic response of nanostructures, the analysis of snap shots or transient states in the mechanical response of nanoobjects, the implementation of nanoscale time-of-flight experiments with high temporal resolution, or the quick transmission of information through nanocontacts by mechanical means. The use of higher beat frequencies opens up the possibility to scan at higher lateral scanning speeds while recording material information. Phase-HFM facilitates straightforward measurements of phase-delays between tip and sample vibrations, with extremely high sensitivity. The opportunities brought about by this technique are still to explored.

Acknowledgments. M.T.C. thanks S.K. Biswas and W. Arnold for carefully reviewing this chapter. Financial support from the Junta de Comunidades de Castilla-La Mancha (JCCM) under project PBI-08-092 is gratefully acknowledged.

References

1. Lemons RA, Quate CE (1974) Acoustic Microscope-scanning version. Appl Phys Lett 24:163
2. Quate CF (1985) Scanning acoustic microscopy. Physics Today, 34
3. Briggs A (1992) Acoustic microscopy. Clarendon Press, Oxford, UK
4. Briggs A, Arnold W (1995) Advances in acoustic microscopy vol. I. Plenum Press, New York.
5. Briggs A, Arnold W (1996) Advances in acoustic microscopy vol. II. Plenum Press, New York.
6. Hirsekorn S, Pangraz S, Weides G, Arnold W (1995) Measurement of elastic impedance with high spatial resolution using acoustic microscopy. Appl Phys Lett 67:745
7. Ngwa W, Luo W, Kamanyi A, Fomba KW, Grill W (2005) Characterization of polymer thin films by phase-sensitive acoustic microscopy and atomic force microscopy: a comparative review. J Microscopy 218:208
8. Grill W, Hillmann K, Würz KU, Wesner J (1996) Scanning ultrasonic microscopy with phase contrast. In: Briggs A, Arnold W (eds) Advances in acoustic microscopy, vol 2. Plenum Press, New York
9. Twerdowski E, Von Buttlar M, Razek N, Wannemacher R, Schindler A, Grill W (2006) Combined surface-focused acoustic microscopy in transmission and scanning ultrasonic holography. Ultrasonics 44:e1301, and ref therein
10. Xu M, Wang LV (2006) Photoacoustic imaging in biomedicine. Rev Sci Instr 77:041101, and ref therein.
11. Rabe U, Arnold W (1994) Acoustic microscopy by atomic force microscopy. Appl Phys Lett 64:1493
12. Rabe U, Janser K, Arnold W (1996) Vibrations of free and surface-coupled atomic force microscope cantilevers: theory and experiments. Rev Sci Instr 67:3281
13. Minne SC, Manalis SR, Atalar A, Quate CF (1996) Contact imaging in the AFM using a high order flexural mode combined with a new sensor. Appl Phys Lett 68:1427
14. Yamanaka K, Nakano S (1998) Quantitative elasticity evaluation by contact resonance in an atomic force microscope. Appl Phys A 66:S313
15. Scherer V, Arnold W, Bhushan B (1999) Lateral force microscopy using acoustic force microscopy. Surf Interface Anal 27:578
16. Reinstädtler M, Rabe U, Scherer V, Turner JA, Arnold W (2003) Imaging of flexural and torsional resonance modes of atomic force microscopy cantilevers using optical interferometry. Surf Sci 532:1152
17. Reinstädtler M, Rabe U, Scherer V, Hartnann U, Goldade A, Bhushan B, Arnold W (2003) On the nanoscale measurement of friction using atomic-force microscopy cantilever torsional resonances. Appl Phys Lett 82:2604
18. Cretin B, Sthal F (1993) Scanning microdeformation microscopy. Appl Phys Lett 62:829
19. Variac P, Cretin B (1996) Scanning microdeformation microscopy in reflection mode. Appl Phys Lett 68:461
20. Burnham NA, Kulik AJ, Gremaud G, Gallo PJ, Oulevey F (1996) Scanning local-acceleration microscopy. J Vac Sci Technol B 14:794
21. Rohrbeck W, Chilla E. (1992) Detection of surface acoustic waves by scanning force microscopy. Phys Stat Sol (a) 131:69
22. Kolosov O, Yamanaka K (1993) Nonlinear detection of ultrasonic vibrations in an Atomic Force Microscope. Jpn J Appl Phys 32:L1095
23. Hesjedal T, Chilla E, Froehlich H-J (1995) Scanning acoustic force microscopy measurements in grating-like electrodes. Appl Phys A 61:237

24. Bheme G, Hesjedal T, Chilla E, Fröhlich H-J (1998) Transverse surface acoustic wave detection by scanning acoustic force microscopy. Appl Phys Lett 73:882

25. Yamanaka K, Ogiso H, Kolosov O (1994) Ultrasonic Force Microscopy for nanometer resolution subsurface imaging. Appl Phys Lett 64:178

26. Cuberes MT, Martínez JJ (2007) Mechanical diode ultrasonic friction force microscopy. J Phys: Conf Ser 61:224

27. Cuberes MT, Assender HE, Briggs GAD, Kolosov OV (2000) Heterodyne force microscopy of PMMA/rubber nanocomposites: nanomapping of viscoelastic response at ultrasonic frequencies. J Phys D: Appl Phys 33:2347

28. San Paulo A, Black JP, White RM, Bokor J (2007) Detection of nanomechanical vibrations by dynamic force microscopy in higher cantilever eigenmodes. Appl Phys Lett 91:053116

29. Cuberes MT (2007) Ultrasonic machining at the nanometer scale. J Phys: Conf Ser 61:219

30. Shekhawat GS, Dravid VP (2005) Nanoscale imaging of buried structures via Scanning Near-Field Ultrasound Holography. Science 310:89

31. Cantrell SA, Cantrell JH, Lillehei PT (2007) Nanoscale subsurface imaging via resonant difference-frequency atomic force ultrasonic microscopy. J Appl Phys 114324m.

32. Chilla E, Hesjedal T, Fröhlich H-J (1997) Nanoscale determination of phase velocity by scanning acoustic force microscopy. Phys Rev B 55:15852

33. Safar H, Keliman RN, Barber BP, Gammel PL, Pastalan J, Huggins H, Fetter L, Miller R (2000) Imaging of acoustic fields in bulk acoustic-wave thin-film resonators. Appl Phys Lett 77:136

34. Dinelli F, Castell MR, Ritchie DA, Mason NJ, Briggs GAD, Kolosov OV (2000) Mapping surface elastic properties of stiff and compliant materials on the nanoscale using ultrasonic force microscopy. Philosophical Mag A 80:2299

35. Inagaki K, Kolosov O, Briggs A, Wright O (2000) Waveguide ultrasonic force microscopy at 60 MHz. Appl Phys Lett 76:1836

36. Cuberes MT, Briggs GAD, Kolosov O (2001) Nonlinear detection of ultrasonic vibration of AFM cantilevers in and out of contact with the sample. Nanotechnology 12:53

37. Cuberes MT (2008) Nanoscale ultrasonics in liquid environments. J Phys: Conf Ser 100:052014

38. Dinelli F, Biswas SK, Briggs GAD, Kolosov OV (2000) Measurements of stiff material compliance on the nanoscale using ultrasonic force microscopy. Phys Rev B 61:13995

39. Huey BD (2007) AFM and Acoustics: Fast, Quantitative Nanomechanical Mapping. Annu Rev Mater Res 37:351

40. Inagaki K, Matsuda O, Wright OB (2002) Hysteresis of the cantilever shift in ultrasonic force microscopy. Appl Phys Lett 80:2386

41. Szoszkiewicz R, Huey BD, Kolosov OV, Briggs GAD, Gremaud G, Kulik AJ (2003) Tribology and ultrasonic hysteresis at local scales. Appl Surf Sci 219:54

42. Szoszkiewicz R, Kulik AJ, Gremaud G (2005) Quantitative measure of nanoscale adhesion hysteresis by ultrasonic force microscopy. J Chem Phys 122:134706

43. Szoszkiewicz R, Bhushan B, Huey BD, Kulik AJ, Gremaud G (2005) Correlations between adhesion hysteresis and friction at molecular scales. J Chem Phys 122:144708

44. Szoszkiewicz R, Bhushan B, Huey BD, Kulik AJ, Gremaud G (2006) Adhesion hysteresis and friction at nanometer and micrometer lengths. J Appl Phys 99:014310

45. Szoszkiewicz R, Kulik AJ, Gremaud G, Lekka M (2005) Probing local water contents of in vitro protein films by ultrasonic force microscopy. Appl Phys Lett 86:123901

46. Hirsekorn S, Rabe U, Arnold W (1997) Theoretical description of the transfer of vibrations from a sample to the cantilever of an atomic force microscope. Nanotechnology 8:57

47. Dinelli F, Biswas SK, Briggs GAD, Kolosov OV (1997) Ultrasound induced lubricity in microscopic contact. Appl Phys Lett 71:1177

48. Cuberes MT (2007) Nanoscale friction and ultrasonics, Chapter 4. In Gnecco E, Meyer E (eds) Fundamentals of friction and wear on the nanometer scale. Springer, Berlin Heidelberg New York, pp 49–71, and ref therein
49. Cuberes MT (2007) Ultrasonic machining at the nanometer scale. J Phys: Conf Ser 61:219
50. Tsuji T, Yamanaka K (2001) Observation by ultrasonic atomic force microscopy of reversible displacement of subsurface dislocations in highly oriented pyrolitic graphite. Nanotechnology 12:301
51. Yaralioglu GG, Deggertekin FL, Crozier KB, Quate CF (2000) Thin film characterization by atomic force microscopy at ultrasonic frequencies. J Appl Phys 87:7491
52. Sarioglu AF, Atalar A, Degertekin FL (2004) Modeling the effect of subsurface interface defects on contact stiffness for ultrasonic atomic force microscopy. Appl Phys Lett 84:5368
53. Geisler H, Hoehn M, Rambach M, Meyer MA, Zschech E, Mertig M, Romanov A, Bobeth M, Pompe W, Geer RE (2001) Elastic mapping of sub-surface defects by ultrasonic force microscopy: limits of depth sensitivity. Inst Phys: Conf Ser 169:527
54. McGuigan AP, Huey BD, Briggs GAD, Kolosov OV, Tsukahara Y, Yanaka M (2002) Measurement of debonding in cracked nanocomposite films by ultrasonic force microscopy. Appl Phys Lett 80:1180
55. Nalladega V, Sathish S, Brar AS (2006) Nondestructive evaluation of submicron delaminations at polymer/metal interface in flex circuits. AIP Conf Proc, 820 II, 1562.
56. Geer RE, Kolosov OV, Briggs GAD, Shekhawat GS (2002) Nanometer-scale mechanical imaging of aluminium damascene interconnect structures in a low-dielectric-constant polymer. J Appl Phys 91:4549
57. O. V. Kolosov, M. R. Castell, C. D. Marsh, G. A. D. Briggs, T. I. Kamins, and R. S. Williams (1998) Imaging the elastic nanostructure of Ge islands by ultrasonic force microscopy. Phys Rev Lett 81:1046
58. Cuberes MT, Stegemann B, Kaiser B, Rademann K (2007) Ultrasonic Force Microscopy on strained antimony nanoparticles. Ultramicroscopy 107:1053
59. Cuberes MT (2007) Ultrasonic nanofabrication with an atomic force microscope. Imaging & Microscopy 9:36
60. Cuberes MT (2007) Sonolubrication and AFM nanoparticle manipulation. Proceedings of the 30th Annual Meeting of the Adhesion Society, 430
61. Cuberes MT (2008) Manipulation of nanoparticles using ultrasonic AFM. J Phys: Conf Ser 100:052013
62. Gnecco E, Bennewitz R, Gyalog T, Meyer E (2001) Friction experiments on the nanometer scale. J Phys: Condens Mat 13:R619
63. Gao J, Luedtke WD, Landman U (1998) Friction control in thin-film lubrication. J Phys Chem B 102:5033
64. Scherer V, Bhushan B, Rabe U, Arnold W (1997) Local elasticity and lubrication measurements using atomic force and friction force microscopy at ultrasonic frequencies. IEEE Trans Magnetics 33:4077
65. Reinstädtler M, Rabe U, Goldade A, Bhushan B, Arnold W (2005) Investigating ultra-thin lubricant layers using resonant friction force microscopy. Tribol Int 38:533
66. Huang L., Su C (2004) A torsional resonance mode AFM for in-plane tip surface interactions. Ultramicroscoy 100:277
67. Kasai T, Bhushan B, Huang L, Su C (2004) Topography and phase imaging using the torsional resonance mode. Nanotechnology 15:731
68. Reinstädtler M, Kasai T, Rabe U, Bhushan B, Arnold W (2005) Imaging and measurement of elasticity and friction using the TRmode. J Phys D: Appl Phys 38:R269
69. Song Y, Bhushan B (2005) Quantitative extraction of in-plane surface properties using torsional resonance mode of atomic force microscopy. J Appl Phys 97:083533

70. Song Y, Bhushan B (2006) Simulation of dynamic modes of atomic force microscopy using a 3D finite element model. Ultramicroscopy 106:847
71. Misawa M, Ono M (2006) Nanotribology with torsional resonance operation. Jap J Appl Phys 45:1978
72. Kawai S, Kitamura S, Kobayashi D, Kawakatsu H (2005) Dynamic lateral force microscopy with true atomic resolution. Appl Phys Lett 87:173195
73. Carpick RW, Ogletree DF, Salmeron M (1997) Lateral stiffness: a new nanomechanical measurement for the determination of shear strengths with friction force microscopy. Appl Phys Lett 70:1548
74. Johnson KL (1985) Contact mechanics. Cambridge University Press, Cambridge, UK
75. Lantz MA, O'Shea SJ, Hoole ACF, Welland ME (1997) Lateral stiffness of the tip and tip–sample contact in frictional force microscopy. Appl Phys Lett 70:970
76. Riedo E, Gnecco E, Bennewitz R, Meyer E, Brune H (2003) Interaction potential and hopping dynamics governing sliding friction. Phys Rev Lett 91 084502-1
77. Cuberes MT, to be published
78. French AP (1998) Vibration and waves. Chapman and Hall, London

4 Contact Atomic Force Microscopy: A Powerful Tool in Adhesion Science

Maurice Brogly · Houssein Awada · Olivier Noel

Abstract. Adhesion between two objects appears confusing or ambiguous, because the term is employed generally for two things: first, the formation of the interface between a pair of materials, i.e. the establishment of interfacial bonds through forces at the interface which cause materials to attract one another and second, the breaking stress or energy required to break the formed assembly. One can easily see that both interfacial forces and mechanical properties of adherents in the vicinity of the interface and in the bulk contribute to the global mechanical response of the assembly. Such a fundamental issue reflects a paradox that has stimulated intensive research for decades: what is the interplay between surface forces, surface rheology, and adhesive strength? In recent years, Atomic Force Microscopy (AFM) has become a powerful tool, sensitive enough, to detect small surface forces and to study adhesion at the nanoscale. Precise analysis of adhesion forces and surface mechanical properties of model polymer surfaces can be achieved with such a nanometer probe. The purpose and scope of this chapter is to highlight the experimental methods that enable one to dissociate the different contributions (chemical and mechanical) included in an AFM force-distance curve in order to establish quantitative relationships between interfacial tip–polymer interactions and surface viscoelastic properties of a polymer surface. New relationships are proposed that provide a complete understanding of how the adhesion separation energy depends on both surface chemistry and rheological behavior of the surface and thus at a local scale.

Key words: Adhesion, Atomic force microscopy, Self assembly monolayer, Polymer surfaces, Nano-indentation, Contact mechanics, Nanoscale

4.1 Introduction

Advances in nanotechnology and miniaturization of devices has led to the requirement of an increased understanding of the interactions and adhesion phenomena present at nanoscale contacts. It is well known that adhesion plays a major role in the tribological behavior and contact mechanics of many modern nano-devices. Adhesion of materials is ruled by interactions that extend down to the atomic scale of the contributing materials. The forces involved include Van der Waals forces, hydrogen bonds, Coulomb forces as well as chemical covalent bonds in some cases. Although Van der Waals forces are only short range, they often rule the adhesion between two materials and are therefore responsible for contact interactions. In recent years, Atomic Force Microscopy (AFM) has become a powerful tool, sensitive enough, to detect small surface forces and to study adhesion at the nanoscale. Precise analysis of adhesion forces and surface mechanical properties of model polymer surfaces can be achieved with such a nanometer probe. Retraction profiles from force-distance (FD) AFM measurements provide quantitative, chemically sensitive, adhesion information at the nanoscale between probe tips and sample surfaces. Information about local adhesion [1–3], friction [4–6], or surface stiffness [7, 8] is readily obtainable. The purpose and scope of this chapter is to highlight the experimental methods that enable one to dissociate the different contributions (chemical and mechanical) included in an AFM force-distance curve in order to establish quantitative relationships between interfacial tip–polymer interactions and surface viscoelastic properties of a polymer surface. Measurements of local adhesive forces between a silicon nitride tip (Si_3N_4) and model substrates are performed by using the AFM contact mode, at ambient temperature, in the air. Considering that the main technical uncertainties have been listed and minimized, surface force measurements are, in a first step, detailed on chemically modified silicon substrates (grafted with hydroxyl, amine, methyl, and ester functional groups). In order to investigate the effects of mechanical or viscoelastic contributions, force measurements on model polymer networks, whose surfaces are chemically controlled with the same functional groups as before (silicon substrates) were achieved. Such model experiments show that the viscoelastic contribution is dominating in the adhesion force measurement. At the end of the chapter, new relationships are proposed that express the local adhesion force versus the energy dissipated in the tip–polymer contact, and the surface properties of the materials (thermodynamic work of adhesion). Moreover, for crosslinked polymers the local adhesion forces depends on Mc, the mass between crosslinks of the network.

4.2
Adhesion Science

4.2.1
Adhesion and Adhesive Strength

The term "adhesion" between two objects appears confusing or ambiguous, because it is employed generally for two things: first, the formation of the interface between a pair of materials, i.e. the establishment of interfacial bonds through forces at the interface which cause materials to attract one another and second, the breaking stress or energy required to break the formed assembly. One can easily see that both interfacial forces and mechanical properties of adherents in the vicinity of the interface and in the bulk contribute to the global mechanical response of the assembly. Such a fundamental issue reflects a paradox that has stimulated intensive research for decades: if adhesion is only a matter of thermodynamics the work needed for separation would be close to the surface energies of adherents, which are typically 10^{-2} to 10^{-1} J/m^2. The practical work of separation can be 10 to 10^6 greater! On the other hand, separation has often been viewed as involving processes at the molecular level. The force needed for separation should then be as large as $10^{-2}/10^{-10}$ or 10^8 N/m^2 to $10^{-1}/10^{-10}$ or 10^9 N/m^2, where 10^{-10} m is the typical atomic size! Therefore, probing adhesion at a very local scale relates more precisely to the question: What, precisely, is the role of surface chemistry in determining the adhesive strength to separate a nanometer size tip from a surface? In other words, do interfacial forces determine, by their "weakness" or "strength," the weakness or strength of adhesive contact?

The development of intimate molecular contact at an interface is a necessary condition for good adhesion. That is why polymers are good candidates for adhesives. Made of long flexible molecules, they flow under stress like viscous liquids, at long timescales, whereas they deform like soft elastic solids, at short timescales. Bulk properties are mainly governed by entanglements. Without intimate molecular contact, interfacial attraction will be weak, and the applied stress that can be transmitted from one phase to the other, through the interface will accordingly be very low. Therefore, numerous adhesives are applied in a rather liquid-like state, leading to the formation of an assembly through a liquid–solid contact step. Criteria of good adhesion become criteria of good wetting. However, intimate molecular contact alone is not sufficient. The nature and magnitude of the interfacial forces are also important as will be discussed. The most common interfacial forces result from van der Waals and Lewis acid–base interactions. The magnitudes of these forces can generally be related to fundamental thermodynamic quantities, such as surface free energies.

From a thermodynamic point of view, the work required to separate reversibly the interface between two phases 1 and 2 from their equilibrium interacting distance to infinity is termed the "work of adhesion" and is equal to:

$$W_{12} = \gamma_1 + \gamma_2 - \gamma_{12} \tag{4.1}$$

where γ_1 and γ_2 are the surface free energies of phases 1 and 2, respectively and γ_{12} is the interfacial free energy between phases 1 and 2. The work of adhesion is the decrease of Gibbs free energy per unit area when an interface is formed from two individual surfaces. Thus, the greater the interfacial attraction, the greater the work of adhesion will be and the smaller the interfacial free energy between phases 1 and 2 will be. The quantitative determination of the work of adhesion give rise to numerous experimental and theoretical approaches which were published [9–15]. Nevertheless, all these methods are mainly macroscopic and not made "in situ." As a consequence, characterization of local properties have stimulated widespread interest.

4.2.2
Adhesion at a Local Scale

It is important to mention recent works that have contributed to the understanding of adhesion mechanisms at a local scale. Burns et al. [16] demonstrated the influence of adhesion on sliding and on friction forces on chemically modified surfaces. Jones et al. [17] studied the effects of relative humidity, roughness, and surface treatment on adhesive properties of a glass–glass contact. Rabinovich [18] considered the influence of roughness at the nanometer scale. But none of these studies has been devoted to the contact between a local probe and a chemically modified polymer surface. The main reason is the difficulty of quantitative interpretation of the results obtained on viscoelastic materials [19, 20]. However, Basire and Fretigny [21] have done important work that allow them to determine static and dynamic modulus of polymers even if the Dupre energy they found appears to be rather small ($10 \, \text{mJ} \, \text{m}^2$). Nysten and Cuenot [22, 23] have conducted oscillatory experiments of an AFM probe under an electrostatic field. The resonant frequency of the probe is directly related to the mechanical properties of the studied material. Finally, Aimé and his team [24, 25] have studied dissipative phenomena in the adhesive contact by using dynamic AFM. They propose an original model [26] for the determination of polymer viscosity on the basis of noncontact resonant AFM mode. Nevertheless, none of these studies were done on chemically modified polymer surfaces in order to investigate the interplay between surface chemistry and mechanical properties at a local scale.

4.3
Force vs. Distance Measurements with an AFM

Force measurements with AFM, in the contact mode, consist of detecting the deflection of a spring (or cantilever) bearing a silicon nitride tip at its end, when interacting with the sample surface. The deflection of the cantilever is detected by an optical device (four quadrant photodiode) while the tip is vertically moved forward and backward thanks to a piezoelectric ceramic (or actuator). Thus, provided that the spring constant of the cantilever is known, one can obtain a deflection-distance (DD) curve and then a force-distance curve, by using Hooke's law. The DD curves presented in this chapter were performed in air with an available commercial apparatus (Nanoscope IIIa D3000, Digital Instruments). A schematic representation of a DD curve obtained when probing a hard surface is shown in Fig. 4.1

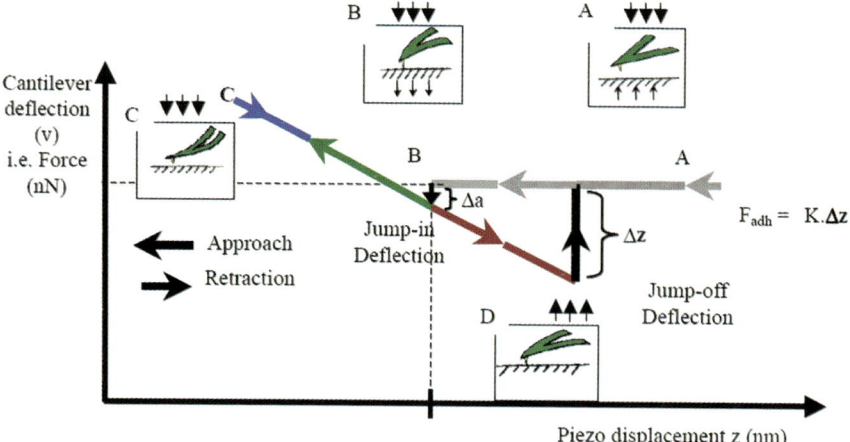

Fig. 4.1. Schematic representation of a DD curve. The slope of the "contact zone" is equal to unity when considering the contact between the tip and a rigid surface, whereas it is lower than 1 when considering the contact between the tip and a soft material (polymer)

In zone A (Fig. 4.1), the cantilever is far from the surface and stays in a state of equilibrium (no interaction with the surface). The cantilever deflection is zero. During the approach toward (or withdrawal from) the surface, the tip interacts with the sample and a jump-in (or jump-off) contact occurs [zones B (for loading) and E (for unloading)]. These instabilities take place because the cantilever becomes mechanically unstable. Usually, for underfomable surfaces, because of mechanical instabilities, the jump-in contact is not significant for determining attractive Van der Waals forces. When in contact, the cantilever deflection is equal to the piezoelectric ceramic displacement provided no indentation of the substrate occurs [zones C (for loading) and D (for unloading)]. An underfomable reference sample (cleaned silicon wafer) is used to scale the DD curve in deflection by fixing to unity the slope value of the contact line. Because of adhesion forces during contact, the jump-off is more interesting than the jump-in and occurs in position D. Considering the cantilever like a spring, knowing its spring constant, one can obtain the adhesion force between the tip and the sample by using Hooke's law:

$$F = k_{tip}. \Delta z \qquad (4.2)$$

where k_{tip} represents the spring constant of the cantilever and Δz represents the jump-off during retraction of the tip from the surface.

In order to carry out a quantitative analysis, different experimental points should be taken into consideration such as: selection of tips, cantilever spring constant, tip radius, linearity of the photodiodes, piezodriver hysteresis, cantilever and piezo thermal stabilities, and tip contamination [27].

4.4
AFM Calibration

One of the fundamental points to obtain reproducible, quantitative, and reliable data is the calibration procedure, which should be rigorous and systematic for all measurements.

4.4.1
Selection of Tips

For force-distance experiments a triangular cantilever terminated by a pyramidal tip (Si_3N_4) has been used. The length of the cantilever is about 100 μm and the height of the tip is about 3 μm. Tips, fabricated from the same Silicon wafer, have an identical spring constant but do not have the same radius, at the apex, which depends on an anisotropic basic attack during fabrication [28]. In order to choose tips having the same radius, force-distance curves between different tips and a reference surface (silicon wafer) must be produced. As all experiments are carried out under the same experimental conditions (temperature, humidity, surface), the adhesion between the tip and the surface only depends on the contact area, i.e. on the tip radius. The tips that give the same adhesion force (i.e. the same radius) are then selected. In a further step, estimation of each tip radius has to be done on the basis of Scanning Electron Microscopy (SEM) images.

4.4.2
Determination of the Spring Constant of the Cantilever

Several methods have been reported [29–32] to measure the spring constant of a cantilever. We have chosen the method proposed by Torii [32] for its simplicity. This nondestructive method refers to the use of rectangular reference cantilevers (calibrated with the resonant frequency method and supplied by NanoMetrology). As mentioned, the cantilevers used for probing nano-adhesion were triangular-shaped cantilevers (supplied by Nanosensor, Germany) and have an effective spring constant equal to $0.30 \pm 0.03\,\mathrm{N\,m^{-1}}$ (the value specified by the supplier is $0.58\,\mathrm{N\,m^{-1}}$!). Cantilevers having the same spring constant were systematically selected.

4.4.3
Determination of the Tip Radius

To deduce quantitative data from a force-distance curve in order to access work of adhesion from experimental values, it is necessary to have a good estimation of the radius of the probe and thus of the tip–sample contact area. SEM images reveal that the tip shape can be represented as a cone ended by a sphere. Figure 4.2 shows a typical SEM image of a tip.

The tip radius can easily be estimated.

Fig. 4.2. Scanning electron microscopy image of an AFM tip

4.4.4
Nonlinearity of the Quadrant of Photodiodes

The nonlinearity of the optical detector is the consequence of a nonhomogeneous spreading of the laser spot on the detector. This nonlinearity has been studied by reporting the slope of the contact line (zones C or D) of the DD curve (obtained on a hard surface and considering that there is no nonlinearity at the middle of the photo detector) versus the tension (V) measured by the detector. The domain of linearity of the detector lies between $\pm 2V$. If nonlinearity is not taken into account, the error on the quantitative results can be significative, because the slope of the contact line determines the Y-scale.

4.4.5
Scan Rate of the Cantilever

The piezodriver used to move the cantilever vertically and forward shows hysteresis and nonlinearity in its vertical displacement. This hysteresis can be studied by reporting the slope of the contact zones (zones C and D) versus the amplitude of the contact zone and the scan rate. During the experiments, the actuator is considered as thermally stable. We observed that a discrepancy appears for very low scan rates ($60 \, nm \, s^{-1}$). For higher scan rates($18 \, \mu m \, s^{-1}$), the viscosity of the environment could be significant (damping effect). A rate of about $6 \, \mu m \, s^{-1}$ is a good compromise. In order to stabilize thermally the piezodriver, the machine is turned on 12 hours before. Contamination of the tip was checked by measuring adhesion force between the tip and a reference surface after 20 force-distance curves were realized on polymer samples.

4.4.6
Systematic Check

In addition, checking regularly (and also randomly) the adhesion force on a reference silicon wafer verifies contamination of the tip during the measurements. In that way contamination of the tip was checked by measuring adhesion force between the tip and a reference surface after 20 force-distance curves were realized on polymer samples. When the tip is contaminated, a new tip is used and characterized. It is important to mention that tip contamination occurs rarely in comparison to the great number of realized DD curves. Finally, the reported results are an average of about 100 DD curves for each substrate. Relative uncertainty is equal to 8% in all cases [33].

4.4.7
Estimation of the Uncertainties Related to the Experimental Pull-Off Force Measurements

If we consider that the main uncertainties in a pull-off force measurement with AFM are due to the determination of the spring constant and the radius of the tip, we can approximate an error of 10% on the measurement of an adhesion force (due to the uncertainty of the method of calibration of the tip), and a relative error of 20% on the determination of the work of adhesion and surface tensions (due to a 10% error on the determination of the tip radius on the basis of SEM images). Nevertheless, even if 20% of error is high one must consider this error as a systematic error. Indeed, all measurements (about 100 for a given surface) were performed with the same tip of the same radius. Then experimental results are significant.

4.5
Adhesion Forces and Surface Energies

In order to quantitatively and "in situ" measure the work of adhesion between an AFM tip and a given surface one has to proceed according to a proper procedure that enables one to separate the chemical contribution from the mechanical contribution in an adhesion force. In that way the following strategy is suggested. In order to distinguish the chemical and the mechanical contribution, two kinds of materials were used: the first one represents a rigid silicon wafer surface with different controlled surface chemical end groups and the other one represents a soft polymer surface with variable Young's modulus but which exhibits the same controlled surface chemistry.

4.5.1
Materials

Si(100) silicon wafers (supplied by Mat Technology, France) polished on one side were used as the substrate for grafting self-assembly monolayer (SAM) thin films. "As received silicon ($Si_{as\ received}$)" refers to a silicon wafer previously cleaned with ethanol in an ultrasonic bath. That means that a contaminated layer still remains on the surface. Four organosilane grafts (supplied by ABCR, Karlsruhe, Germany) were used for the elaboration of homogeneous model surfaces on the substrate. Two hydrophobic model surfaces were prepared by using hexadecyltrichlorosilane ($C_{16}H_{42}O_3Si$ or Si_{CH3}) and 1H,1H,2H,2H-perfluorodecylmethyldichlorosilane ($C_{11}H_7Cl_2F_{17}Si$ or Si_{CF3}) and two hydrophilic model surfaces by using (6-aminohexyl)-aminopropyltrimethoxysilane ($C_{12}H_{30}N_2O_3Si$ or Si_{NH2}) and 2(carbomethoxy)ethyltrichlorosilane ($C_4H_7Cl_3O_2Si$ or Si_{ester}). Polydimethylsiloxane (PDMS) polymers were supplied by ABCR (Karlsruhe, Germany). All other chemicals used in chemical handling (cleaning, synthesis) were of reagent grade or better (supplied by Aldrich).

4.5.2
Preparation of Oxidized Silica Surface

Before coating, the substrates surface must be chemically modified in order to become highly hydrophilic. The silicon surface is, first, cleaned with ethanol and dried with nitrogen before oxidation. Then, oxidized surfaces are obtained after cleaning the substrate in a warm (60 C) oxidative solution (3:7 v/v, 30% H_2O_2 and H_2SO_4 mixture) for about 30 minutes in order to keep a smooth surface, and then, thoroughly rinsed with deionized and twice distilled water. Just before being grafted with organosilane the wafers are dried with nitrogen. This treatment produces a high hydroxyl group density on the surface (SiOH groups), to which functional silanes will adsorb upon hydrolysis [34].

Silicon wafers covered with hydroxyl end-groups (Si_{OH}) were synthesized with this method and immediately probed in order to avoid contamination of the surface by the environment due to the high reactivity of SiOH groups.

4.5.3
Grafting of Functionalized SAMs on Silicon Wafer

Three different techniques are frequently used to obtain SAMs: Langmuir–Blodgett techniques, involving an air–water interface to transfer the assembled film to a solid substrate, solution adsorption of film molecules onto the substrate, and a vapor-phase molecular self-assembling technique [34], which uses vapor deposition of the film onto the substrate. Our functionalized SAMs were prepared with the last technique slightly improved in the laboratory. Through the lack of solvent contamination by small solvent molecules and defects are prevented. Moreover, a previous study [35, 36] showed that the molecular films prepared with this method are homogeneous, stable, and resistant.

The silicon wafers are placed above a previously de-aired organosilane (100 μl) / paraffin (3 ml) mixture. The vapor-phase deposition of the molecular film on the substrate is performed in a vacuum chamber (50 min at 5×10^{-3} Torr) at room temperature.

4.5.4
Characterization of the SAMs

Figure 4.3 represents, as an example, AFM topographic and phase contrast images of the Si–CH$_3$ wafer. Tapping-mode images confirm that no aggregates were formed and show a complete homogeneous recovery of the grafts. Average roughness ranges between 0.1 nm (Si–OH) and 0.3 nm (Si–CH$_3$), whatever the grafted substrate. These values confirm that the grafts are well ordered and packed at nanometer scale.

Contact angle measurements with water droplets, surface energy (determined by wettability) and ellipsometric results obtained on wafer and PDMS-grafted substrates are gathered in Table 4.1.

Contact angle measurements (Table 4.1) show that grafting is effective and values obtained correspond to expected values [9]. The same comments are valid for surface energy values. Thicknesses of organic SAM films were determined by ellipsometry (Sopra ES4M ellipsometer) measurements by fitting the refractive index and the thickness of the organic film. Experimental values are in good agreement (Table 4.1) with the estimated theoretical values [37, 38]. Values reported in the literature correspond to untilted grafts relative to the surface plane. In fact the graft

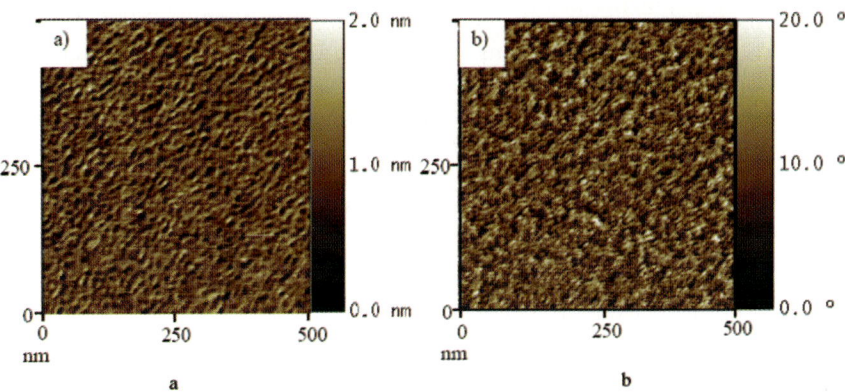

Fig. 4.3. Topographic (**a**) and phase contrast (**b**) images (500 × 500 nm) of a CH_3-grafted silicon wafer

Table 4.1. Water contact angles, surfaces energies, and thicknesses of SAM on silicon wafer substrates

Substrates	Contact angle of water (°)	Surface energy (mJ m^{-2})	Experimental SAM thickness (Å)	Theoretical SAM thickness (Å)
Si–CF$_3$	106 ± 2	21 ± 1	12 ± 1	14
Si–CH$_3$	103 ± 2	22 ± 1	21 ± 1	22.5
Si$_{asreceived}$	78 ± 2	32 ± 1	–	–
Si–COOR	71 ± 2	43 ± 1	5 ± 1	6
Si–NH$_2$	57 ± 2	53 ± 1	9 ± 2	15
Si–OH	6 ± 2	76 ± 1	–	–

molecules are tilted with a characteristic angle between the long-molecular axis and the surface normal [39]. This angle explains the difference between experimental and theoretical values. Nevertheless ellipsometric results confirm that only a monolayer is formed whatevever the substrate and the grafted molecule.

All the measurements performed on the SAMs (AFM imaging, ellipsometric and contact angle) show that homogeneous and well-packed grafting is obtained.

4.5.5
Force-Distance Curves on Rigid Systems Having Controlled Surface Chemistry

The AFM tip–substrate pull-off or adhesion force was first measured on chemically modified SAMs obtained on purely rigid substrates (silicon wafers) and was compared with that for as-received silicon wafers. Figure 4.4 shows that AFM force-distance measurements under our conditions are sensitive to a chemical modification of the wafer surface by adsorption of SAM.

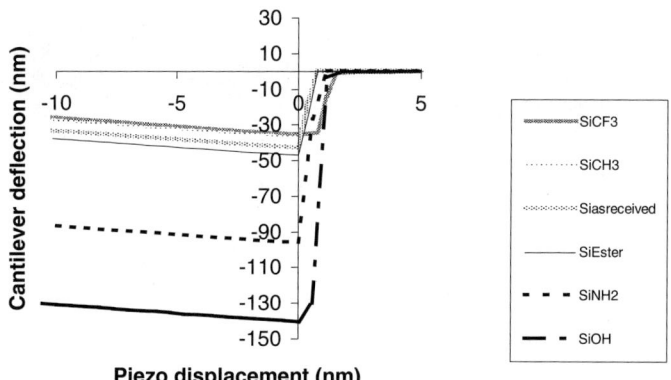

Fig. 4.4. Experimental force-distance curves (retraction) on functionalized silicon wafers

When jump-out of contact occurs, the corresponding pull-off deflection is measured. Pull-off deflection values ($D^{\text{pull-off}}$) increase in the following order:

$$D^{\text{pull-off}}_{\text{SiCF3}} < D^{\text{pull-off}}_{\text{SiCH3}} < D^{\text{pull-off}}_{\text{Siasreceived}} < D^{\text{pull-off}}_{\text{SiCOOR}}$$
$$< D^{\text{pull-off}}_{\text{SiNH2}} < D^{\text{pull-off}}_{\text{SiOH}}$$

Knowing the pull-off deflection one can easily deduce the adhesion force, knowing the cantilever spring constant k_{tip}:

$$F_{\text{adh}} = k_{\text{kip}} . D^{\text{pull-off}} \qquad (4.3)$$

k_{tip} has been determined according to the method described in [31]. The pull-off deflection and thus the adhesion force value increases with the hydrophilicity of the surface. The measured adhesion force depends strongly on the tip radius in the case of purely rigid substrates. Sugawara et al. [40] suggested that the adhesion force is proportional to the tip radius.

The DMT theory [41], also establishes a relationship between the adhesion force (F_{adh}), the tip radius (R), and the thermodynamic work of adhesion (W_0):

$$F_{\text{adh}} = 2.\pi.R.W_0 \qquad (4.4)$$

On the basis of this relation one can deduce W_0 from experimental adhesion forces. Figure 4.5 shows that the thermodynamic work of adhesion, W_0, is proportional to the surface energy deduced from classical wettability measurements.

In the case of hydrophilic surfaces (Si–NH$_2$, Si–OH), the AFM tip interacts with the water layer adsorbed on the surface and the measured adhesion force results from the adhesion between water adsorbed on both surfaces (tip and sample) and from the adhesion force due to the Van der Waal force between the NH$_2$ or OH sites existing at the grafted wafer surface. On the contrary, during wettability experiments the water droplet deposed on the surface does not feel the OH sites existing at the hydroxylated wafer surface.

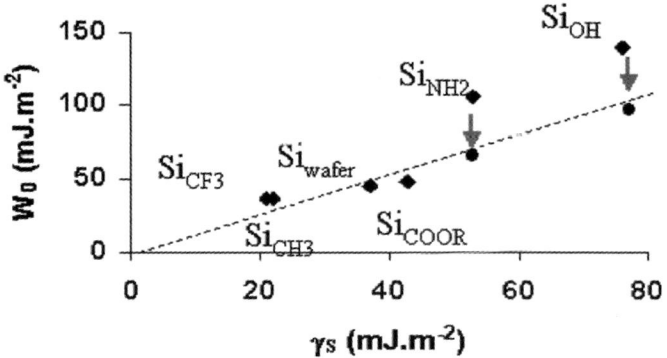

Fig. 4.5. Thermodynamic work of adhesion deduced from AFM (DMT theory) versus surface energy of SAM grafted wafers

4.5.6
Influence of Capillary Forces on Adhesion Forces

For technical reasons, all the AFM measurements were carried out in air at about 20 °C. Under these conditions, the influence of the capillary forces cannot be neglected. Indeed, all the experiments were carried out at a constant relative humidity of 30%. Many studies have reported the effect of humidity on the separation force. Xiao et al. [42] and Sedin et al. [43] have shown that a water meniscus cannot be formed between the tip and the surface if relative humidity is below 20%. He and co-workers [44] have observed meniscus formation only in the case of hydrophilic contact but not for hydrophobic contact. These results are confirmed by Zhang et al. [45]. Salmeron [46], for a hydrophilic tip–mica contact has observed that water capillary effects are significant if relative humidity is higher than 40%. As in our experiments the humidity is 30% and one can suspect the existence of a capillary bridge. On the theoretical point of view, the minimum thickness of the water capillary film is expressed as [47, 48]:

$$e - a_0 \left(\frac{\gamma_W}{S} \right) \tag{4.5}$$

Where e is the film thickness, γ_W the surface tension of water, a_0 a capillary length, and S the spreading coefficient ($S = \gamma_S - \gamma_{SL} - \gamma_L$).

As a consequence, the force measured by AFM includes the contribution of Van der Waals and capillary forces. The total adhesion force is given by the following expression:

$$F_{adh} = F_{cap} + F_{VDW} \tag{4.6}$$

The capillary force depends on a meniscus formed between the two surfaces and is given by Israelachvili [36] and by Riedo et al. [49], in the case of a sphere-plane contact, by:

$$F_{cap} = 2\pi R \gamma_W (\cos \theta_{w/tip} + \cos \theta_{w/wafer}) \tag{4.7}$$

Table 4.2. Respective contribution of capillary forces and Van der Waals forces to F_{adh} and W_0

Substrates	F_{adh} (nN)	F_{cap} (nN)	F_{vdw} (nN)	W_0 (mJ m^{-2})
Si–NH$_2$	33	17	16	52
Si–OH	45	27	18	60

where γ_w is the surface tension of water, R the tip radius, and θ the contact angle between water and the tip (determined on the back side of the cantilever) or water and the wafer. In a recent study, Weeks et al. [50] have used environmental scanning electron microscopy to image water meniscus formation between an AFM tip and a surface. Values of W_0 obtained after correction from capillary forces effects on Si–NH$_2$ and Si–OH hydrophilic surfaces are represented by black circles in Fig. 4.5. Values of capillary forces and van der Waals forces in the case of hydrophilic contacts are gathered in Table 4.2.

Taking into account the contribution of capillary forces shows (Fig. 4.5) that a good correlation is observed between the work of adhesion (W_0) and the surface energy of the chemically modified solids (γs). As a consequence, these results demonstrate that force-distance experiments may be used to determine quantitative and representative data at a local scale on a rigid substrate.

4.6
Adhesion Forces Measurements on Polymers

4.6.1
Cross-Linking and Functionalization of PDMS Networks

PDMS samples were cross-linked under nitrogen in a glove box using tetrakis(dimethylsiloxy)silane as a cross-linker and a platinum-based catalyst. All the chemicals were supplied by ABCR (Karlsruhe, Germany). All the size and mechanical characteristics of the different PDMS are gathered in Table 4.3.

Mc values represent the average molecular masses between crosslinks after crosslinking. Mc values range betwwen 0.8 and 34 kg/mol. Then, PDMS 0.8 k is the hardest substrate, whereas 34 k refers to the softest one. Mc values were determined by swelling experiments. Flory's law of rubber elasticity [51], which represents the reciprocal tensile modulus versus Mc, the mass between crosslinks, is satisfied for the synthesized networks. This proves that a good control of macroscopic mechanical properties is achieved. Moreover, the Young's modulus is independent of the strain rate in the range of strain rate corresponding to that used during AFM experiments (at ambient temperature for all the networks). PDMS networks are then treated by water plasma and functionalization with the same grafts as for silicon wafers using the vapor deposition technique. Data gathered in Table 4.3 show that the grafting technique leads to the formation of a homogeneous monolayer of controlled hydrophilcity or hydrophobicity.

Table 4.3. Average mass between crosslinks, mechanical properties and surface chemistry of the cross-linked PDMS

PDMS	Average mass between crosslinks M_C (g/mol)	Elasticity domain (%)	Deformation at break (%)	Young's modulus (MPa)
0.8 k	800	40	196	2.24
8.5 k	8,500	46	210	0.64
13 k	13,000	47	250	0.30
23 k	23,000	50	245	0.19
34 k	34,000	52	250	0.13
Substrates	Water contact angle (°)	Surface energy (mJ m^{-2})	SAM thickness (Å)	Theoretical thickness (Å)
PDMS	104 ± 4	28 ± 1	/	/
PDMS–CH$_3$	108 ± 4	22 ± 1	22 ± 2	22.5
PDMS–NH$_2$	50 ± 4	53 ± 1	10 ± 3	15

4.6.2
Force-Distance Curves on Soft Polymer Surfaces

A comparison between the force-distance curves obtained on a silicon substrate and on a PDMS substrate shows important differences probably due to the specific mechanical behavior of PDMS polymer chains (Fig. 4.6). First compared to silicon wafer, in the tip–polymer separation process, the jump-off contact occurs over a large piezo displacement scale and could correspond to a progressive dewetting of the tip by polymer chains during tip retraction. Second, the jump-off amplitude is higher than for silicon wafer. Finally, the loading and unloading slope in the tip deflection-piezo displacement representation is much lower than unity in the case of soft polymer systems. The beginning of the indentation is assumed to be at the minimum of the force-distance curve.

Creep experiments have also been performed with the AFM. Considering the experimental contact time ($t_{exp} < 0.1$ s), the creep effect is neglected in our force curve measurements. Nevertheless, a crucial question concerns the beginning of indentation in the specific case of soft polymers. Indeed, for very low modulus materials, the jump-in contact deflection appears to be very important and somehow comparable to the jump-off contact (Fig. 4.6). Such an amplitude (60–80 nm) could not be only due to mechanical instability of the cantilever, estimated to be equal to 4–5 nm. In order to explain the important deflection when contact occurs, we propose the hypothesis of the formation of a nanoprotuberance at the PDMS surface. This protuberance comes into contact with the AFM tip and relaxes till zero deformation of the surface. This relaxation induces a high deflection of the tip. Aime et al. [52] have proposed such a model that takes into account the polymer viscoelasticity. As our PDMS present very low modulus and high eleasticity we have used and adapted a model based on linear elasticity [53]. Detailed information can be found in [33].

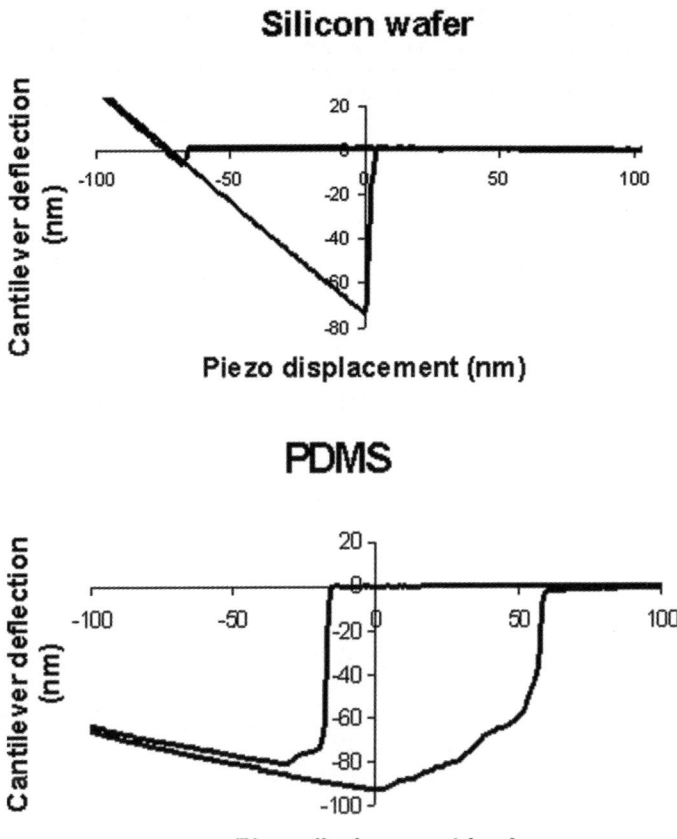

Fig. 4.6. Comparison between a force-distance curve obtained on a silicon substrate (*top*) and a PDMS substrate (*bottom*)

Briefly, at a given time t, the AFM tip is at a distance $d(t)$ from the polymer surface. The tip–surface interaction force is $F(t)$. This force produces the formation of a nanoprotuberance of height $h(t)$. At time $t + \Delta t$, the tip–sample distance is:

$$d(t + \Delta t) = d(t) - h(t) - p(\Delta t) - \delta(t) \tag{4.8}$$

where $p(\Delta t)$ is the piezo displacement and $\delta(t)$ is the cantilever deflection. Then the force $F(t + \Delta t)$ and height $h(t + \Delta t)$ are calculated. Table 4.4 gathers the protuberance heights obtained by simulation and the experimental values.

These results show that the deflection of the cantilever is negligeable when contact occurs. As a consequence, the jump-in amplitude at the approach is due to the relaxation of the nanoprotuberance till zero deformation. Because of adhesion forces, the AFM tip is pulled down in the relaxation process. Good agreement is observed and indicates that real indentation of the polymer surface occurs at the minimum of the jump-in curve. Then, during retraction of the tip, adhesion forces are determined in the same manner as previously for rigid substrates.

Table 4.4. Nanoprotuberance heights obtained by simulation and jump-in contact amplitude

PDMS	Nanoprotuberance height (simulation) (nm)	Jump-in contact amplitude (average over 100 DD curves) (nm)
0.8 k	24	20–25
13 k	50	50–65
34 k	73	70–85

4.6.3
Real Indentation of the AFM Tip Inside a Soft Polymer

Before monitoring force-distance curves on a soft polymer surface, the actuator and the cantilever were thermally stabilized. The laser spot in contact with the tip must be positioned in such a way that tangential forces, due to frictional forces, were minimized. We carried out these measurements on the basis of the above prerequisites so that force measurements could be liable and compared. Force vs. indentation (F-I) curves are deduced from tip deflection-piezo displacement curves by assuming that for a given force, the indentation depth is the difference between the experimental deflection value and the one that should be observed if the material was underfomable (deduced from the slope of 1 for underfomable materials). To deduce work of adhesion from our experimental values, it is necessary to have a good estimation of the radius of the probe and thus of the tip–sample contact area. The SEM analysis reveal that the tip shape can be represented as a cone ended by a sphere. Figure 4.7 shows that the F-I curve evolves according to two regimes.

In our case, the transition between the two modes corresponds to the change in the contact geometry (sphere-plane contact to cone-plane contact). The observed transition shows that our tip radius can be estimated to 50 nm ± 5 nm.

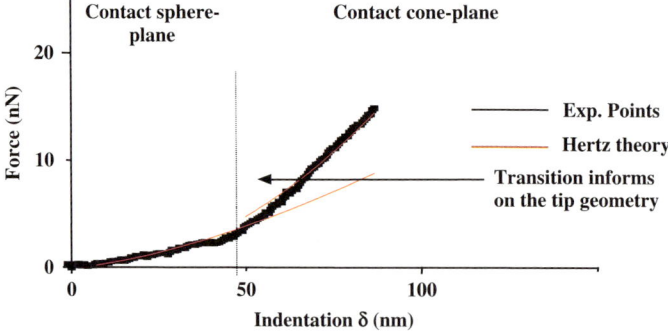

Fig. 4.7. Force vs. real indentation on a soft polymer: two regimes are observed, one for the penetration of the spherical part of the AFM tip (*right*) and the second when the conical part of the tip penetrates the polymer (*left*)

Moreover one has to ask if the stiffness of the cantilever is adapted. Considering the normal stiffness of the cantilever, it is possible to determine the maximal indentation depth (δ_{max}), using the following relation:

$$k_{tip} = 2 \times E^* \times \sqrt{(R\delta_{max})} \tag{4.9}$$

where k_{tip} is the spring constant of the cantilever, E^* is the reduced modulus, and R is the tip radius. δ_{max} represents the maximum penetration depth of the tip in a given PDMS sample. Once this indentation is obtained, even if the loading of the tip is increased, the tip does not penetrate deeper in the sample. Calculation of δ_{max} gives values ranging between 89 nm and 27 μm respectively for the hardest (0.8 k) and the softest (34 k) PDMS. In order to perform experiments at constant indentation for all the PDMS and considering that for our substrates the minimum value of δ_{max} is 89 nm, we have decided to perform constant nanoindentation experiments up to 80 nm.

4.7
AFM Nano-Indentation Experiments on Polymer Networks

4.7.1
Force-Indentation Curves

Force vs. indentation (F-I) curves are deduced from tip deflection-piezo displacement curves by assuming that for a given force, the indentation depth is the difference between the experimental deflection value and the one that should be observed if the material was underfomable deduced from the slope of 1 for underfomable materials. All F-I curves obtained on PDMS and grafted PDMS have been performed at a 80-nm indentation depth, as discussed previously. We now consider PDMS of different Young's modulus grafted with identical molecules than for the silicon wafers (CH_3 and NH_2 SAMs). We show in Fig. 4.8 representative tip deflection-indentation curves on nongrafted PDMS soft polymer networks.

Figure 4.8 clearly shows the influence of the network mechanical properties on F-I response when surface chemistry is constant (CH_3 grafts). For a low Mc value (0.8 k) loading and unloading are quite similar and no dissipation of energy is observed. On the contrary when Mc increases (34 k) unloading is very different from loading and dissipation and adhesive contact occurs. The loading part obeys classical Hertz contact for all the PDMS samples. Such a behavior is observed when the loading force (i.e. deflection during loading) depends on the real indentation according to a power law. More precisely the power law exponent is equal to 1.5 for a Hertzian sphere-plane contact. Thus, it is possible to extract from the loading curve the surface modulus of the soft polymer substrate. Indeed the following relation is used:

$$F = \frac{4}{3} \times E_{AFM} \times (1 - \nu^2) \times \sqrt{R} \times \delta^{\frac{3}{2}} \tag{4.10}$$

Where F is the loading force, E_{AFM} is the polymer surface Young's modulus, ν is the Poisson coefficient (0.5 for elastomers), R is the AFM tip radius, and δ is the real

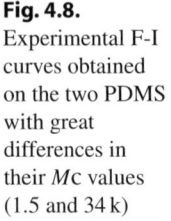

Fig. 4.8.
Experimental F-I curves obtained on the two PDMS with great differences in their *M*c values (1.5 and 34 k)

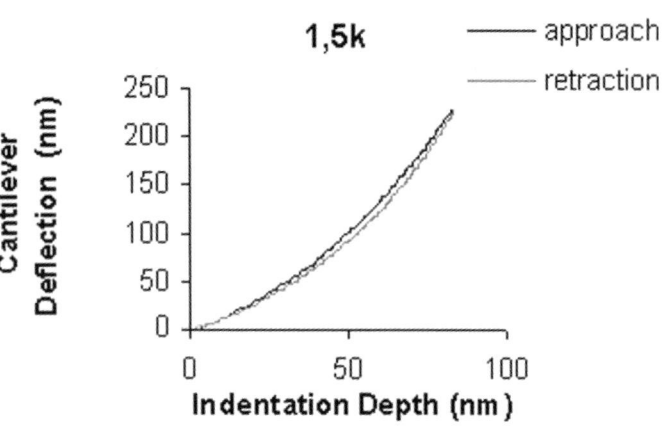

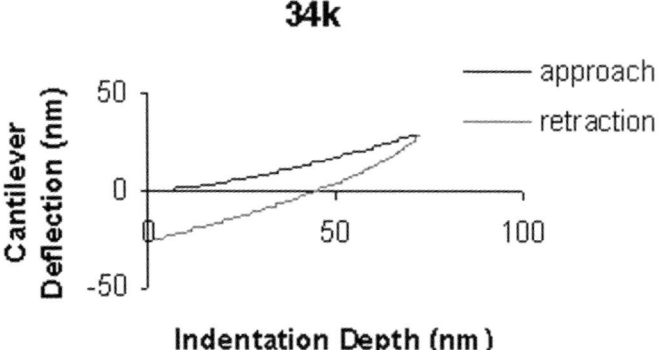

indentation. The calculated values of the polymer surface modulus are gathered in Table 4.5.

The values gathered in Table 4.5, indicate that the surface Young's modulus increases when *M*c decreases. This result is fully consistent with the expected one. Futhermore, the reciprocal of E_{AFM} varies linearly with *M*c, indicating that even at the nanoscale Flory's law of rubber elesticity is verified.

Table 4.5. PDMS surface Young's modulus, determined on the basis of nano-indentation experiments. Average mass between crosslinks (*M*c) is also reported

PDMS	E_{AFM} (Mpa)	$1/E_{AFM}$ (Mpa)	*M*c (g mol^{-1})
0.8 k	4.43	0.23	800
8.5 k	2.00	0.50	8,500
13 k	0.64	1.56	13,000
34 k	0.25	4.00	34,000

4.7.2
Nano-Indentation and Nano-Adhesion on Soft Polymers Having Controlled Surface Chemistry and Mechanical Properties

All nano-indentation curves obtained on PDMS and grafted PDMS have been performed at an 80-nm indentation depth, as discussed previously. We now consider PDMS having different Young's modulus grafted with molecules identical to those for the SAMs on silicon wafers (CH_3 and NH_2 SAMs). The idea is to compare the adhesion forces of two systems of identical surface chemistry but of different mechanical properties in order to determine the influence of viscoelastic dissipation in a separation process at the nanoscale. We have shown in Fig. 4.9 typical tip deflection-piezo displacement curves during tip retraction, for CH_3-grafted PDMS of various mass between crosslinks (0.8, 13, and 34 k).

One can see that the adhesion force significantly increases when the network becomes softer (i.e. when Mc increases). As the surface chemistry is identical for the three PDMS (Fig. 4.9), this increase of the adhesion force reflects the increase of viscous dissipation during tip retraction. Quantitative information can be extracted from such experimental curves.

The following ratio $\left(\frac{Fadh_{PDMSX}}{Fadh_{SiX}} \right)$, where $Fadh_{PDMS}$ and $Fadh_{Si}$ represents the adhesion force measured on PDMS and silicon substrates, respectively, and X represents the surface chemistry of the grafted SAM (i.e. CH_3 or NH_2), is calculated for all the PDMS substrates (Table 4.6).

It appears that for a given PDMS substrate (for example, 1.5 k), the ratios are independent of the surface chemistry, whereas for a given grafting, the ratios depend on the mechanical properties of the substrate, with the same values for all the grafts.

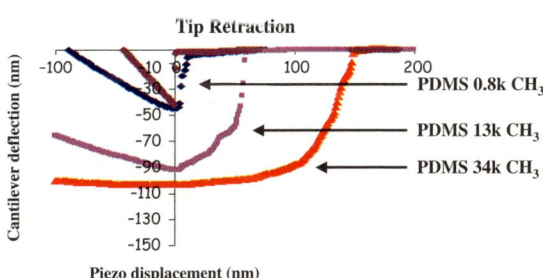

Fig. 4.9. Tip deflection vs. piezo displacement curves (during tip retraction) for CH_3-grafted PDMS of various mass between crosslinks (0.8, 13, and 34 k)

Table 4.6. Ratio $\left(\frac{Fadh_{PDMSX}}{Fadh_{SiX}} \right)$ for CH_3, NH_2, and ungrafted substrates and for three PDMS of different mechanical properties

PDMS	F_{PDMS}/F_{Si} ratio	$F_{PDMS-CH3}/F_{Si-CH3}$ ratio	$F_{PDMS-NH2}/F_{Si-NH2}$ ratio
0.8 k	1.1 ± 0.1	1.0 ± 0.1	1.2 ± 0.1
13 k	1.8 ± 0.1	1.5 ± 0.1	1.5 ± 0.1
34 k	2.0 ± 0.1	1.9 ± 0.1	2.6 ± 0.1

Thus, for a given substrate:

$$Fadh_{PDMSX} = Fadh_{SiX} \times C_1 \tag{4.11}$$

Where C_1 is a constant.

However, for an undeformable substrate, the DMT [41] theory gives:

$$Fadh_{SiX} = 2.\pi.R.W_0 \tag{4.12}$$

(where R is the tip radius and W_0 is the thermodynamic work of adhesion).

As a consequence:

$$Fadh_{PDMSX} = 2.\pi.R.W_0.C_1 \tag{4.13}$$

and when introducing a dimensional constant $(D_1 = (2.\pi.R)^{-1})$ this relationship becomes:

$$Gadh_{PDMSX} = W_0.C_1.2.\pi.R.D_1 \tag{4.14}$$

which can be rewritten as:

$$Gadh_{PDMSX} = W_0 \times f(Mc, \, v, \, T) \tag{4.15}$$

where G is the separation energy and $f(Mc, \, v, \, T)$ is called a "viscoelastic dissipation function" which depends on the network molecular structure (Mc), temperature T, and separation rate v.

This relationship clearly expresses the respective part of the mechanical contribution from the chemical one in an AFM force-indentation experiment.

As D_1 is a dimensional constant equal to $(2.\pi.R)^{-1}$, from Eqs. (4.14) and (4.15) we propose that:

$$f(Mc, \, v, \, T) = \{C_1\} \tag{4.16}$$

where $\{C_1\}$ is the average of the values of C_1 obtained for PDMS, PDMS–CH_3, and PDMS–NH_2, and thus for a given Mc (mass between crosslinks). According to Eq. (4.16), $f(Mc, v, T)$ is determined for each substrate for a given rate $(6\,\mu m\ s^{-1})$ and a given temperature (293 K). $f(Mc, v, T)$ does not depend on the surface chemistry and $f(Mc, v, T)$ cannot theoretically be lower than 1 (which corresponds to a zero separation rate) and should increase, while the Young's modulus decreases (which means energy dissipation in the bulk is higher when the network is softer). The values of the dissipative function $f(Mc, v, T)$ obtained (at room temperature and a separation rate of $6\,\mu m\ s^{-1}$) and thermodynamic work of adhesion W_0 are gathered in Table 4.7.

Values obtained are coherent with the previous assumptions. Moreover, even if the dependance of $f(Mc, v, T)$ on the separation rate v is not dominating in the range of the available separation rates with the AFM, it is obvious (Figs. 4.8 and 4.9), that dissipation occurs for the 34-k sample and that this dissipation is included in the function f. The dissipative function f is constant for a given grafting and increases as Mc does, i.e. as PDMS Young's modulus decreases. This is an expected result as viscoelastic dissipation should increase when the network becomes softer. According to Table 4.7, $f(Mc, v, T)$ is determined for each substrate for a given rate and a given temperature and is independent of the surface chemistry.

Table 4.7. Average viscoelastic dissipation function (f) and calculated thermodynamic work of adhesion (W_0)

PDMS	Dissipation function (f)	Work of adhesion W_0 (CH$_3$ grafts) (mJ m^{-2})	Work of adhesion W_0 (NH$_2$ grafts) (mJ m^{-2})
0.8 k	1.1 ± 0.1	35 ± 4	107 ± 15
13 k	1.6 ± 0.2	36 ± 4	105 ± 15
34 k	2.2 ± 0.4	32 ± 4	106 ± 15

Assuming that the separation of the tip from the polymer sample could be described as a Johnson, Kendall and Roberts (JKR) process, it is also possible to determine W_0, the thermodynamic work of adhesion. According to JKR formalism [54]:

$$W_0(X) = (2. Fadh_{PDMSX})/(3. \pi. R) \tag{4.17}$$

where $W_0(X)$ is the tip–PDMS thermodynamic work of adhesion for a given graft (X). Values of W_0 are gathered in Table 4.7, for each PDMS (0.8, 13, and 34 k) and for each grafting. Values of W_0 are in agreement with values obtained on silicon wafers (Fig. 4.5) that have the same surface chemistry and also with quoted values in the literature. Indeed, usual work of adhesion values lie between 40 and 70 mJ m^{-2} for organic–organic contacts such as between two silinated silica [55], and between 40 and 145 mJ m^{-2} for a contact between a raw material such as silica and a silinated silica. Finally, we have reported in Fig. 4.10 the evolution of the dissipative function f versus Mc, the mass between crosslinks.

A linear relationship is obtained. Therefore, a modified expression of G, the separation energy is proposed:

$$f(Mc, v, T) = 1 + Mc.f'(v, T) \tag{4.18}$$

thus:

$$Gadh_{PDMSX} = W_0 \times (1 + Mc.f'(v, T)) \tag{4.19}$$

where $f'(v, T)$ represents a dissipative viscoelastic function which depends only on the temperature T and separation rate v. As Mc varies linearly with the surface

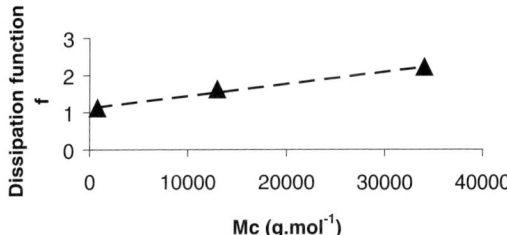

Fig. 4.10. Dissipation function f versus Mc, the average mass between crosslinks

modulus of the substrate, a complementary equation can be proposed:

$$Gadh_{PDMSX} = W_0 \times (1 + C_M \times (1/E_{AFM}) \times f'(v,\ T)) \tag{4.20}$$

Where C_M is a constant that links the reciprocal surface modulus (E_{AFM}) to the average mass between crosslinks. Equation (4.20) gives a complete understanding of how the adhesion separation energy depends on both surface chemistry and rheological behavior of the surface [E_{AFM} and $f'(v,\ T)$] and thus at a local scale.

4.8
Conclusion

Analysis of adhesive and mechanical properties of model polymer surfaces can be achieved with an Atomic Force Microscope, in contact mode. This chapter represents an attempt to develop new relationships for adhesion at the nanoscale using atomic force microscopy. The purpose was to dissociate the different contributions (chemical and mechanical) included in an AFM force-distance curve in order to establish relationships between interfacial tip–polymer interactions and surface viscoelastic properties of the polymer. Discussion was first focused on experimental and technical considerations that must be checked in order to properly conduct force-distance measurements. Then surface force measurements are, in a first step, conducted on chemically modified silicon substrates (grafted with hydroxyl, amine, methyl, and ester functional groups). Quantitative data can easily be extracted and give access to in situ determination of the thermodynamic work of adhesion. In order to investigate the effects of mechanical or viscoelastic contributions, force measurements on model PDMS networks, whose surfaces are chemically modified with the same functional groups as for silicon substrates, were made. New relationships are proposed between the local adhesion force and both the dissipation energy in the tip–polymer contact and the surface properties of the materials (thermodynamic work of adhesion, surface Young's modulus, dissipation function). Moreover, it is clearly shown that the dissipation function is related to M_C, the mass between crosslinks of the polymer network, which varies linearly with the reciprocal surface modulus of the soft polymer substrate.

References

1. Dubourg F, Aimé JP (2000) Surface Sci 466:137
2. Noel O, Awada H, Castelein G, Brogly M, Schultz J (2006) J Adhesion 82:649
3. Brogly M, Noel O, Awada H, Castelein G, Schultz J (2006) Comptes Rendus de l'Académie des Sciences, Chimie 9:99
4. Gauthier S, Aimé JP, Bouhacina T, Attias AJ, Desbat B (1996) Langmuir 12:5126
5. Paiva A, Sheller N, Forster MD, Crosby AJ, Shull KR (2000) Macromolecules 33:1878
6. Bushan B, Sundararajan S (1998) Acta Mater 46:3793
7. Beake BD, Leggett GJ, Shipway PH (1999) Surface Interface Anal 27:1084
8. Tomasetti E, Legras R, Nysten B (1998) Nanotechnologies 9:305
9. Israelachvili J (1991) Intermolecular and surface forces, 2nd ed. Academic Press, New York
10. Gent AN, Schultz J (1972) J Adhesion 3:281

11. Gutmann V (1977) The donor-acceptor approach to molecular interaction. Plenum Press, New York
12. Drago RS, Wayland BJ (1965) Am Chem Soc 87:3571
13. Cain SR (1991) In: Mittal KL, Anderson H Jr (eds) Acid-base interactions: relevance to adhesion science and technology. VSP, Zeist
14. Fowkes FM (1987) J Adhesion Sci Technol 1:7
15. Brogly M, Nardin M, Schultz J (1996) J Adhesion 58:263
16. Burns AR, Houston JE, Carpick RW, Michalske TA (1999) Langmuir 15:2922
17. Jones R, Pollock HM, Cleaver JAS, Hodges CS (2002) Langmuir 18:8045
18. Rabinovich YI, Adler J, Ata A, Singh RK, Moudgil BM (2000) J Colloid Interface Sci. 232:10
19. Unertl WN (2000) J Adhesion 74:195
20. Basire C (1998) Ph. D. Thesis, Université Paris VI, Paris, France
21. Basire C, Fretigny C (2001) Tribol Lett 10:189
22. Cuenot S, Duwez AS, Martin P, Nysten B (2002) Chimie Nouvelle 79:89
23. Cuenot S (2003) Ph. D. Thesis, Université Catholique de Louvain, Louvain, Belgium
24. Aime JP, Boisgard R, Nony L, Couturier G (2001) J Chem Phys 114:4945
25. Boisgard R, Aime JP, Couturier G (2002) Surface Sci 511:171
26. Dubourg F (2002) Ph. D. Thesis, Université de Bordeaux I, Bordeaux, France
27. Noel O, Brogly M, Castellein G, Schultz J (2004) Langmuir 20:2707
28. Albrecht TR, Akamine S, Carver, TE.,Quate CF (1990) J Vac Sci Technol 8:3386
29. Sader JE, White LR (1993) J Appl Phys 74:1
30. Hutter JL, Bechhoefer J (1993) Rev Sci Instrum 64:1868
31. Sader JE, Larson I, Mulvaney P, White LR (1995) Rev Sci Instrum 66:3789
32. Torrii A, Sasaki M, Hane K, Okuma S (1996) Meas Sci Technol 7:179
33. Noel O (2003) Ph. D. Thesis, Université de Haute Alsace, Mulhouse, France
34. Chaudhury MK, Whitesides GM (1992) Science 255:1230
35. Noel O, Brogly M, Castelein G, Schultz J (2004) Eur Polym J 40:965
36. Awada H, Castelein G, Brogly M (2005) Surface Interface Anal 37:755
37. Ruhe J, Novotny VJ, Kanazawa KK, Clarke T, Street GB (1993) Langmuir 9:2383
38. Wasserman SR, Whitesides GM, Tidswell IM, Ocko BM, Pershan PS, Axe JD (1989) J Am Chem 111:5852
39. Allara DL, Parikh AN, Rondelez F (1995) Langmuir 11:2357
40. Sugawara Y, Ohta M, Konishi T, Morita S, Suzuki M, Enomoto Y (1993) Wear 168:13
41. Derjaguin BV, Muller VM, Toporov YP (1975) J Colloid Interface Sci 53:314
42. Xiao X, Oian L (2000) Langmuir 16:8153
43. Sedin DL, Rowlen KL (2000) Analytical Chem 72:2183
44. He M, Szuchmacher Blum A, Aston DE, Buenviaje C, Overney RM, Luginbuhl R (2001) J Chem Phys 114:1355
45. Zhang L, Li L, Chan S, Jiang S (2002) Langmuir 18:5448
46. Hu J, Xiao XD, Ogletree DF, Salmeron M (1995) Surface Sci 344:221
47. Bruinsma R (1990) Macromolecules 23:276
48. de Gennes PG (1985) Rev Mod Phys 57:827
49. Riedo E, Levy F, Brune H (2002) Phys Rev Lett 88:185505
50. Weeks BL, Vaughn MW, DeYoreo JJ (2005) Langmuir 21:8096
51. Flory PJ (1944) Chem Rev 35:51
52. Aime JP, Michel D, Boisgard R, Nony L (1999) Phys Rev B 59:2407
53. Pethica JB, Sutton AP (1988) J Vac Sci Technol A 6:2490
54. Johnson KL, Kendall K, Roberts AD (1971) Proc Royal Soc A324:301
55. Papirer E, Balard H, Sidqi M (1993) J Colloid Interface Sci 159:238

5 Contact Resonance Force Microscopy Techniques for Nanomechanical Measurements*

Donna C. Hurley

Abstract. Contact resonance force microscopy (CR-FM) methods such as atomic force acoustic microscopy show great promise as tools for nanoscale materials research. However, accurate and reliable CR-FM measurements require the simultaneous optimization of a large number of experimental conditions. Among these variables are cantilever spring constant, applied static load, reference material, and resonant mode (mode type and order). In addition, results depend on the models used for data analysis and interpretation (e.g., choice of contact-mechanics model). All of these parameters are linked in numerous ways that are not straighforward to classify. In this chapter, we present a "user's guide" to quantitative measurements of nanomechanical properties with CR-FM methods. The discussion emphasizes the experimental methods and their practical implementation, providing a snapshot of the current state of the art. We discuss the basic physical principles involved and show how they can be used to make informed choices about experimental parameters and operating conditions. Experimental data and the results of theoretical models are provided as specific examples of the abstract concepts. Ideas for future work are also discussed, including ways to simplify the measurement process or improve measurement accuracy. The objective is not only to enable readers to perform their own CR-FM measurements, but also to optimize experimental conditions for a given material system. By gaining a better understanding of the underlying measurement principles, more researchers will be encouraged to further extend the technique and use it for an ever-wider range of applications for the nanoscale characterization of materials.

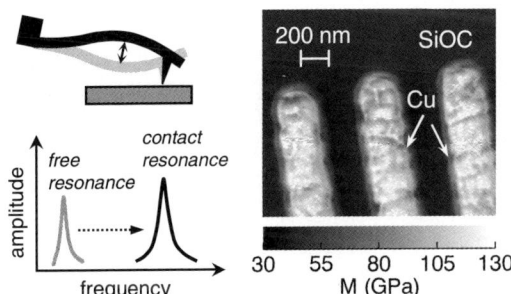

Key words: Atomic force acoustic microscopy, Contact resonance force microscopy, Elastic properties, Nanomechanical properties

Abbreviations

a	Contact radius between the tip and the sample
α	Tilt angle of the cantilever
b	Thickness of the cantilever
β	Exponent to describe the dependence of contact stiffness on applied force
c_B	Characteristic parameter of the cantilever
d	Deflection of the cantilever
d_0	"Jump to contact" deflection of the cantilever
E	Young's modulus
E^*	Reduced elastic modulus
F_N	Static force applied normal to the surface by the tip
f_n	Contact-resonance frequency of the nth flexural mode
f_n^0	Natural (free) frequency of the nth flexural mode
γ	Relative position of the cantilever tip
h	Height of the tip
k_{lever}	Flexural stiffness or spring constant of the cantilever
k, κ	Normal, lateral contact stiffness
L	Length of the cantilever
L_1	Distance from the fixed end of the cantilever to the tip
L'	Distance from the free end of the cantilever to the tip
M	Indentation (plane strain) modulus
m	Exponent to describe the tip–sample contact
n	Mode number
ν	Poisson's ratio
p_0	Normal component of the compressional stress applied at the sample surface by the tip
R	Radius of curvature of the tip
r	Radial distance from the axis of the tip
ρ	Mass density
S	Sensitivity of the mode response; derivative of the frequency versus contact stiffness curve
s	Sensitivity of the cantilever
σ_r	Radial component of the tensile stress in the sample
σ_z	Normal component of the compressional stress in the sample beneath the axis of the tip
w	Width of the cantilever
$x_n^{\ 0}$	Wavenumber of the nth flexural free resonance
z	Normal depth from the surface of the sample
AFM	Atomic force microscopy
AFAM	Atomic force acoustic microscopy
CR-FM	Contact resonance force microscopy
NI	Instrumented (nano-) indentation
SAWS	Surface acoustic wave spectroscopy
UAFM	Ultrasonic AFM
UFM	Ultrasonic force microscopy

5.1
Introduction

"The ability to accurately and reproducibly measure the properties and performance characteristics of nanoscale materials, devices, and systems is a critical enabler for progress in fundamental nanoscience, in the design of new nanomaterials, and ultimately in manufacturing new nanoscale products [1]." This quotation from the US National Nanotechnology Initiative emphasizes the need for measurement tools in emerging nanomaterial applications, a field predicted to generate a multibillion-dollar market within 10 years. One specific measurement need is for nanomechanical information—knowledge on the nanoscale of mechanical properties such as elastic modulus, adhesion, and friction. Accurate information is essential not only to predict the performance of a system before use, but also to evaluate its reliability during or after use. The measurement need is motivated partly by the fact that new applications often involve structures with nanoscale dimensions (e.g., nanoelectromechanical systems, nanoimprint lithography). Measurements of such structures by necessity must provide nanoscale spatial resolution. Other new structures have larger overall dimensions, but integrate disparate materials on the micro- or nanoscale (e.g., electronic interconnect, nanocomposites). In such cases, nanoscale information is needed in order to differentiate the properties of the various components.

Many methods to measure small-scale mechanical properties have been devised, including ones based on indentation [2–4], on ultrasonics [5,6], and on other physical phenomena [7,8]. Such methods often have drawbacks: they are not sufficiently quantitative, are limited to specialized geometries, and so forth. For instance, instrumented or "nano-" indentation (NI) [2] is inherently destructive, creating indents hundreds to thousands of nanometers wide. Conventional NI techniques may also provide insufficient spatial resolution as dimensions shrink further. A promising approach combines low-load NI techniques with force modulation and scanning [3]. However, the lateral resolution is still restricted by the radius (a few hundred nanometers) of the Berkovich diamond indenter employed.

Atomic force microscopy (AFM) methods present an attractive alternative for measuring mechanical properties. The small radius of the cantilever tip ($\sim$5–50 nm) provides nanoscale spatial resolution. Furthermore, the scanning capability of the AFM instrument enables rapid, in-situ imaging. AFM was originally developed to measure surface topography with atomic spatial resolution [9]. Since then, several AFM techniques have been demonstrated to sense mechanical properties [10–13]. Methods to measure mechanical properties that are based on force-displacement curves have also been extensively developed (see Ref. [14] for a review). Force-displacement methods work best when the compliance of the cantilever is roughly comparable to that of the test material. Therefore, these methods are better suited to very compliant ("soft") materials, and lose effectiveness as the material stiffness increases. The most promising AFM methods for quantitative measurements of relatively stiff materials such as ceramics or metals are dynamic approaches in which the cantilever is vibrated at or near its resonant frequencies [15]. These methods are often labeled "acoustic" or "ultrasonic," due to

the frequency of vibration involved (~100 kHz–3 MHz). Among them are ultrasonic force microscopy (UFM) [16, 17], heterodyne force microscopy [18], ultrasonic atomic force microscopy (UAFM) [19], and atomic force acoustic microscopy
(AFAM) [20].

Of these methods, AFAM (and to a lesser extent, UAFM) has achieved the
most progress in quantitative measurements. A general name for approaches such
as AFAM and UAFM is "contact resonance spectroscopy AFM" or more simply
"contact resonance force microscopy" (CR-FM). The key concepts of CR-FM are
illustrated in Fig. 5.1. Resonant vibrational modes of the cantilever are excited
either by an external actuator, as shown in the figure, or by an actuator attached
to the AFM cantilever holder. When the tip of the cantilever is out of contact
with the sample (Fig. 5.1a), the resonant modes occur at specific frequencies that
depend on the geometry and material properties of the cantilever. When the tip
is placed in contact (Fig. 5.1b), the frequencies of the resonant modes increase
due to tip–sample forces (Fig. 5.1c). CR-FM involves measuring the frequencies
at which the free and contact resonances occur. The mechanical properties of
the sample are then deduced from these frequencies with the help of appropriate models. Quantitative measurements have also been demonstrated with UFM
methods [21].

In this chapter, we present a "user's guide" to quantitative measurements of
nanomechanical properties with CR-FM methods. An earlier chapter in this series
by Rabe [22] provided a comprehensive review focusing on the technique's theoretical foundation. Here, the discussion emphasizes the experimental methods and their
practical implementation, providing a snapshot of the state of the art. Current best
practices for data acquisition and analysis are not merely stated, but are explained
and justified in terms of the physical principles involved. Practical examples are provided to illustrate the concepts discussed. Our objective is to enable readers not only
to perform CR-FM measurements, but also to optimize experimental conditions for
their particular needs. By gaining a better understanding of the underlying measurement principles, more researchers will be encouraged to further extend the technique
and use it for an ever-wider range of applications.

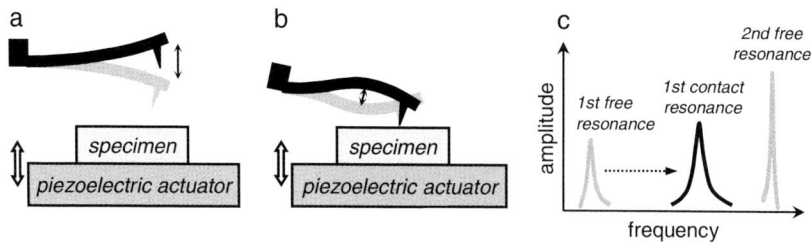

Fig. 5.1. Concepts of contact resonance force microscopy (CR-FM). Resonant modes of the
cantilever are excited by a piezoelectric actuator when the tip is **a** in free space and **b** in contact
with a specimen under an applied static force. **c** Resonant spectra. The lowest-order contact
resonance occurs at a higher frequency than the first free-space resonance, but is lower than the
second free-space resonance

5.2
Cantilevers for Contact Resonance Force Microscopy

Accurate CR-FM measurements begin by choosing a suitable cantilever. Measurements on stiff materials (modulus greater than approximately 50 GPa) involve relatively stiff cantilevers—ones with a spring constant k_{lever} of approximately 30 to 50 N/m. A variety of such cantilevers are available commercially and are usually sold as "noncontact" or "intermittent contact" probes. As seen below, interpretation of CR-FM data requires a model for the vibrating cantilever. The model assumes that the cantilever is a uniform rectangular beam, so that analytical expressions can be derived. For this reason, the cantilevers typically used in CR-FM have a simple rectangular shape and are micromachined from single-crystal silicon (Si). The long axis of the cantilever is oriented in the <110> crystalline direction, and the axis of the tip is <001>-oriented.

Table 5.1 contains information about several AFM cantilevers. The table shows the nominal length L, width w, thickness b, and spring constant k_{lever} for each cantilever, as given by the manufacturer. For a simple rectangular cantilever beam, k_{lever} may be determined from the relationship $k_{lever} = Eb^3w/(4L^3)$, where E is Young's modulus. The table also gives measured values of the frequency f_n^0 of the nth order flexural free resonance for each cantilever. Cantilever #1 is the type most frequently used for CR-FM experiments in our laboratory. Cantilever #2 has dimensions and spring constant similar to those of #1, but was made by a different manufacturer. Note that these cantilevers are relatively long. Shorter cantilevers with similar values

Table 5.1. Properties of cantilevers used in CR-FM experiments. Shown are the nominal length L, width w, thickness b, and nominal spring constant k_{lever} cited by the manufacturer. Also given are the measured values of the four lowest free resonant frequencies and the frequency ratio f_n^0/f_1^0 of the nth free frequency to the first free frequency. f_4^0 could not be measured for cantilevers #1 and #2 due to bandwidth limitations of the AFM photodiode. The "theory" column shows the values of f_n^0/f_1^0 predicted by the analytical model discussed in the text

Property	Cantilever #				Theory
	1	2	3	4	
L (μm)	225	230	230	450	
w (μm)	38	40	40	55	
b (μm)	7	7	3	4	
k_{lever} (N/m)	48	40	3.5	1.6	
f_1^0 (kHz)	171.66	167.68	72.76	20.99	
f_2^0 (kHz)	1,066.5	1,071.2	501.57	131.40	
f_3^0 (kHz)	2,937.3	3,011.7	1,464.2	368.24	
f_4^0 (kHz)	—	—	2,956.4	721.09	
f_2^0/f_1^0	6.21	6.39	6.89	6.26	6.27
f_3^0/f_1^0	17.11	17.96	20.12	17.54	17.55
f_4^0/f_1^0	—	—	40.63	34.35	34.39

of k_{lever} are available (e.g., $L \approx 125\,\mu m$), but their resonant frequencies will be higher and could be difficult to detect, given the finite bandwidth of the AFM photodiode detector. Cantilever #3 is similar to #1 and #2 in length and width, but is thinner and therefore has a lower k_{lever}. The spring constant of cantilever #4 is even lower, because it is thinner and longer.

To perform measurements, cantilevers are mounted in the standard holder provided by the AFM manufacturer. Recent research indicates that the mounting or clamping conditions of the cantilever in the holder can affect the measurements in some cases [23]. Depending on the specific holder used, it may therefore be advisable to develop an improved mounting method (e.g., gluing the cantilever in the holder, using a firmer clamp). Further work is needed to better understand the practical implications of these effects on measurements.

Before performing contact experiments, the cantilever's free (natural) frequencies when the tip is out of contact must be measured. The values of the free frequencies are used to characterize the cantilever's properties, as discussed below. Typically, the free frequencies f_n^0 of the lowest two or three flexural modes are measured. One way to measure the free frequencies is with the AFM's "tuning" software intended for intermittent-contact operation. The subroutine controls a small piezoelectric actuator at the clamped end of the cantilever. Depending on the frequency characteristics of the actuator, however, it may be difficult to excite the higher-order modes in this way. Alternatively, the free frequencies can be measured with a piezoelectric actuator mounted beneath the specimen. The cantilever is brought close to, but not in contact with, the specimen. Driving the actuator at relatively high voltages creates acoustic vibrations that are large enough to excite the cantilever's free resonances via air coupling. Measured values of f_n^0 for cantilevers #1 – #4 are given in Table 5.1.

Determining the free frequencies is necessary for data analysis, but also serves a second purpose. The ratio f_n^0/f_1^0 of the nth free mode to the lowest mode f_1^0 is an indicator of measurement quality. The closer the measured values of f_n^0/f_1^0 are to those predicted by the analytical model, the more likely it is that the model accurately describes the dynamics of the actual cantilever. Table 5.1 contains the measured ratios f_n^0/f_1^0 for the four cantilevers. The predicted and measured ratios differ by less than 3% for #1 and #2, and less than 0.2% for #4. The discrepancies are probably due to small variations in the micromachining process, for instance small irregularities in the cross section or a thickness variation along the length of the cantilever [24]. In the case of cantilever #3, the measured and predicted values of f_n^0/f_1^0 differ by more than 10%. Use of such cantilevers should be avoided if possible. One possible explanation for the discrepancy is that the top side of cantilever #3 contains a thin reflective film to boost the detected signal amplitude. In practice, we are able to achieve adequate signal amplitude with uncoated cantilevers.

The scanning electron microscope (SEM) can be used to obtain additional information about cantilevers. Figure 5.2 shows SEM images of two cantilevers that are candidates for CR-FM experiments. Figure 5.2a–c were acquired with a cantilever similar to #1 in Table 5.1, while Fig. 5.2d–f were acquired with cantilever #2. Several observations can be made from the images. First, the cross section of each cantilever is actually trapezoidal, not rectangular. This is more pronounced in the first cantilever. The cross section is uniform throughout most of the length of the cantilever,

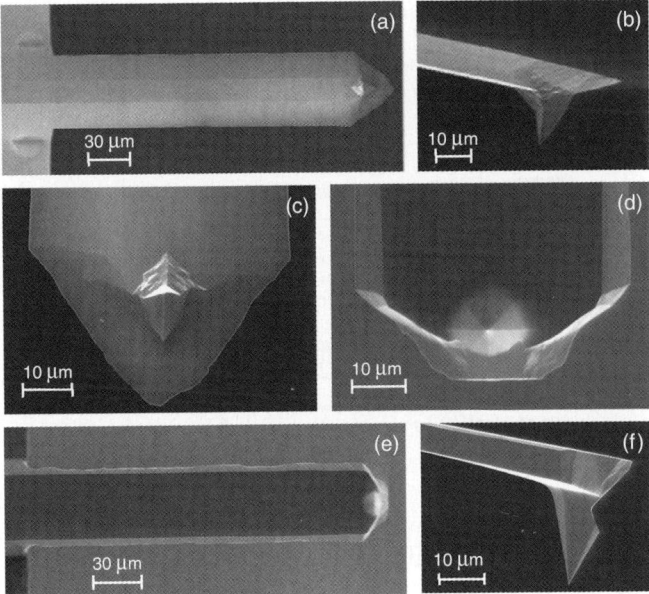

Fig. 5.2. SEM images of single-crystal silicon cantilevers. **a** Plan view of the underside of a cantilever similar to #1 in Table 5.1. **b** Side view and **c** plan view of the tip end of the cantilever. **d** Closeup of the tip end of cantilever #2 in Table 5.1. **e** Plan view of the underside of cantilever #2. **f** Side view of the tip end of the cantilever

but varies in both width and thickness at the very end of the cantilever. In addition, the tip is not located at the very end of the cantilever. The distance between the tip and the end of the cantilever differs from cantilever to cantilever. Finally, the tip of cantilever #2 is noticeably longer.

5.3
Data Acquisition Techniques

In this section, we describe current practices for quantitative measurements of elastic modulus with CR-FM. A "recipe" is presented based on standard practices of our own and other groups. Following this recipe will enable readers to perform similar measurements themselves. Familiarity with the recipe is also necessary to understand subsequent discussions about improving and optimizing measurements.

Figure 5.3 shows a block diagram of the apparatus used in our laboratory for CR-FM experiments. The constituent components, all of which are commercially available, include (a) an AFM instrument, (b) a function generator and piezoelectric actuator to excite the cantilever resonances, (c) a lock-in amplifier for frequency-selective signal detection, and (d) a computer for instrument control and data acquisition. The equipment used in our laboratory is specified here, but other equipment with similar characteristics could also be used [25].

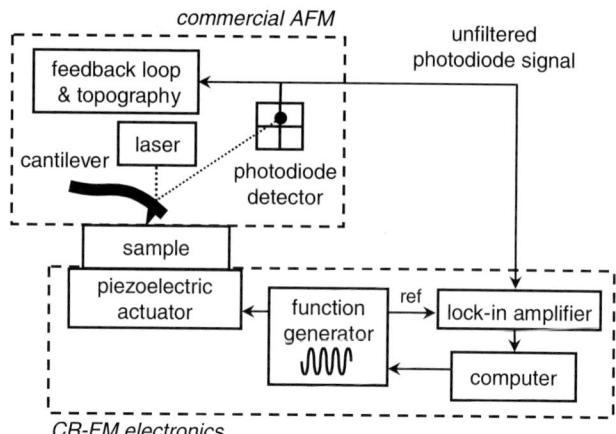

Fig. 5.3. Block diagram of apparatus for CR-FM experiments

Contact-resonance experiments have been performed with several different AFM instruments, including the Digital Instruments Dimension 3000 (Veeco Metrology, Inc., Santa Barbara, CA) and the Asylum MFP-3D (Asylum Research, Santa Barbara, CA). The AFM instrument must meet two main requirements. First, it must provide access to the unfiltered, high-frequency photodiode detector signal. Second, the bandwidth of this signal must be sufficient to detect frequencies up to at least 2 MHz (preferably 3 MHz). Access is provided in some instruments by an accessory unit. The ability to input an external signal for image acquisition is also useful, although not strictly required for modulus measurements.

As a piezoelectric actuator, we find it convenient to use ultrasonic contact transducers designed for nondestructive testing (Panametrics, Olympus NDT, Waltham, MA). Longitudinal transducers with the excitation motion normal to the transducer surface are suitable for experiments involving flexural modes. The piezoelectric element in these transducers is heavily damped for a broad frequency response (typically, -6 dB rolloff at $\pm 50\%$ of the center frequency). Thus, a single transducer with a center frequency of 1 MHz or 2.25 MHz can excite all of the cantilever resonances from approximately 100 kHz to 3 MHz. Another advantage of these transducers is their relatively large diameter (approximately 1–3 cm). This means that reasonably large specimens can be accommodated, and that the vibration amplitude varies fairly slowly across the specimen. The simplest way to mount the sample is to bond it directly to the top of the transducer. Various glues provide a rigid bond, but sample removal is easier if an acoustic couplant such as glycerin is used. Samples mounted in this way should be allowed to "settle" for several hours or overnight to ensure stable measurements.

The piezoelectric transducer is driven with a continuous sine wave signal from a function generator (33120A, Agilent Technologies, Santa Clara, CA). The primary requirements are a programmable output with adjustable frequency ($\sim$10 kHz–3 MHz) and amplitude (0–5 V peak-to-peak). The required voltage amplitude depends on the actuator used. The actuator vibration must be sufficiently strong to excite the resonant modes of the cantilever when the tip is in contact, yet the vibration amplitude must be sufficiently small that the tip and the sample remain in contact at all times. Continuous contact ensures a linear elastic regime, so that the assumptions of the data analysis model are valid. With the transducers described above, a driving amplitude of approximately 100 mV or less is usually adequate. Optical measurements with a Michelson interferometer on similar transducers suggest that the peak vibration amplitude at the surface of the transducer is much less than 0.1 nm at this level of excitation voltage.

The photodiode detector signal from the AFM instrument is connected to the input of a lock-in amplifier (SR844, Stanford Research Systems, Sunnyvale, CA). The lock-in must have sufficient bandwidth to detect the contact-resonance frequencies. The sync signal from the function generator provides the reference signal for the lock-in amplifier. With this arrangement, the output amplitude of the lock-in gives the amplitude of cantilever vibration only at the excitation frequency.

Data acquisition software is straightforward and can be created with commercial tools such as LabVIEW (National Instruments Corp., Austin, TX). In the data acquisition routine, the function generator is programmed to output a continuous sine wave with a specific frequency and amplitude. The resulting output signal of the lock-in amplifier is then recorded. The excitation frequency is incrementally increased, and the sequence is repeated until the desired maximum frequency is reached. In this way, a spectrum of the cantilever vibration amplitude versus frequency is recorded. Both the amplitude and phase of the lock-in amplifier signal are recorded. To date, the phase information has not been formally used in the data analysis. However, it can be useful in identifying the exact resonant frequency if the amplitude peak is very small.

Examples of experimental contact-resonance spectra are shown in Fig. 5.4. The data were acquired with cantilever #1 in Table 5.1. Spectra of the first, second, and third free-space flexural resonances are shown in Fig. 5.4a, c, e, respectively. The corresponding contact-resonance spectra are shown in Fig. 5.4b, d, f. The sample was a <102> SnO_2 nanobelt [26]. Spectra are shown for two values of the static cantilever deflection d. Using the relation $F_N = k_{lever} d$ between the deflection d and the resulting static force F_N and assuming $k_{lever} = 48$ N/m, it is found that $F_N = 0.7 \mu N$ and $2.2 \mu N$ for $d = 15$ nm (dashed line) and $d = 45$ nm (solid line). The spectra show that the peak contact-resonance frequency increases with increasing F_N (or d). As discussed below, such behavior is predicted by the contact mechanics models used for data analysis. The amplitude of the signal also tends to increase with increasing F_N, although this is not always true. The signal amplitude can be seen to decrease with increasing mode number or frequency. One explanation for this trend could be rolloff in the frequency response of the AFM photodiode detector.

CR-FM experiments consist of acquiring contact-resonance spectra for two specimens in alternation: the "test" (unknown) specimen and a "reference" specimen whose elastic properties are known. Reasons for this referencing approach are discussed below in Sect. 5.4. The elastic properties of reference specimens can be determined by various means, including pulse-echo ultrasonics [27] and instrumented (nano-) indentation [2, 28]. For accurate measurements, the elastic properties of the reference specimen should be similar to those of the test specimen [24,29,30]. If possible, a means to measure both specimens at the same time should be devised in order to avoid repeated mounting and unmounting. It is important that the alignment of the

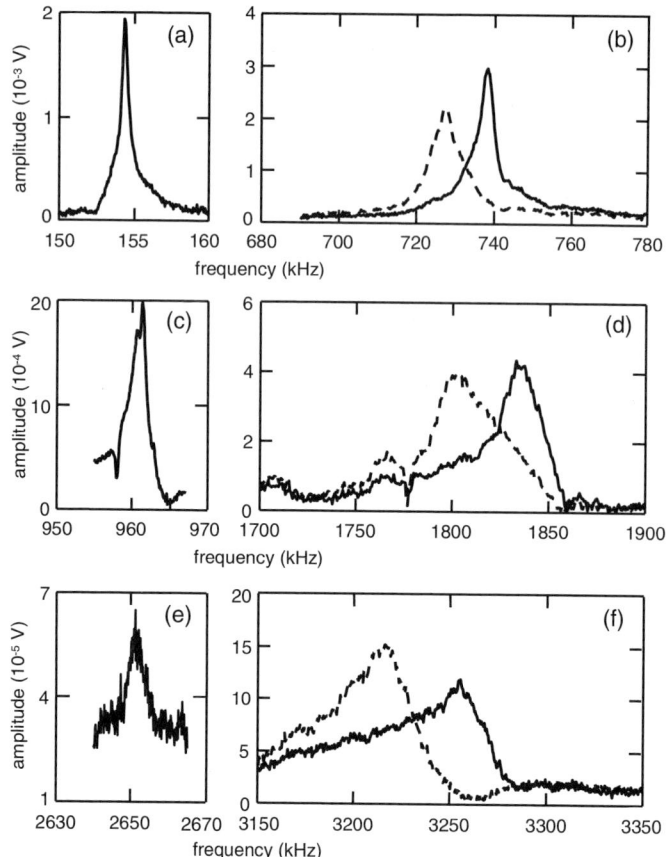

Fig. 5.4. Examples of experimental results. **a, c,** and **e**: Spectra of the first, second, and third free-space flexural resonances. **b, d, f**: Corresponding contact-resonance spectra for $d = 15\,\mathrm{nm}$ (*dashed line*) and $d = 45\,\mathrm{nm}$ (*solid line*)

laser spot on the cantilever remains the same throughout all of the measurements. This ensures that measurements made at the same static deflection d correspond to the same static load F_N, without direct measurement of the spring constant k_{lever}.

The basic measurement procedure consists of the following steps:

1. Measure the free frequencies f_n^0 of the two lowest flexural modes.
2. Bring the cantilever into contact with the reference specimen. The tip must remain stationary (i.e., set the scan size to zero).
3. Acquire force-distance ("force calibration") curves to determine the cantilever sensitivity ("optical lever sensitivity"). The sensitivity s relates the deflection d of the cantilever to the output voltage of the photodiode detector. Thus, if $s = 100\,\mathrm{nm/V}$, increasing the setpoint voltage by 0.1 V will increase d by 10 nm. Also note the "pull-on" or "jump to contact" deflection d_0, which indicates the adhesion.
4. Set the static deflection d of the cantilever to a specific value. Typically, d ranges from 10 to 60 nm. Elastic contact between the tip and the sample is needed to

ensure that the analysis model is valid. Therefore, d should be at least tentimes larger than the deflection d_0 due to adhesion. However, d should be kept as low as possible to minimize potential damage to both the tip and the sample.

5. Acquire contact-resonance spectra for the lowest two flexural modes. Typically, spectra are acquired for three or more values of d at a given location on the specimen. This not only increases the number of data points, but also provides insight into the tip–sample contact.

6. Repeat steps 2–5 on the test specimen, at the same values for d. The setpoint voltages may differ slightly from those for the reference specimen.

7. Repeat steps 2–5 on the reference and test specimens in alternation, ending with the reference specimen. Perform sets of measurements several times on both specimens, in order to achieve a statistically sufficient number of data.

5.4
Data Analysis Methods

Data analysis consists of two distinct steps, each involving a separate model. First, the measured frequencies are related to the tip–sample interaction force by means of a model for the dynamic motion of the cantilever. Next, the interaction force—the contact stiffness—is used to determine the elastic properties of the sample through a model for the contact mechanics between the tip and the sample. Here, we present without proof the equations needed for both models. For further discussion, the reader is referred to Ref. [22]. The development here is intended to facilitate software implementation. Because the analysis models are based on analytical formulas, calculations are straightforward to implement in commercial software such as LabVIEW or IDL (ITT Visual Information Solutions, Boulder, CO).

5.4.1
Model for Cantilever Dynamics

Figure 5.5a shows the model to describe the cantilever's vibrations in free space. It is important to note that a distributed-mass model is used, not a point-mass (harmonic oscillator) model. It has been shown that the point-mass approximation does

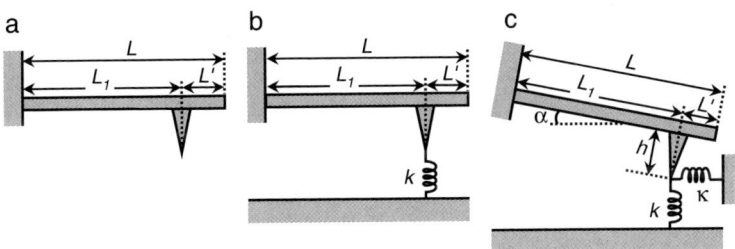

Fig. 5.5. Models for cantilever dynamics. **a** Tip out of contact. **b** Tip in contact. Only normal (vertical) elastic forces are considered. **c** Tip in contact, with both normal and tangential elastic forces included

not produce accurate results under the conditions described here [31, 32]. Because estimates of the tip mass are typically less than 0.5% of the cantilever mass, it is neglected. The cantilever is modeled as an elastically isotropic beam of uniform cross section with length L, width w, thickness b, density ρ, and Young's modulus E. The tip is located at a distance $L_1 < L$ from the clamped end of the cantilever. The remaining distance to the free end of the cantilever is L', so that $L = L_1 + L'$. The flexural spring constant of the cantilever is $k_{lever} = Eb^3 w/(4L_1^3)$. The frequency f_n^0 of the nth free flexural resonance is related to the wavenumber x_n^0 by

$$\left(x_n^0 L\right)^2 = 4\pi f_n^0 \frac{L^2}{b} \sqrt{\frac{3\rho}{E}} = f_n^0 (c_B L)^2. \tag{5.1}$$

$x_n^0 L$ is also a root of

$$1 + \cos x_n^0 L \cosh x_n^0 L = 0. \tag{5.2}$$

The first four roots of Eq. (5.2) are $[x_1^0 L, \ x_2^0 L, \ x_3^0 L, \ x_4^0 L] = [1.8751,\ 4.6941,\ 7.8548,\ 10.996]$. Experimental values for the cantilever parameter $c_B L$ for each mode can therefore be calculated from the measured values of the free frequencies f_n^0. This approach means that cantilever properties such as L and E need not be determined directly. Equation (5.1) also shows that

$$\frac{f_n^0}{f_{n-1}^0} = \left(\frac{x_n^0 L}{x_{n-1}^0 L}\right)^2. \tag{5.3}$$

Inserting the values of $x_n^0 L$ given above into Eq. (5.3) yields the ratios listed in Table 5.1.

Figure 5.5b depicts the simplest model for cantilever dynamics if the tip is in contact. In this case, the tip–sample interaction is entirely elastic and acts in a direction normal (vertical) to the sample surface. The tip–sample interaction is represented by a spring with spring constant k, also known as the contact stiffness. By considering the dynamics of this system [32], it is found that the normalized contact stiffness k/k_{lever} is given by

$$\begin{aligned}
\frac{k}{k_{lever}} & \left[\left(\sin x_n L' \cosh x_n L' - \cos x_n L' \sinh x_n L'\right)\left(1 - \cos x_n L_1 \cosh x_n L_1\right)\right. \\
& \left. - \left(\sin x_n L_1 \cosh x_n L_1 - \cos x_n L_1 \sinh x_n L_1\right)\left(1 + \cos x_n L' \cosh x_n L'\right)\right] \\
& = \frac{2}{3}(x_n L_1)^3 \left(1 + \cos x_n L \cosh x_n L\right)
\end{aligned} \tag{5.4}$$

Here, x_n is the wavenumber of the nth flexural contact resonance. Rewriting Eq. (5.4) in terms of the relative tip position ratio $\gamma = L_1/L$ and rearranging terms, we find

$$\frac{k}{k_{lever}} = \frac{2}{3}(x_n L \gamma)^3 \frac{\left(1 + \cos x_n L \cosh x_n L\right)}{D}, \tag{5.5}$$

where

$$D = [\sin x_n L(1 - \gamma) \cosh x_n L(1 - \gamma) - \cos x_n L\gamma \sinh x_n L\gamma][1 - \cos x_n L\gamma \cosh x_n L\gamma]$$
$$- [\sin x_n L\gamma \cosh x_n L\gamma - \cos x_n L\gamma \sinh x_n L\gamma][1 + \cos x_n L(1 - \gamma) \cosh x_n L(1 - \gamma)].$$

In the data analysis, Eq. (5.5) is used to calculate k/k_{lever} for each measured contact-resonance frequency f_n. First, the values of f_n are used to calculate $x_n L$:

$$x_n L = c_B L\sqrt{f_n} = x_n^0 L\sqrt{\frac{f_n}{f_n^0}}, \tag{5.6}$$

where Eq. (5.1) has been used. From the experimental values of $x_n L$, Eq. (5.5) is then used to plot k/k_{lever} as a function of the relative tip position $\gamma = L_1/L$ for each mode n. Figure 5.6 is an example of the resulting plot. The graph shows that the values of k/k_{lever} are the same for the two flexural modes at only one value of γ. This value of k/k_{lever} where the two modes intersect or "cross" is taken as the solution. An online calculator that performs this operation for $n = 1$ and 2 is available [33].

To aid the reader in developing data analysis software, Table 5.2 contains experimental values of the contact-resonance frequencies and the corresponding values of k/k_{lever} and L_1/L. It can be seen that for these data, the calculated values of $L_1/L \approx 0.97$. Direct measurement of the cantilever dimensions in SEM and optical micrographs suggests that the actual tip location $L_1/L \approx 0.91$–0.92 for this particular cantilever. Discrepancies between the calculated and measured values of L_1/L have been observed previously [24, 34]. The most likely explanation is the deviation of the cantilever's actual behavior from that predicted by the idealized beam model. For instance, Fig. 5.2 reveals that the cantilever cross section near the tip end is not uniform for this type of cantilever. A mass or volume analysis of the actual cantilever shape suggests a higher value of L_1/L than that obtained from simple length analysis.

If lateral elastic forces are included, the model shown in Fig. 5.5c is used. The cantilever is tilted by an angle α with respect to the sample surface. The tip has height h. The elastic interaction between the tip and the sample is represented by two springs: a vertical spring with stiffness k and a horizontal (tangential) spring with stiffness κ. The vertical contact stiffness k normalized by the flexural cantilever stiffness k_{lever} is given by [34–36] the positive root of

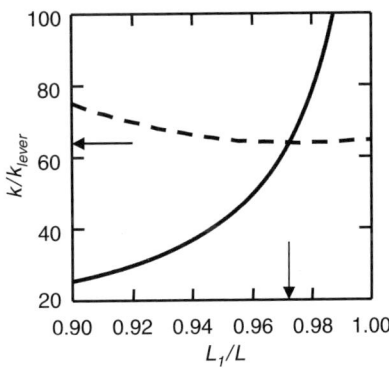

Fig. 5.6. Normalized contact stiffness k/k_{lever} as a function of the relative tip position L_1/L. Results are shown for $n = 1$ (*solid line*) with $f_1 = 734.5$ kHz and $n = 2$ (*dashed line*) with $f_2 = 1,806.4$ kHz (see Table 5.2). The measurements were made with cantilever #1 in Table 5.1. The *arrows* indicate the values of k/k_{lever} and L_1/L where the two curves intersect or "cross"

Table 5.2. Examples of experimental results. The test specimen was a $<102>$ SnO$_2$ nanobelt (NB) and the reference material was $<100>$ Si. Measurements were performed with cantilever #1 in Table 5.1. Shown are the contact-resonance frequencies f_1 and f_2 of the lowest two flexural modes versus the cantilever static deflection d. Also shown are the values of the relative tip position L_1/L for which the normalized contact stiffness k/k_{lever} is the same for both modes

Specimen	Trial #	d (nm)	f_1 (kHz)	f_2 (kHz)	L_1/L	k/k_{lever}
Si $<100>$ (reference)	1	15	732.0	1,803.4	0.974	63.7
		30	734.5	1,806.4	0.973	64.1
		45	737.3	1,832.8	0.973	67.5
SnO$_2$ NB (test)	1	15	727.3	1,784.2	0.976	61.3
		30	732.5	1,802.8	0.974	63.7
		45	734.5	1,817.8	0.973	65.5
Si $<100>$ (reference)	2	15	735.0	1,818.0	0.973	65.6
		30	737.8	1,827.5	0.972	66.8
		45	739.3	1,840.5	0.972	68.5

$$\frac{k}{k_{lever}} = \frac{-B \pm \sqrt{B^2 - 4AC}}{6A}, \tag{5.7}$$

where

$$A = \left(\frac{\kappa}{k}\right)\left(\frac{h}{L_1}\right)^2 (1 - \cos x_n L \cosh x_n L_1)(1 + \cos x_n L' \cosh x_n L'),$$

$$B = B_1 + B_2 + B_3,$$

$$C = 2(x_n L_1)^4 (1 + \cos x_n L_1 \cosh x_n L),$$

with

$$B_1 = \left(\frac{h}{L_1}\right)^2 (x_n L_1)^3 \left(\sin^2 \alpha + \frac{\kappa}{k}\cos^2 \alpha\right)$$
$$\times [(1 + \cos x_n L' \cosh x_n L')(\sin x_n L_1 \cosh x_n L_1 + \cos x_n L_1 \sinh x_n L_1)$$
$$- (1 - \cos x_n L_1 \cosh x_n L_1)(\sin x_n L' \cosh x_n L' + \cos x_n L' \sinh x_n L')],$$

$$B_2 = 2\left(\frac{h}{L_1}\right)(x_n L_1)^2 \left(\frac{\kappa}{k}\cos \alpha \sin \alpha\right)$$
$$\times \left[(1 + \cos x_n L' \cosh x_n L')(\sin x_n L_1 \sinh x_n L_1)\right.$$
$$+ (1 - \cos x_n L_1 \cosh x_n L_1)(\sin x_n L' \sinh x_n L')\Big],$$

$$B_3 = (x_n L_1)\left(\cos^2 \alpha + \frac{\kappa}{k}\sin^2 \alpha\right)$$
$$\times [(1 + \cos x_n L' \cosh x_n L')(\sin x_n L_1 \cosh x_n L_1 - \cos x_n L_1 \sinh x_n L_1)$$
$$- (1 - \cos x_n L_1 \cosh x_n L_1)(\sin x_n L' \cosh x_n L' - \cos x_n L' \sinh x_n L')].$$

The inclusion of lateral forces greatly increases the number of variables involved. It is therefore desirable to use the simpler model in Fig. 5.5b when possible. The conditions under which this model is valid are discussed in Sect. 5.6. More detailed discussions of this model and its application to CR-FM analysis have been presented elsewhere [22, 34, 36].

Damping (inelastic or dissipative) interactions are beyond the scope of this discussion. In the model for cantilever dynamics, damping is included by means of one or more dashpots in parallel with the springs. This model is discussed in detail elsewhere [22, 36]. Analysis of CR-FM data with the inclusion of damping effects has been presented [37, 38]. Further work is needed to develop a practical data analysis procedure that includes damping terms.

5.4.2
Model for Contact Mechanics

After obtaining values for the normalized contact stiffness k/k_{lever} from the measured contact-resonance frequencies, we can use these values to determine elastic properties of the specimen. This step of the analysis requires a second model, namely one for the contact mechanics between the tip and the sample. A complete discussion of contact mechanics is given elsewhere [39]. The analysis is similar in many respects to that used to interpret instrumented (nano-) indentation data [2, 28].

The key parameters of two different models for contact mechanics are shown in Fig. 5.7. The indentation moduli of the tip and the sample are M_{tip} and M_s, respectively. Figure 5.7a represents Hertzian contact between a hemispherical tip with radius of curvature R and a flat sample. A vertical (normal) static load F_N is applied to the tip, creating a circular contact of radius a. Figure 5.7b shows flat-punch contact between a flat tip and a flat sample. In this case, the contact area is constant with F_N and is determined by the tip diameter $2a$. For both cases, the vertical contact stiffness k between the tip and the sample is given by

$$k = 2aE^*, \tag{5.8}$$

where E^* is the reduced elastic modulus for the tip–sample system:

$$\frac{1}{E^*} = \frac{1}{M_{tip}} + \frac{1}{M_s}. \tag{5.9}$$

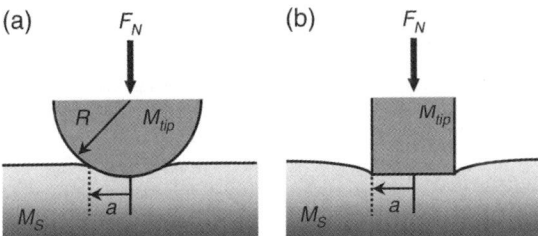

Fig. 5.7. Models for **a** Hertzian contact between a hemispherical tip and a flat sample and **b** flat-punch contact between a flat tip and a flat sample

For elastically isotropic materials, the indentation or plane strain modulus $M = E/(1 - v^2)$, where E is Young's modulus E and v is Poisson's ratio. Then

$$\frac{1}{E^*} = \frac{1 - v_{tip}^2}{E_{tip}} + \frac{1 - v_s^2}{E_s}. \tag{5.10}$$

To determine E^* from Eq. (5.8), it is necessary to know the contact radius a. For a flat indenter ("flat punch"), a is constant. For Hertzian contact, a is given by

$$a = \sqrt[3]{\frac{3RF_N}{4E^*}}. \tag{5.11}$$

In principle, one could measure a directly (for a flat punch), or determine experimental values for R and F_N and calculate a. Equation (5.8) or (5.11) could then be used to determine E^* and hence M. In practice, a referencing or comparison approach is used [24, 29] in which measurements are performed on the test (subscript s) and reference (subscript ref) samples at the same values of F_N. The approach assumes that the values of M_{ref} and thus E^*_{ref} are known. Then it can be shown [29] that

$$E^* = E^*_{ref} \left(\frac{k_s}{k_{ref}} \right)^m = E^*_{ref} \left(\frac{\frac{k_s}{k_{lever}}}{\frac{k_{ref}}{k_{lever}}} \right)^m, \tag{5.12}$$

where $m = 3/2$ for Hertzian contact and $m = 1$ for a flat punch. From the values of E^* obtained by Eq. (5.12) with the experimental values of k_s/k_{lever} and k_{ref}/k_{lever}, the modulus M of the sample can be calculated with Eq. (5.9). As discussed in Sect. 5.7, the true shape of the tip is usually intermediate between a hemisphere and a flat [40]. Thus, $m = 3/2$ and $m = 1$ represent upper and lower limits on M. Values for E^* obtained with this approach for the data in Table 5.2 are shown in Table 5.3. Also included are the values of the sample modulus M calculated from E^* assuming Hertzian and flat-punch contact.

The referencing approach invented by Rabe and coworkers [29] represents an important measurement innovation for CR-FM methods. Accurate determination

Table 5.3. Results of contact-mechanics analysis for the data in Table 5.2. The contact-stiffness ratio k_{test}/k_{ref} is given for each combination of test and reference measurements. The corresponding values of the reduced elastic modulus E^* and the indentation modulus M of the test specimen are shown, assuming either Hertzian ($m = 3/2$) or flat-punch ($m = 1$) contact. The values $M_{tip} = M_{ref} = 165.1$ GPa were used

Data pair	d(nm)	k_{test}/k_{ref}	E^* (GPa) $m = 1$	M (GPa) $m = 1$	E^* (GPa) $m = 3/2$	M (GPa) $m = 3/2$
test 1/ref 1	15	0.963	79.5	153.4	78.1	148.0
	30	0.993	82.0	162.9	81.7	161.8
	45	0.971	80.2	155.9	79.0	151.6
test 1/ref 2	15	0.936	77.2	145.1	74.7	136.4
	30	0.953	78.7	150.2	76.7	143.5
	45	0.956	78.9	151.3	77.2	145.0

of properties such as R, a, and F_N are extremely difficult due to their small size. Moreover, the Si tip often changes size and shape over the course of the measurements. This means that direct characterization approaches such as the "area function" method used in nanoindentation [28] are fraught with error for AFM tips. The referencing method avoids these issues. It also avoids direct measurement of the cantilever spring constant k_{lever}, which can be difficult to do with sufficient accuracy (see, for instance, Ref. [41]). As long as the laser alignment on the cantilever remains constant, a given cantilever deflection always yields the same applied static force F_N. The only assumption made about the tip shape is that of axial symmetry.

5.5
Survey of Contact Resonance Force Microscopy Measurements

Table 5.4 contains a survey of modulus measurements reported in the literature [24, 26, 29, 30, 37, 42–56]. All of the listed experiments used CR-FM methods, although some [50, 55] used a data analysis approach different from that described above. The body of work has grown significantly in recent years. The table indicates the wide scope of materials that have been examined. In some cases, "model" materials such as bulk single crystals or blanket thin films were investigated in order to better understand certain measurement issues. In other cases, the spatial resolution of CR-FM made it possible to examine micro- to nanoscale features such as piezoelectric domains and grain boundaries. Some measurements were made with the point-by-point approach described above, while others involved quantitative images (modulus maps).

Also shown in Table 5.4 are values for the approximate measurement uncertainty σ for each experiment. The values were estimated from information given in the references. In many cases, σ was determined by scatter in the individual measurements and not by an uncertainty analysis. Strictly speaking, these values correspond to measurement precision or repeatability, not accuracy. Previous discussions of accuracy [44, 48] gave a conservative estimate of 40%. It can be seen from Table 5.4, however, that many experiments achieve better measurement uncertainty—as low as $\pm 1\%$ in some cases. There are many factors that account for the variation in σ from experiment to experiment, for instance, sample smoothness or uniformity, the amount of tip wear during measurements, and assumptions made in the data analysis concerning tip shape. It should be noted that the lowest values of σ have usually been achieved in imaging experiments [30, 46, 51], which yield a very large number of measurements for statistical analysis.

Other work besides that listed in Table 5.4 has been performed to characterize materials with CR-FM methods. Martensitic phase transformations in nickel–titanium alloys were studied by measuring the cantilever vibration amplitude near resonance as a function of temperature [57]. The dependence of contact-resonance frequencies on the thickness of tungsten and polymer films was examined [58]. Measurements of contact-resonance frequencies were used to determine Young's modulus of polypyrrole polymer nanotubes [59]. Contact-resonance measurements

Table 5.4. Survey of experimental results for the indentation modulus M measured by CR-FM. The first two columns give the material and the type of specimen, including the thickness t if appropriate. Values in the "σ" column represent the estimated measurement uncertainty

Material	Specimen	M (GPa)	σ (%)	Comments	Ref.
Al	Film, $t = 1.09\,\mu m$	54–81	±3–5	Compared to SAWS & NI	[42]
Au	Film, $t = 300\,nm$	102–104	±1–5	Dual reference method; values depend on which reference specimens used	[30]
CaF_2	<100> single crystals	121–123			
MgF_2		164–180			
Si		169–181			
$BaTiO_3$	<100> grain	(a) 75–350	±9–23	Values for (a) modulus map; (b) a-domains;	[43]
		(b) 318 ± 30		(c) c-domains	
		(c) 220 ± 50			
Clay (dickite)	c-axis particles	6.8 ± 2.7	±15–40	Multiple reference materials	[44]
Diamond-like carbon (DLC)	Films, $t = 5\,nm$	107 ± 12	±10–30	"Preliminary values"	[45]
	$t = 20\,nm$	210 ± 63			
DLC	Thin film	35–300	±4	Values from modulus map	[46]
Epoxy/silica	Nanocomposite plate	7–80	N/A	Modulus maps	[47]
Nanocrystalline (nc) ferrites	Films, $t = 200\,nm$	70–190	±8–10	M vs. oxidation temperature	[48]
Glass (FSG)	Film, $t = 3.08\,\mu m$	57–64	±5	Humidity effects observed	[37]

Table 5.4. (continued)

Material	Specimen	M (GPa)	σ (%)	Comments	Ref.
GaAs InP	<100> single crystals	112–125 88–96	±1–15	Values depend on model and cantilever used	[49]
Lead zirconate titanate (PZT)	Plate, $t = 7\,mm$	137 ± 18 147 ± 19	±13	Inhomogeneous sample; values for two positions	[29]
PZT	Plate, $t = 300\,\mu m$	(a) 132 (b) 116	N/A	Values (a) within domain, (b) at domain boundary	[50]
Nb	Film, $t = 280\,nm$	86–127	±1–18	Values depend on models, cantilever, reference(s)	[24, 42]
Nb nc Ni	Film, $t = 200\,nm$ Films, $t = 53\,nm$ $t = 204\,nm$ $t = 772\,nm$	119 ± 7 223 ± 28 220 ± 19 210 ± 26	±6 ±9–13	Modulus map Nanocrystalline effects; consistent with NI and SAWS results	[51] [52, 53]
Si	<111> single crystal $t = 44\,nm$	171–174	±1	Compared to NI	[29]
SnO$_2$ nanobelt		154 ± 18	±9–12	Agrees with UFM results	[26]
nc SnSe	Film, $t = 20\,nm$	20–50	N/A	Modulus map	[54]
WC-Co	Cermet plate	730 ± 50 (WC) 260 ± 40 (Co)	±7–15	Analysis by fit to full spectrum; no reference	[55]
ZnO nanowires	26–134 nm diameter	115–218	±7–13	Size effects observed	[56]

and images were used along with other methods to characterize nanocrystalline chromium nitride films [60].

So far, we have discussed measurements of *flexural* contact-resonance frequencies to determine the elastic modulus M. Other work has utilized *torsional* cantilever modes to obtain additional information. Various CR-FM methods have been used to investigate the tribological properties of materials (see, for instance, Refs. [16, 61–65]). Detailed reviews of this topic can be found elsewhere [66, 67]. It was also recently shown that by measuring the contact-resonance frequencies of both flexural and torsional modes, shear elastic properties such as Poisson's ratio or shear modulus could be determined independently from Young's modulus or M [34].

5.6
Theoretical Principles for Optimizing Experiments

As discussed above, CR-FM involves two separate models: one for the cantilever dynamics, and one for the tip–sample contact mechanics. To achieve good measurements, the experimental parameters must be chosen to satisfy *both* models simultaneously. In this section, we address the key issues of each model that are involved in optimizing these parameters.

5.6.1
Cantilever Dynamics: Which Modes?

In the procedure described above, the frequencies of the two lowest-order flexural resonances are measured. Why are these particular modes used? The answer is found

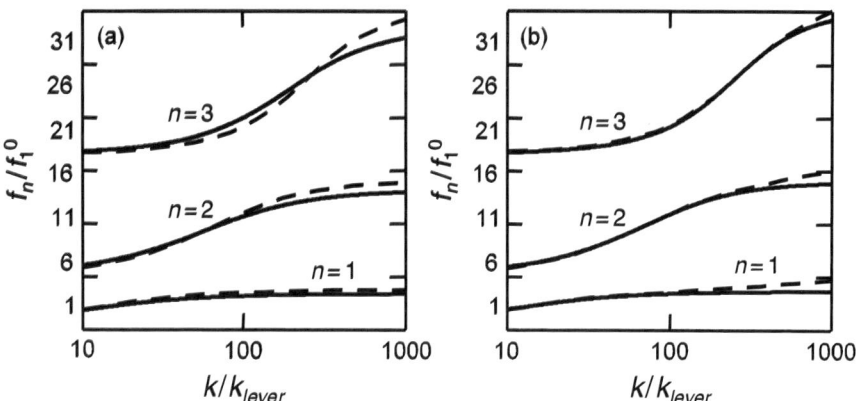

Fig. 5.8. Normalized contact-resonance frequency f_n/f_1^0 vs. normalized contact stiffness k/k_{lever} for the three lowest-order flexural modes ($n = 1, 2, 3$). The curves in **a** were calculated assuming normal forces only for $L_1/L = 1.0$ (*solid lines*) and $L_1/L = 0.97$ (*dashed lines*). In **b**, the *solid lines* indicate the calculated response for $L_1/L = 0.97$ assuming normal forces only. The *dashed lines* were calculated including both normal and lateral forces with $L_1/L = 0.97$, $\alpha = 11°$, $h/L_1 = 0.05$, and $\kappa/k = 0.9$ (*dashed line*)

by considering the relation between the contact-resonance frequency f_n and the normalized contact stiffness k/k_{lever}. We will call this relation the "response curve." Figure 5.8 shows f_n versus k/k_{lever} for the first three flexural modes ($n = 1, 2, 3$). The values of f_n are normalized to the first free flexural frequency f_1^0. Such curves are calculated with Eq. (5.5) or Eq. (5.7) or with an online calculator [33]. The figure shows that the response is qualitatively similar for each flexural mode. When k/k_{lever} is relatively small, f_n remains close to its free-space value f_n^0. As k/k_{lever} increases, the change in frequency for a given change in contact stiffness gradually increases. Figure 5.8a shows that the relative tip position L_1/L affects the overall shape of the response curve.

The region of the response curve with the highest sensitivity is where the frequency change is the greatest (i.e., with the largest slope or derivative). In this region, small changes in contact stiffness produce measurable changes in the contact-resonance frequency. It can be seen from Fig. 5.8 that this region occurs at higher values of k/k_{lever} as the mode number n increases. To ensure accurate measurements, experimental parameters should be chosen to maximize the sensitivity of at least one mode. When k/k_{lever} increases beyond the region of highest sensitivity for a given mode, the mode approaches a "pinned" state, where even relatively large changes in contact stiffness result in only small frequency shifts. Here, the contact-resonance frequency of the nth mode approaches the free-space frequency of the $(n + 1)$th mode. Experimental conditions should be chosen to avoid mode pinning. Not only is this the least sensitive part of the response curve, but in addition measurements made under pinned conditions could be misleading. Measurements made at different static forces F_N will yield very similar values of f_n and thus very similar values of k/k_{lever}. This could lead to the incorrect conclusion that the tip shape is flat.

In addition, lateral effects become increasingly significant as the tip approaches a pinned state [68]. This can be seen in Fig. 5.8b, which compares the response curve for normal forces alone (solid lines) to that for normal and lateral forces combined (dashed lines). The difference between the two curves for each mode increases with k/k_{lever}. Operating under conditions of high sensitivity therefore has the added benefit that the simpler (normal-forces only) model for data analysis is valid.

More insight into mode sensitivity can be gained from Fig. 5.9. Like Fig. 5.8, Fig. 5.9a shows how f_n varies with k/k_{lever}. Figure 5.9b shows the sensitivity S of each mode as a function of contact stiffness. S is defined as the derivative of f_n/f_1^0 with respect to k/k_{lever} [69]. Analytical expressions for S for the flexural and torsional modes of cantilevers with both uniform and nonuniform cross sections have been published elsewhere [70]. Other theoretical work that includes lateral and damping forces in the calculation of S has been published [71, 72]. The values of S in Fig. 5.9b were calculated numerically from the curves in Fig. 5.9a. The circles indicate the values of k/k_{lever} at which S is a maximum for each mode, while the thick lines show where S is at least 80% of the maximum value. (There is no thick line or circle shown for $n = 1$ because the region of greatest sensitivity occurs for $k/k_{\text{lever}} < 1$.) Figure 5.9b clearly shows that the absolute mode sensitivity is greatest for $n = 1$ and decreases as n increases. This leads to the generalization that the lowest-order modes possible should be used. The figure also indicates that the value of k/k_{lever} at which S is a maximum increases with increasing n.

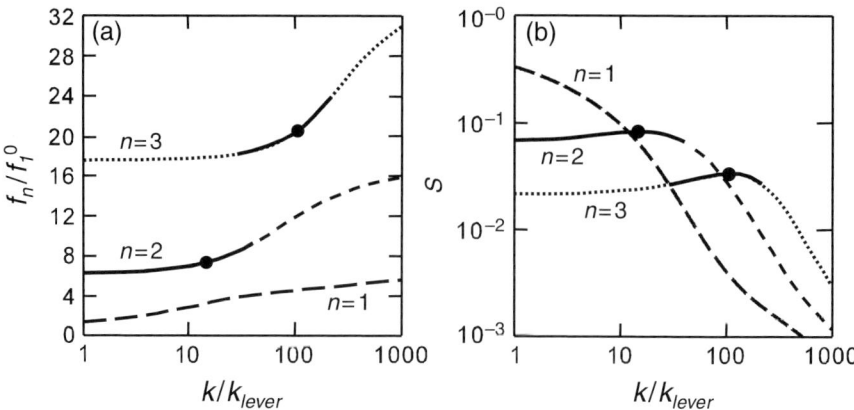

Fig. 5.9. a Normalized contact-resonance frequency f_n/f_1^0 as a function of normalized contact stiffness k/k_{lever} for $n = 1$ (*dash-dotted line*), $n = 2$ (*dashed line*), and $n = 3$ (*dotted line*). The curves were calculated including both normal and lateral forces with $L_1/L = 0.97$, $\alpha = 11°$, $h/L_1 = 0.05$, and $\kappa/k = 0.9$. (**b**) Sensitivity S as a function of normalized contact stiffness k/k_{lever} for $n = 1$ (*dash-dotted line*), $n = 2$ (*dashed line*), and $n = 3$ (*dotted line*). The *circles* indicate the point of maximum sensitivity for each mode. The *thick lines* indicate where S is at least 80% of its maximum value. The region of maximum sensitivity for $n = 1$ occurs for $k/k_{\text{lever}} < 1$

The information in graphs such as these help to establish rough guidelines for mode selection. The values used in Fig. 5.9 for the parameters L_1/L, κ/k, α, and h/L_1 are representative of typical experimental conditions. With these values, we calculate that adding lateral forces changes k/k_{lever} by approximately 5% or less for the following conditions: for $n = 1$, $k/k_{\text{lever}} < 40$; for $n = 2$, $k/k_{\text{lever}} < 200$; for $n = 3$, $120 \leq k/k_{\text{lever}} \leq 600$. The most sensitive range of k/k_{lever} (80% or more of maximum) is approximately as follows: for $n = 1$, $k/k_{\text{lever}} < 1$; $n = 2$, $0 \leq k/k_{\text{lever}} \leq 40$; and for $n = 3$, $30 \leq k/k_{\text{lever}} \leq 210$. In many experiments in the literature, k/k_{lever} ranges from approximately 50 to 100. From a sensitivity perspective, the second and third flexural modes seem the best choice for these measurements. Lateral effects are also more significant for the first mode than for the third mode in this range of k/k_{lever}. However, from a practical standpoint, the third mode is usually more difficult (if not impossible) to detect than the first mode. The absolute sensitivity of the first and third modes is also approximately the same for this range of k/k_{lever}. These facts may explain why the first and second modes have typically been used in experiments on stiff materials. It is important to recognize that the guidelines will change somewhat depending on the values of L_1/L, κ/k, α, and h/L_1 used. Readers are therefore encouraged to perform their own calculations.

The effect of including lateral forces in the data analysis is further illustrated in Fig. 5.10, which shows the normalized contact stiffness k/k_{lever} versus relative tip position L_1/L for the three lowest-order flexural modes. The curves in Fig. 5.10 were calculated with contact-resonance frequency data from Table 5.2. If only normal forces are included (solid lines), the different mode pairs cross at different values of k/k_{lever} and L_1/L. Data analysis with this approach will yield a greater measurement

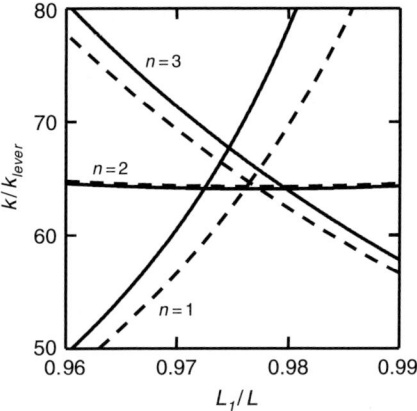

Fig. 5.10. Normalized contact stiffness k/k_{lever} as a function of relative tip position L_1/L for data obtained with cantilever #1 in Table 5.1. The results are shown for the measured values $f_1 = 734.5 \, \text{kHz}, f_2 = 1,806.4 \, \text{kHz}$, and $f_3 = 3,226.0 \, \text{kHz}$ for the contact-resonance frequencies of the three lowest flexural modes. The *solid lines* were calculated assuming vertical forces only, while the *dashed lines* include both normal and lateral forces with $\alpha = 11°$, $h/L_1 = 0.03$, and $\kappa/k = 0.9$

uncertainty, because the variation in the crossing-point value of k/k_{lever} for different mode pairs will result in a larger scatter in M. If lateral effects are included (dashed lines), it is possible for all three modes to converge at a single value of k/k_{lever}. (The value of L_1/L also converges, and in this case increases to a value somewhat closer to that estimated from measurements of the cantilever's dimensions.) It might be possible to exploit this convergence effect to experimentally determine κ/k or h/L_1. For the data given here, the values of k/k_{lever} for the first and third modes change more than those for the second mode when lateral effects are included. This conclusion can also be reached from a detailed inspection of response curves like those in Fig. 5.8b. However, it is the value of k/k_{lever} at which the mode pairs cross that is important. The figure shows that the crossing-point value of k/k_{lever} is virtually the same for the mode pairs $n = 1, 2$ and $n = 2, 3$ whether lateral effects are included or not. Because k/k_{lever} changes much more for the mode combination $n = 1, 3$ when lateral forces are included, measurements with this pair are less reliable than with either mode pair $n = 1, 2$ or $n = 2, 3$ in this case.

5.6.2
Contact Mechanics: What Forces?

Section 5.6.1 shows that conditions are optimized for certain values of k/k_{lever}. How do we select the experimental parameters to ensure these values? To do so, one must understand and correctly apply the principles of contact mechanics. Here, the discussion is limited to Hertzian contact mechanics. The corresponding discussion for flat-punch contact is simpler because the contact area does not depend on force.

The cantilever spring constant k_{lever} directly affects the ratio k/k_{lever}, but it also influences the contact stiffness k through the applied force F_N. F_N is determined by

k_{lever} and the static cantilever deflection d through $F_N = k_{lever}d$. For Hertzian contact, the relationship between k and F_N is found by combining Eqs. (5.8) and (5.11):

$$k = 2aE^* = 2E^* \left[\sqrt[3]{\frac{3RF_N}{4E^*}} \right] = \sqrt[3]{6E^{*2}RF_N}. \tag{5.13}$$

Further considerations affect the choice of k_{lever} and F_N. F_N must be sufficiently greater than any adhesion forces present to ensure that an elastic model for contact mechanics is valid. Moreover, the applied force determines the stress field in the sample, which determines the volume of the sample that is measured. For this discussion, it is simpler to work in terms of the contact radius a. Equation (5.11) gives the relation between a and F_N. For Hertzian contact, the compressional stress σ_z directly beneath the indenter as a function of depth z into the sample is given by [39]

$$\sigma_z = p_0 \left(1 + \frac{z^2}{a^2} \right)^{-1}, \tag{5.14}$$

where $p_0 = 3F_N/(2\pi a^2)$ is the maximum stress applied at the surface, $z = 0$. Equation (5.14) shows that σ_z decreases rapidly with increasing distance into the sample. For $z = 3a$, $\sigma_z = 0.1p_0$; thus for $z > 3a$, σ_z is considered negligible. This leads to the general guideline that measurements probe to a depth z $\sim 3a$ [39,73].

This guideline can be used to tailor the experimental conditions to a particular sample. For instance, measurements of bulk specimens or relatively thick films should probe sufficiently deeply to minimize the contribution of any damage or contamination layers. Such layers are typically a few to several nanometers thick. For $k_{lever} \sim 40$ N/m on materials with $M > 50$ GPa, the probed depth exceeds 10 nm at an applied force of a few hundred nanonewtons, so this condition is easily met. In other cases, it may be important to minimize the measurement depth. In thin-film systems, for instance, the properties of the substrate will affect the measurements if the stress field extends too deeply. By keeping the applied force, and thus the depth of measurement sensitivity, sufficiently low, the film properties can be measured directly without the added complication of incorporating substrate effects in the data analysis. For nanocrystalline nickel films on silicon, a film only ~ 50 nm thick was directly measured with CR-FM methods [53]. Utilizing this effect to determine film thickness has also been investigated [58,74].

Figure 5.11 provides an example of depth effects. CR-FM experiments were performed on a sample containing a nanocrystalline nickel (Ni) film ~ 15 nm thick on a <001 > Si substrate. The reference sample was a single crystal of <001> Ni with indentation modulus $M_{Ni} = 219$ GPa. Contact-resonance spectra were acquired at values of the static cantilever deflection $d = 10, 20, 30, 40, 50,$ and 60 nm. With a cantilever spring constant $k_{lever} = 11 \pm 1$ N/m, the corresponding range of applied static force F_N was approximately 110–660 nN. It can be seen in Fig. 5.11 that the measured value of M depends on F_N. At low applied forces, the stress field does not significantly penetrate into the Si substrate, and the values of M are similar to those measured in thicker nanocrystalline Ni films ($\sim$200–230 GPa) [53]. As the applied force increases, the stress field extends further into the substrate, and the measured response depends on the properties of both the Ni film and the Si substrate. At the

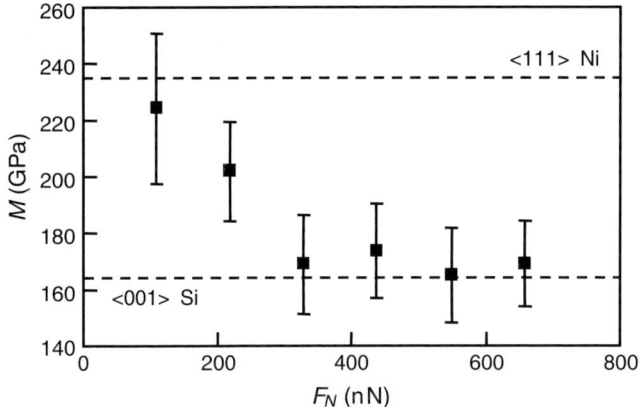

Fig. 5.11. Experimental values of the indentation modulus M as a function of applied static force F_N for a nanocrystalline Ni film $\sim$15 nm thick on a <001> Si substrate. The *dashed lines* show the values of M for <111> Ni and <100> Si. The *error bars* represent the standard deviations of six separate measurements

highest values of F_N, the volume of material probed in the measurement consists more of the substrate than of the thin film. The response is therefore dominated by the properties of the substrate, and the measured values of M approach that of <001> Si (165 GPa). The cantilever used in these experiments was more compliant than those typically used ($k_{lever} \sim 40$ N/m). In the case of stiffer cantilevers, even very low deflections result in forces that are too high to sense only the properties of the film. Cantilevers with even lower values of k_{lever} could also be used to ensure suitable values of F_N. In that case, it might be necessary to utilize higher-order resonant modes to minimize lateral effects and maximize sensitivity.

The contact radius a also determines the lateral extent of the measurement sensitivity. At $z = 0$, the radial tensile stress σ_r as a function of distance r from axis of the indenter is given by [39]

$$\sigma_r = p_0(1 - 2v^2)\frac{a^2}{3r^2} \quad (r > a),\tag{5.15}$$

where v is Poisson's ratio. For $v \sim 0.2$–0.3, σ_r falls to 10% of its maximum value at $r \sim 1.1a$–$1.4a$. The lateral spatial resolution is approximately twice this value, or $\sim 2a$–$3a$, and not simply a. Thus, a balance must be maintained between achieving sufficient depth of penetration and high lateral spatial resolution. For the representative values $F_N = 1\,\mu$N, $R = 25$ nm, and $E^* = 82.5$ GPa, we obtain $a = 6.1$ nm. Thus, $3a \approx 20$ nm is a conservative estimate of both lateral and depth resolution.

5.6.3
An Example

The following example illustrates the principles involved in selecting experimental conditions. A sample of interest has an indentation modulus $M = 80$ GPa. Using

Eq. (5.9) and assuming $M_{tip} = 165$ GPa for the <100> Si tip, we find $E^* = 54$ GPa. The radius R of the cantilever tip typically ranges from less than 10 nm (new) to approximately 50 nm (used) [40]. To probe to a depth $z \approx 3a = 10$ nm, $F_N \approx$ 200–500 nN from Eq. (5.11). With these values of E^*, R, and F_N from Eq. (5.12) we obtain $k \approx 400$–800 N/m at the minimum load. At a maximum load of $2\,\mu$N, $k \approx 600$–1,200 N/m. For $k_{lever} = 40$ N/m, this means that $k/k_{lever} \approx 10$–30. For these values of k/k_{lever}, modes $n = 1$ and $n = 2$ are relatively sensitive and have minimal lateral effects. Therefore, this combination of k_{lever}, F_N, and d is a suitable choice for this specimen. Note that for $k_{lever} = 1$ N/m, deflections of $d = 200$ nm and greater would be needed to achieve $3a \approx 10$ nm. Some AFM instruments do not permit such large values of d. This is one reason why relatively stiff cantilevers are often used.

The referencing approach described in Sect. 5.4.2 avoids the need to determine absolute values of k_{lever} and F_N. The approach presents a distinct practical advantage, because accurate measurements of k_{lever} and AFM forces remain a challenge [14, 41]. However, this discussion shows that the values of k_{lever} and F_N critically affect the experimental conditions. Without direct measurements of k_{lever}, it is possible only to estimate properties such as F_N and a. Therefore, calculations such as the ones in the previous paragraph can only estimate k/k_{lever}. The actual values of k/k_{lever} may also differ from those calculated for other reasons. For instance, the actual contact mechanics are likely to differ from Hertzian behavior. It is advisable to compare the values of k/k_{lever} obtained in the first few measurements to the predicted values. The operating conditions, particularly F_N, can then be adjusted to enhance the response.

5.7
Practical Issues for Optimizing Experiments

The previous sections have discussed how some of the underlying theoretical principles affect measurement sensitivity. There are also some purely practical issues that must be considered when choosing the experimental conditions. In this section, we discuss some of these issues.

5.7.1
Exciting and Detecting the Cantilever's Resonant Modes

In the method described above, the resonant modes of the cantilever are excited by a separate actuator (transducer) mounted beneath the specimen. One might wonder why this additional transducer is used, when almost all AFM instruments contain a built-in actuator. Intended for use in intermittent-contact mode, this actuator is located in the cantilever holder and excites vibrations from the clamped end of the cantilever. If the electrical drive signal to the actuator can be externally controlled, CR-FM measurements can be performed with this "internal" actuator instead of an "external" transducer. This approach may be useful in certain applications, for instance if the sample is curved or otherwise prohibits access from below.

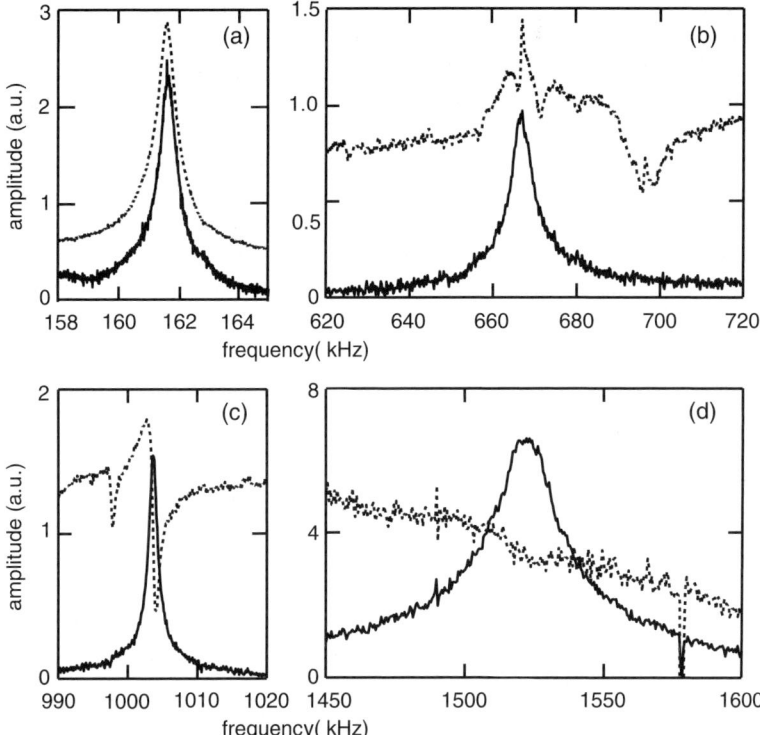

Fig. 5.12. Spectra excited by an "external" transducer mounted beneath the sample (*solid lines*) and by the AFM's "internal" piezoelectric element in the cantilever holder (*dashed lines*). The sample was a glass slide. Spectra for the first and second free resonances are shown in **a** and **c**, while the corresponding contact-resonance spectra are shown in **b** and **d**, respectively

Figure 5.12 compares spectra obtained with an internal actuator and an external transducer. Figure 5.12a, c show the spectra of the first and second free resonances, while Fig. 5.12b, d contain the spectra of the corresponding contact resonances. All of the contact-resonance spectra were acquired without changing the contact between the tip and the sample. Although the free-resonance spectra appear similar, the contact resonances obtained by the different excitation means are significantly different. Spectra obtained with the internal actuator contain background signals and extraneous peaks that hinder identification of the contact-resonance frequencies. These signals may be due to resonant modes of the actuator itself, or else the silicon plate on which the cantilever is fabricated [23]. Figure 5.12 also shows that the frequency response of the internal actuator may not be adequate to sufficiently excite the higher-order resonant modes. Because results depend strongly on the specific AFM instrument used, general conclusions should not be drawn from the spectra in Fig. 5.12.

Alternatively, a custom actuator can be implemented to excite the cantilever resonances. This has usually been accomplished by attaching the actuator to the base of the cantilever [19, 75]. However, cantilevers with integral actuation elements have recently been developed [76]. Excitation by means of a transducer beneath the sample was first developed as the AFAM method by Rabe and coworkers [22, 31]. Excitation with an actuator at the clamped end of the cantilever was first developed as the UAFM method by Yamanaka and coworkers [19, 75]. Mathematical analyses of both systems have been published [22, 77]. The analyses show that as long as damping terms are small, the contact-resonance peaks occur at the same frequencies regardless of excitation method. However, the amplitude of the cantilever vibration and other details of the response spectrum differs between methods.

Practical issues also affect the ability to experimentally detect the contact resonances. In particular, the size and relative position of the laser spot can affect the detected signal amplitude [31, 78, 79]. This effect can be better understood by considering the vibration amplitude versus position on the cantilever that is predicted for a given mode and given contact stiffness [31]. Such considerations suggest that for optimum detection sensitivity, the best position of the laser spot is not always at the very end of the cantilever. The exact behavior depends strongly on the specific experimental conditions. Under typical experimental conditions, our experience has generally been that the amplitude of the first mode drops relatively sharply as the laser is moved away from the end of the cantilever towards its clamped base. The amplitude of the second mode drops more gradually with distance, and can sometimes actually increase as the spot is moved away from the tip end towards the clamped base. In this way, it may be possible to adjust the relative amplitudes of the detected modes. The finite diameter of the laser spot ($\sim$50 μm) further complicates matters, especially for the higher-order modes. If the laser spot is sufficiently large compared to the wavelength of a given mode, the detected signal may be virtually zero in some positions. The lateral (side-to-side) position of the laser spot may also change the detected signal. Poor lateral positioning is one possible reason that spectra can sometimes appear asymmetrical.

5.7.2
Tip Shape and Tip Wear

Measurements are also affected by practical issues concerning the tip shape. One issue is the nonideal shape of the real cantilever tip and how it affects the nanoscale contact mechanics. As seen in SEM images such as those shown in Figs. 5.2, 5.13a, and 5.14a, the cantilever tip usually does not conform to an idealized shape such as a hemisphere. Several studies have been performed to better understand the effect of the true tip shape on measurements [40, 45, 49, 80].

In our studies of tip shape [40], CR-FM experiments were performed with several cantilevers with similar characteristics. The tips of the cantilevers were imaged with high-resolution SEM techniques before and after each set of experiments. The SEM images were used to determine directly the tip shape and dimensions such as the tip radius R. These values were then compared to those inferred from the CR-FM data by means of contact-mechanics models. The results showed that none of the standard contact-mechanics models describe the actual behavior of the tips. Figure 5.13b shows the measured contact stiffness k as a function of static applied force F_N for three cantilevers. Although the dimensions and free frequencies of the cantilevers

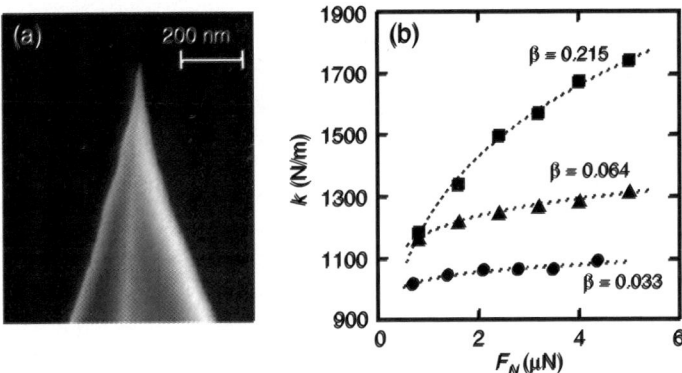

Fig. 5.13. a SEM image of a new cantilever tip. The cantilever was similar to cantilever #1 in Table 5.1. b Vertical contact stiffness k as a function of the applied static load F_N for three cantilevers obtained from the same manufacturer at the same time. The *dotted lines* indicate fits to the form $k \propto F_N{}^\beta$, where $\beta = 0$ corresponds to a flat tip and $\beta = 1/3$ corresponds to a hemispherical tip

were virtually identical, the dependence of k on F_N was different in each case. On the basis of the results from ten different cantilevers, we found that the force dependence was best described by $k \propto F_N{}^\beta$. The exponent β was varied to allow for tip shapes intermediate between a flat punch ($\beta = 0$) and a hemisphere ($\beta = 1/3$).

Results such as these could lead to improved measurement procedures. The experimental uncertainty in the modulus M is often inflated in order to accommodate the range of values that results from assuming $m = 1$ and $m = 3/2$ in Eq. (5.12). Developing a tip characterization procedure might allow a more exact value of m to be used for each set of measurements, and thus reduce uncertainty. The procedure could be based on the dependence of k on F_N [40, 80], or might take a different approach.

Another practical issue is tip wear—changes in tip shape and size during use. When used in contact at loads up to a few micronewtons, sharp tips such as those shown in Figs. 5.2, 5.13a, and 5.14a gradually grow blunter, suddenly break, or plastically deform [40, 45, 49]. Gradual tip wear is illustrated in Fig. 5.14b, which shows values for k/k_{lever} measured in alternation on fused silica ($M \approx 75\,\text{GPa}$) and an aluminum film ($M \approx 79\,\text{GPa}$). Although the applied load was kept the same throughout the measurements, it can be seen that k/k_{lever} gradually increased due to tip wear. Tip wear means that the radius R increases, causing the contact radius a and thus k to increase. However, each subsequent measurement of k/k_{lever} for the aluminum film was greater than the previous one for fused silica (with one exception). Therefore, the referencing approach described above will yield accurate values for M despite gradual tip wear [29]. This motivates the practice of measuring the reference specimen before and after each measurement on the unknown specimen. Figure 5.14b shows that k/k_{lever} approaches an equilibrium value after a few dozen measurements. The exact behavior depends on the specific materials involved as well as the load history. Such results suggest that it might be desirable to "wear in" tips before use. Although this procedure could degrade the spatial resolution slightly, it

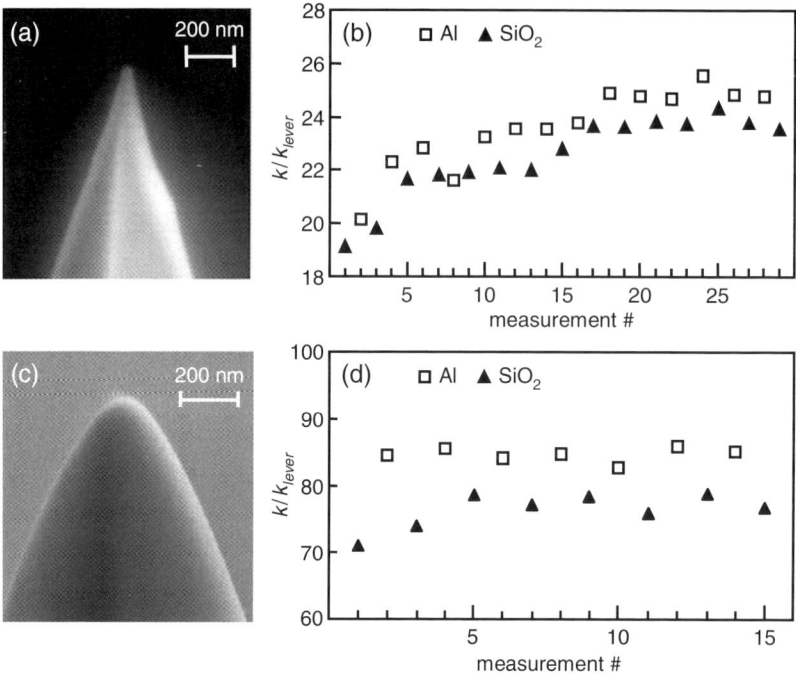

Fig. 5.14. Examples of tip wear in CR-FM experiments. **a** SEM image of a sharp tip in side view. **b** Measured values of the normalized contact stiffness k/k_{lever} for fused silica (SiO_2, *triangles*) and an aluminum film (Al, *squares*). The film was approximately 1 μm thick on a sapphire substrate. The first measurement corresponds to a new tip similar to the one shown in **a**. The measurements were made at a static deflection $d = 30$ nm, corresponding to an applied static force $F_N \approx 1.2$ μN. **c** SEM image of a deliberately shaped tip in side view. **d** Normalized contact stiffness k/k_{lever} for SiO_2 (*triangles*) and an Al film (*squares*) using the tip shown in **c**. The first measurement corresponds to a new tip. The measurements were made at a static deflection $d = 35$ nm

could yield more reproducible results and lead to less data scatter. Systematic studies to investigate methodologies for such a "tip wear procedure" have not been reported.

A possible solution to tip wear, breakage, and deformation is the use of tips with a larger radius of curvature R. Figure 5.14 shows the results of CR-FM experiments to investigate this idea. As seen in Fig. 5.14c, the cantilever had a deliberately rounded tip (Nanosensors, Neuchâtel, Switzerland). The nominal value of R ranged between 90 nm (end view) and 160 nm (side view). Figure 5.14d shows that the normalized contact stiffness k/k_{lever} rapidly approaches an equilibrium value, indicating very little wear. Similar results have been observed by another group [81]. Although further measurements are needed, these initial results suggest that larger tips may be useful for quantitative modulus measurements, if a decrease in spatial resolution can be tolerated.

A final issue involving practical considerations is the choice of reference specimen. It has been found that the best quantitative results are obtained if the indentation modulus M_{ref} of the reference material is similar to the modulus M_s of the unknown specimen. One reason is possible uncertainty in the value M_{tip} used in Eq. (5.9) for the modulus of the tip [30]. This can be better understood by rewriting the equations in Sect. 5.4.2 as

$$\frac{1}{M_s} = \frac{1}{M_{ref}} \left(\frac{k_{ref}}{k_s} \right)^m + \frac{1}{M_{tip}} \left[\left(\frac{k_{ref}}{k_s} \right)^m - 1 \right]. \tag{5.16}$$

If the unknown and reference materials have similar modulus, $k_{ref}/k_s \approx 1$. This means that the quantity in square brackets in Eq. (5.16) is small, and the impact of inaccuracy in the value of M_{tip} is minimized. The use of two reference materials has also been demonstrated as a way to overcome the difficulty of accurately knowing M_{tip} [22, 24, 30].

Another consideration in choosing reference materials is the presence of any surface layers such as oxides or adsorbed water. These layers cause the real system to deviate from the idealized one used in the data analysis. An oxide layer means that the sample is a two-layer system instead of a uniform half space. Capillary layers due to adsorbed water create significant adhesive forces, so that pure elastic contact cannot be assumed. Surface roughness is also a consideration. Smooth samples provide greater uniformity in the contact conditions across the surface, and therefore reduce data scatter.

5.8
Imaging with Contact Resonance Force Microscopy

So far in this chapter, we have described how CR-FM is used for measurements at a fixed sample position. However, the two-dimensional scanning capability of the AFM instrument makes it ideal for creating images of the spatial distribution in properties. One reason for the growing demand for spatial visualization of properties is the increasing integration of multiple materials on micro- and nanometer scales. The cause of failure in such systems is often a localized variation in properties. Because measurements of "average" sample properties are simply not sufficient in such cases, images are required.

Initially, images obtained by CR-FM methods gave only qualitative information about elastic properties [12, 20, 75, 82]. Such "amplitude images" are nonetheless useful in many applications and are straightforward to acquire experimentally. In this approach, the excitation frequency is set close to that of the contact resonance for a particular mode. The output amplitude signal of the lock-in amplifier is connected to an auxiliary input channel of the AFM instrument and is used for image acquisition. The resulting image thus contains the relative amplitude of the cantilever vibration at the excitation frequency for each position on the sample. The contrast in amplitude images may be enhanced or suppressed by the choice of excitation frequency [29]. At relatively low frequencies, the cantilever vibration amplitude will be higher for regions with lower elastic modulus. More compliant regions will therefore appear brighter. Increasing the excitation frequency so that it is close to the contact-resonance frequency of stiffer sample components decreases the vibration amplitude of more compliant regions. Hence, stiffer regions will appear brighter. In this way,

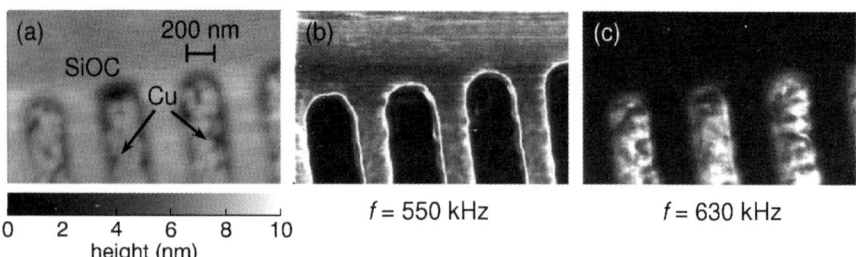

Fig. 5.15. Example of CR-FM or AFAM amplitude imaging. The sample contained copper (Cu) lines in an organosilicate glass (SiOC) film. **a** Topography. The image was acquired in contact mode, at the same time as the image in **c**. **b** Amplitude image of the cantilever vibration for an excitation frequency $f = 550$ kHz. **c** Amplitude image at $f = 630$ kHz. Images were acquired with a cantilever with lowest free-space frequency $f_1^0 = 151.3$ kHz

one can observe a reversal or "inversion" in image contrast by changing the excitation frequency [11].

Examples of amplitude imaging are shown in Fig. 5.15. The sample contained a blanket film of an organosilicate glass (denoted SiOC) approximately 280 nm thick. Copper (Cu) lines were deposited into trenches created in the SiOC blanket film. The sample was etched briefly in a hydrofluoric acid solution to remove any protective surface layers. The topography image of Fig. 5.15a reveals that the sample is very flat, with height variations less than ~5–10 nm. Figures 5.15b, c show amplitude images acquired at two different frequencies. Small features inside the Cu lines can be seen. These are most likely due to shifts in the contact area that arise from small topographical features (e.g., pores, polishing effects). In addition to the SiOC film and the Cu lines, bright regions can be seen at the SiOC/Cu interfaces. This feature corresponds to a thin barrier layer deposited on the sidewall of the trenches, and is not obvious in the topography image. In Fig. 5.15b, the SiOC regions of the image are brighter than the Cu regions. However, the Cu regions are brighter in the image in Fig. 5.15c, which was acquired at a higher excitation frequency. This information suggests that the contact-resonance frequency of the Cu regions is generally higher than that of the SiOC regions. Because higher contact-resonance frequencies imply greater elastic modulus, it can be inferred that the modulus of the Cu lines is greater than that of the SiOC film.

Figure 5.15 shows that components of different elastic stiffness are easily identified by amplitude imaging. However, the figure also illustrates the difficulties involved in trying to evaluate the relative stiffness of different sample components. Instead, quantitative imaging or *mapping* of nanoscale elastic properties is ultimately desired. Quantitative imaging requires detecting the frequency of the contact-resonance peak at each position as the tip moves across the sample. A single such contact-resonance-frequency image can provide more information than an entire series of amplitude images. However, if the sample components differ greatly in their elastic properties, the contact-resonance frequency will vary significantly across the sample, making detection more difficult. Several solutions to this challenge have been demonstrated [17, 30, 43, 46, 83–85]. Typically, there is a tradeoff between the imaging speed and the amount of custom hardware and software required.

We have also developed techniques for nanomechanical mapping with CR-FM [51, 86, 87]. Our approach differs conceptually from other implementations, in that the starting frequency of the frequency sweep window is continuously adjusted to track the contact-resonance peak. In this way, a high-resolution spectrum is acquired with a minimum number of data points, even if the frequency of the peak shifts significantly throughout the image. Other approaches that lack feedback must perform a frequency sweep at every pixel over the same relatively wide range that encompasses all possible peaks. Our approach also utilizes a digital signal processor (DSP) architecture. One advantage of a DSP approach is that it facilitates future upgrades, because changes are made in software instead of hardware.

A schematic of the "frequency-tracking" apparatus is shown in Fig. 5.16. The electronics are explained in greater detail elsewhere [87]. An adjustable-amplitude, swept-frequency sinusoidal voltage is applied to the piezoelectric actuator beneath the specimen. As the cantilever is swept through its resonant frequency by the actuator, the photodiode detects the cantilever's vibration amplitude and sends this signal to the DSP circuit. The circuit converts the signal to a voltage proportional to the root-mean-square (rms) amplitude of vibration and digitizes it with an analog-to-digital (A/D) converter. As each sweep completes, the circuit constructs a complete resonance curve. It finds the peak in the resonance curve and uses this information in a feedback-control loop. The control loop adjusts a voltage-controlled oscillator (VCO) to tune the center frequency of vibration to maintain the cantilever response curve centered on resonance. The control voltage is also routed to an input port of the AFM instrument for image acquisition. Each pixel in the resulting image thus contains a value proportional to the peak (resonant) frequency at that position. A frequency range can be specified in order to exclude all but the cantilever mode of interest. A total of 128 data points are acquired for each resonance curve. At

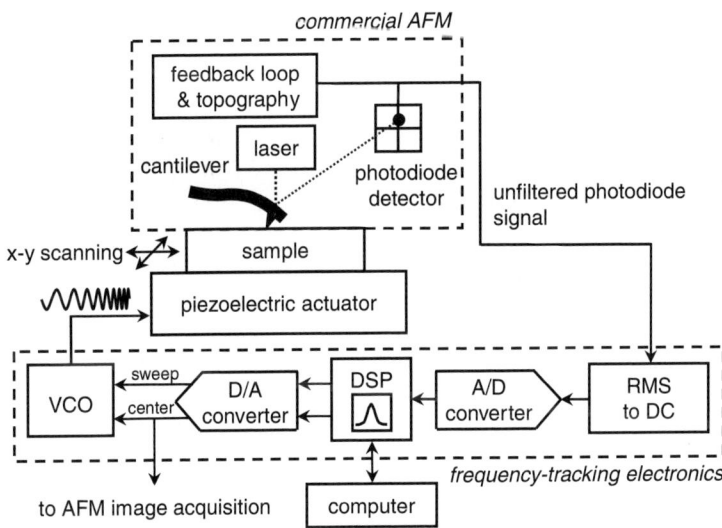

Fig. 5.16. Schematic of experimental apparatus for contact-resonance-frequency imaging

48 kilosamples per second, the system acquires the full cantilever resonance curve 375 times per second. The AFM scan speed must be adjusted to ensure that several spectrum sweeps are made at each image position. For scan lengths up to several micrometers, an image with 256 × 256 pixels is usually acquired in less than 25 min.

An example of quantitative imaging with our frequency-tracking electronics is shown in Fig. 5.17. The images correspond to the same region of the SiOC/Cu structure as that shown in Fig. 5.15. The topography image in Fig. 5.17a shows the SiOC blanket film and the slightly recessed (less than 5 nm) Cu lines. The contact-resonance-frequency images for the two lowest flexural modes of the cantilever are shown in Figs. 5.17b, c, respectively. The frequency images reveal directly that the contact-resonance frequency in the Cu regions is higher than that in the SiOC regions.

Figure 5.17d shows a map of the normalized contact stiffness k/k_{lever} calculated from the frequency images in Fig. 5.17b, c. The image was calculated on a pixel-by-pixel basis with the analysis approach described above for point measurements. Depending on the application, it may be sufficient to evaluate the contact-stiffness map alone. In other cases, a map of the indentation modulus M may be needed. Calculating a modulus map from the contact-stiffness image involves the same models and assumptions used for point measurements. For instance, a specific

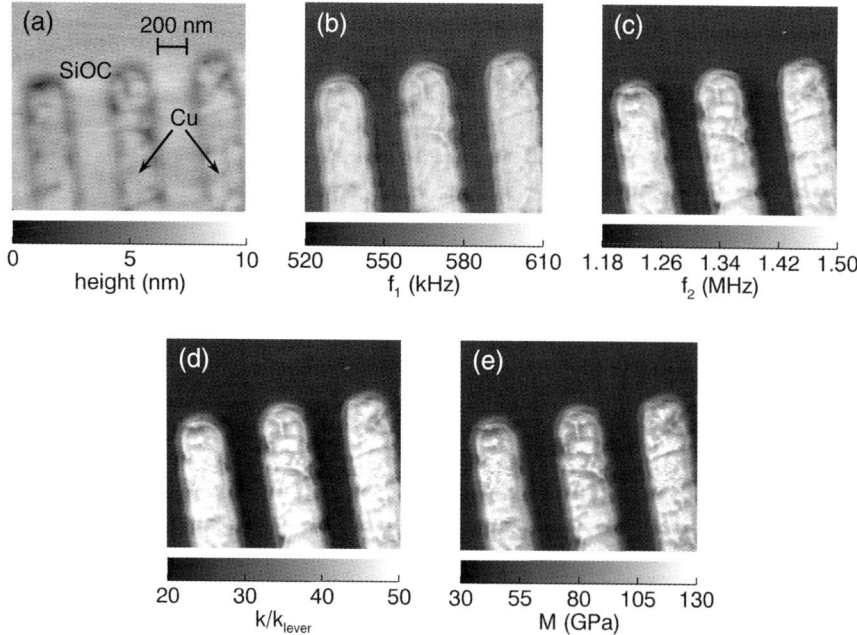

Fig. 5.17. Example of quantitative CR-FM imaging. **a** Topography. **b** and **c** Contact-resonance frequency images of the first (f_1) and second (f_2) flexural modes, respectively. **d** Normalized contact stiffness k/k_{lever} calculated from **b** and **c**. **e** Map of the indentation modulus M calculated from **d**, as described in the text. The free-space frequencies of the cantilever's lowest two flexural modes were $f_1^0 = 151.3\,kHz$, $f_2^0 = 938.0\,kHz$

contact-mechanics model must be chosen. Reference values of E^* and k/k_{lever} are also needed. Here, we calculated the modulus map from the contact-stiffness image with the assumption that the tip was flat. We also assumed that the mean value of E^* in the SiOC region corresponded to $M_{SiOC} = 44.3$ GPa. This value was obtained from independent point measurements directly on the SiOC film with a borosilicate glass as the reference material. The average value of k/k_{lever} in the SiOC region of the image was used as the reference value. The resulting modulus map is shown in Fig. 5.17e. In spite of the assumptions made to obtain the map, it shows how quantitative images of M can be achieved. Other mapping results are listed in Table 5.4.

Imaging techniques might also be used to improve point measurements of M. Recently, values for M in homogeneous samples were obtained from contact-resonance-frequency images [30]. Images for the two lowest flexural modes were acquired by scanning small regions of the sample. The average frequency of each image was determined and used to calculate the contact stiffness k/k_{lever}, from which M could be calculated. An advantage of this approach is that it yields many more data points in a much shorter time than can be obtained from individual point measurements. When combined with the use of two reference materials, the measurement uncertainty of this approach (scatter in data points) was typically less than 1%. Further evaluation of this approach is needed, for instance to establish the variability in images acquired under nominally identical conditions.

Other mechanical properties besides elastic modulus can be imaged with CR-FM methods, if they influence the contact stiffness between the tip and the sample. It was explained in Sect. 5.6.2 how CR-FM probes the sample properties to a depth z roughly three times that of the tip–sample contact radius a. CR-FM can thus be used to sense variations in mechanical properties beneath the surface, for example to investigate subsurface dislocations [88, 89]. In other recent work, amplitude imaging was used to detect very large voids (diameter 50–500 μm) buried several hundred nanometers below the surface [90]. Cracks in buried layers have also been studied with UFM methods [91].

Another property of technological interest is the relative bonding or adhesion between a film and a substrate. To investigate the sensitivity of CR-FM to variations in film adhesion [92], we fabricated a model system of gold (Au) and titanium (Ti) films on (001) Si. Figure 5.18a shows a cross-sectional schematic of the sample. A rectangular grid of Ti 1 nm thick containing 5×5 μm holes (10 μm pitch) was created on Si by standard microfabrication techniques. A blanket film of Au 20-nm thick and a topcoat of Ti 2 nm thick were deposited on top of the grid. A crude scratch test was performed by lightly dragging one end of a tweezer across the sample. Optical micrographs showed that this treatment had removed the film in the scratched regions without a Ti interlayer (squares), and left the gold intact in the scratched regions containing a Ti interlayer (grid). The result confirmed our expectation that the film adhesion was much stronger in regions containing the Ti interlayer. The Ti topcoat was included merely to prevent contamination of the AFM tip by the soft Au film.

Frequency-imaging experiments were performed on the sample. A map of the normalized contact stiffness k/k_{lever} calculated from the resulting contact-resonance frequency images is shown in Fig. 5.18b. The image shows that k/k_{lever} is lower in the region with poor adhesion (no Ti interlayer). A line scan of the average value

Fig. 5.18. Evaluation of film/substrate adhesion with CR-FM imaging. **a** Schematic of sample in cross section. **b** Map of the normalized contact stiffness k/k_{lever} calculated from contact-resonance frequency images. **c** Average stiffness versus position across the center of **b**

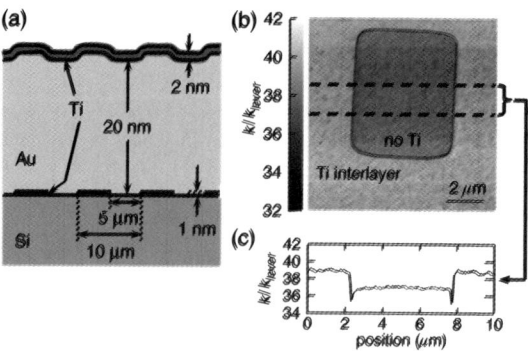

of k/k_{lever} for 40 lines in the center of the image is shown in Fig. 5.18c. The mean value of k/k_{lever} is 39.1 ± 0.6 in the grid regions and 37.1 ± 0.5 in the square, a difference of 5%. Additional images acquired at different sample positions consistently showed a decrease in k/k_{lever} of 4–5% for the regions of poor adhesion that lacked a Ti interlayer.

From Eq. (5.11), we estimate that $a = 6$–8.5 nm for our experimental parameters. The experiments should therefore probe the film interface ($z = 22$–24 nm $\approx 3a$). The observed results are also consistent with theoretical predictions for layered systems with disbonds [74,93]. In that work, the film/substrate system was modeled with an impedance-radiation approach that included a change in boundary conditions at the disbonded interface (i.e., zero shear stress). For a disbond in an aluminum film 20 nm thick on (001) silicon, a reduction of approximately 4% in the contact stiffness was predicted, very similar to our results. Although the parameters used in the theoretical study differed from those in our experiments, the overall combination of conditions (film and substrate modulus, applied force, etc.) was sufficiently similar to ours that we believe a comparison is valid. These results represent progress towards quantitative imaging of adhesion, a goal with important practical implications for many thin-film devices.

5.9
The Road Ahead

In this chapter, we have described contact resonance force microscopy methods and their use for materials characterization on the nanoscale. CR-FM experiments involve measuring the resonant frequencies of a vibrating cantilever when its tip is in contact with a material. Models for the cantilever dynamics and for the tip–sample contact mechanics are then used to relate the contact-resonance frequencies to the near-surface elastic properties. The basic physical principles have been discussed, as well as the experimental apparatus and current best practices for measurements. Principles to optimize experimental conditions were explained and used to motivate the best practices. Some of these principles are of a more theoretical nature, such as the signal response curve. Others are more practical, such as tip shape and wear. Finally, methods for qualitative and quantitative nanomechanical imaging with CR-FM were described. We hope that the discussion stimulates readers to envision applications of CR-FM in their own research.

Although CR-FM methods are sufficiently mature to serve as a useful tool, there is still room for improvement. Extensions or refinements to the basic technique could not only yield better measurements, but could also lead to exciting new applications. Here, we briefly discuss some of the key directions for future work. The ideas can be grouped into three broad themes: improving precision and accuracy, reducing measurement time, and adapting techniques to new material systems.

Measurement accuracy and precision might be improved by using custom-fabricated cantilevers, instead of relying on commercially available ones. For example, optimal sensitivity in combined torsional- and flexural-mode experiments were hampered by the range of available cantilevers [34]. Accuracy could also be improved by fabricating cantilevers whose response more closely matches that predicted by the data analysis model. For instance, a nonuniform or nonrectangular cross section may shift the spacing of the free frequencies. Vibrational behavior can also be affected by asymmetries in the cantilever beam [94], or by the clamping conditions of the cantilever holder [23]. In one case, the resulting spurious modes were reduced by patterning the backside of the cantilever substrate [95]. A more radical approach involves redesigning the cantilever geometry. Examples of this approach in other AFM applications include cantilevers with higher-order modes that occur at an exact integer multiple of the fundamental mode [96], and cantilevers whose free frequencies are adjusted to accommodate the bandwidth of the AFM photodiode [97].

Instead of designing a new cantilever, accuracy might be improved with a better model to predict the dynamic behavior of real cantilevers. One possibility is to replace the relatively simple analytic model with a more detailed model involving finite-element analysis (FEA). FEA models have been developed to evaluate the dynamic behavior of resonant cantilevers in free space [98] and in contact [99–101]. With a FEA approach, information about the actual geometry of the cantilever obtained from SEM or other imaging tools is easily incorporated. This feature is particularly valuable when the cantilever geometry varies significantly from a rectangular shape [100]. In spite of progress, an alternative to the existing analytical model that is suitable for practical measurements has not yet been reported.

Measurements could also be improved by a better understanding of the tip–sample contact behavior. As discussed in Sect. 5.7.2, an improved procedure to characterize the true tip shape could increase precision. Work to characterize tips for other AFM applications might provide insight into this issue [102–104]. Tip wear also adds to measurement uncertainty. As described above and in Ref. [105], deliberately blunted or rounded tips could address this problem. In order to maintain high spatial resolution, however, sharp yet robust tips will ultimately be needed. Tips with hard wear coatings are commercially available and seem ideally suited. In practice, however, the coatings tend to fracture [45], and can be insufficiently smooth for consistent tip–sample contact. As with cantilever geometry, it may be necessary to fabricate custom tips. Carbon-coated tips with a hemispherical shape were created in one study with promising results [106].

Future work to reduce measurement time will contain both experimental and theoretical aspects. From a practical standpoint, minimizing tip wear could reduce measurement time in addition to increasing accuracy. More robust tips might enable measurement redundancy to be minimized and the number of reference measurements to be reduced. From a theoretical standpoint, measurement time could be cut virtually in half by devising a method that requires the frequency of only one resonant mode. Currently, the frequencies of two modes are needed to determine both the normalized contact stiffness k/k_{lever} and the relative tip position L_1/L. Perhaps L_1/L could be determined in another way, for instance through an additional calibration procedure, FEA modeling, or SEM imaging. Rapid or even simultaneous measurement of multiple modes would also reduce measurement time [107]. Increased speed in imaging applications is also important. There are physical limits on the ultimate scanning speed that can be attained. However, current times of ~ 30 min or longer to acquire one frequency image could be reduced considerably before reaching these limits. This is an area of active research not only for AFM in general [108, 109], but also for CR-FM in particular [17].

Extending the range of materials and applications that can be addressed with CR-FM methods is an exciting prospect. For instance, dynamical processes could be studied if faster imaging rates were possible. Measurements on extremely thin films (~ 20 nm or less) also require refinements such as an analysis approach that includes the substrate properties [110]. New applications involving highly compliant materials such as polymers present other challenges, because CR-FM methods were originally developed for use on much stiffer materials. For instance, lower static forces are needed to prevent sample damage, requiring the use of cantilevers with a lower spring constant. To optimize the response of such cantilevers, measurements with higher-order mode pairs (e.g., $n = 2, 3$ or $n = 3, 4$) may be needed. Analysis of low-force data is also likely to involve inelastic or damping effects (see for example Ref. [39] or [111]). Such effects can arise not only from a surface capillary layer, but from the viscoelastic nature of the material itself. It may eventually be possible to quantitatively evaluate extremely compliant ($M < 100$ MPa) materials with CR-FM [112].

The above discussion presents a variety of measurement challenges for CR-FM methods. Successfully meeting these challenges will enable researchers to better address materials problems in emerging applications as well as existing ones. The increased use and applicability of these methods will contribute to the rapid growth of nanoscale materials science. For all of these reasons, the future seems bright for CR-FM methods as a valuable tool for materials characterization.

Acknowledgments. I am grateful for the many contributions made by M. Kopycinska-Müller (now at IZFP/IAVT-TU, Dresden, Germany) to this work while at NIST from 2003 to 2006. The idea for this chapter originated from our discussions together. I also thank colleagues at NIST who contributed in many ways to this work: M. Fasolka, R. Geiss, A. Kos, E. Langlois (now at Analog Devices, Boston, Massachusetts), J. Pratt, P. Rice (now at the University of Colorado-Boulder), D. Smith, C. Stafford, and G. Stan. I value many interactions over the years with researchers from other institutes, especially with J. Turner and students (University of Nebraska-Lincoln), and W. Arnold, U. Rabe, and S. Hirsekorn (Fraunhofer Institute for Nondestructive Testing IZFP, Saarbrücken, Germany). I also value collaborations and discussions with R. Geer (University at Albany, State University of New York), B. Huey (University of Connecticut-Storrs), N. Jennett and M. Monclus (National Physical Laboratory, Teddington, UK), and M. Prasad (Colorado School of Mines, Golden). The data on SnO_2 nanobelts presented in this chapter were the result of a visit to NIST by Y. Zheng (then at the University at Albany).

References

1. http://www.nano.gov/NNI_07Budget.pdf (accessed August 2008)
2. Oliver WC, Pharr GM (1992) J Mater Res 7:1564
3. Syed Asif SA, Wahl KJ, Colton RJ, Warren OL (2001) J Appl Phys 90:1192
4. Li X, Bhushan B (2002) Mater Charact 48:11
5. Every AG (2002) Meas Sci Technol 13:R21
6. Ogi H, Tian J, Tada T, Hirao M (2003) Appl Phys Lett 83:464
7. Cretin B, Sthal F (1993) Appl Phys Lett 62:829
8. Kraft O, Volkert CA (2001) Adv Engng Mater 3:99
9. Binning G, Quate CF, Gerber Ch (1986) Phys Rev Lett 56:930
10. Maivald P, Butt HJ, Gould SAC, Prater CB, Drake B, Gurley JA, Elings VB, Hansma PK (1991) Nanotechnology 2:103
11. Burnham NA, Kulik AJ, Gremaud G, Gallo PJ, Oulevey F (1996) J Vac Sci Technol B 14:794
12. Troyon M, Wang Z, Pastre D, Lei HN, Hazotte A (1997) Nanotechnology 8:163
13. Rosa-Zeiser A, Weilandt E, Hild S, Marti O (1997) Meas Sci Technol 8:1333
14. Cappella B, Dietler G (1999) Surface Sci Repts 34:1
15. Zhong Q, Inniss D, Kjoller K, Elings VB (1993) Surface Sci 290:L688
16. Yamanaka K, Ogiso H, Kolosov OV (1994) Appl Phys Lett 64:178
17. Huey BD (2007) Annu Rev Mater Res (2007) 37:351
18. Cuberes MT, Assender HE, Briggs GAD, Kolosov OV (2000) J Phys D: Appl Phys 33:2347
19. Yamanaka K, Nakano S (1996) Jpn J Appl Phys 35:3787
20. Rabe U, Arnold W (1994) Appl Phys Lett 64:1493
21. Dinelli F, Castell MR, Ritchie DA, Mason NJ, Briggs GAD, Kolosov OV (2000) Phil Mag A 80:2299
22. Rabe U (2006) Atomic force acoustic microscopy. In: Bushan B, Fuchs H (eds) Applied scanning probe methods, vol II. Springer, Berlin Heidelberg New York, p 37
23. Rabe U, Hirsekorn S, Reinstädtler M, Sulzbach T, Lehrer C, Arnold W (2007) Nanotechnology 18:044008
24. Hurley DC, Shen K, Jennett NM, Turner JA (2003) J Appl Phys 94:2347
25. Commercial equipment and materials are identified only in order to adequately specify certain procedures. In no case does such identification imply recommendation or endorsement by the National Institute of Standards and Technology, nor does it imply that the materials or equipment identified are necessarily the best available for the purpose.
26. Zheng Y, Geer RE, Dovidenko K, Kopycinska-Müller M, Hurley DC (2006) J Appl Phys 100:124308
27. Papadakis EP (1990) The measurement of ultrasonic velocity. In: Thurston RN, Pierce AD (eds) Physical Acoustics, vol XIX. Academic Press, San Diego, p 81
28. Oliver WC, Pharr GM (2004) J Mater Res 19:3
29. Rabe U, Amelio S, Kopycinska M, Hirsekorn S, Kempf M, Göken M, Arnold W (2002) Surf Interf Anal 33:65
30. Stan G, Price W (2006) Rev Sci Instr 77:103707
31. Rabe U, Janser K, Arnold W (1996) Rev Sci Instr 67:3281
32. Turner JA, Hirsekorn S, Rabe U, Arnold W (1997) J Appl Phys 82:966
33. http://em-jaturner.unl.edu/AFMcalcs.htm (accessed August 2008)
34. Hurley DC, Turner JA (2007) J Appl Phys 102:033509
35. Wright OB, Nishiguchi N (1997) Appl Phys Lett 71:626
36. Rabe U, Turner J, Arnold W (1998) Appl Phys A 66:S277
37. Hurley DC, Turner JA (2004) J Appl Phys 95:2403

38. Hurley DC, Kopycinska-Müller M, Julthongpiput D, Fasolka MJ (2006) Appl Surf Sci 253:1274
39. Johnson KL (1985) Contact Mechanics. Cambridge University Press, Cambridge UK
40. Kopycinska-Müller M, Geiss RH, Hurley DC (2006) Ultramicroscopy 106:466
41. Langlois ED, Shaw GA, Kramar JA, Pratt JR, Hurley DC (2007) Rev Sci Instr 78:093705
42. Hurley DC, Turner JA, Wiehn JS, Rice P (2002) In: Meyendorf N, Baaklini GY, Michel B (eds) Proc of the SPIE 4703. SPIE Publishers, Bellingham WA, p 65
43. Rabe U, Kopycinska M, Hirsekorn S, Muñoz Saldaña J, Schneider GA, Arnold W (2002) J Phys D: Appl Phys 35:2621
44. Prasad M, Kopycinska M, Rabe U, Arnold W (2002) Geophys Res Lett 29:13–1
45. Amelio S, Goldade AV, Rabe U, Scherer V, Bhusan B, Arnold W (2001) Thin Solid Films 392:75
46. Passeri D, Bettucci A, Germano M, Rossi M, Alippi A, Sessa V, Fiori A, Tamburri E, Terranova ML (2006) Appl Phys Lett 88:121910
47. Preghnella M, Pegoretti A, Migliaresi C (2006) Polymer Testing 25:443
48. Kester E, Rabe R, Presmanes L, Tailhades Ph, Arnold W (2000) J Phys Chem Solids 61:1275
49. Passeri D, Bettucci A, Germano M, Rossi M, Alippi A, Orlanducci S, Terranova ML, Ciavarella M (2005) Rev Sci Instr 76:093904
50. Tsuji T, Saito S, Fukuda K, Yamanaka K, Ogiso H, Akedo J, Kawakami K (2005) Appl Phys Lett 87:071909
51. Hurley DC, Kopycinska-Müller M, Kos AB, Geiss RH (2005) Meas Sci Technol 16:2167
52. Hurley DC, Geiss RH, Jennett NM, Kopycinska-Müller M, Maxwell AS, Müller J, Read DT, Wright JE (2005) J Mater Res 20:1186
53. Kopycinska-Müller M, Geiss RH, Müller J, Hurley DC (2005) Nanotechnology 16:703
54. Passeri D, Rossi M, Alippi A, Bettucci A, Manno D, Serra A, Filippo E, Lucci M, Davoli I (2008) Superlattices and Microstructures, in press, doi:10.1016/j.spmi.2007.10.004
55. Dupas E, Gremaud G, Kulik A, Loubet J-L (2001) Rev Sci Instr 72:3891
56. Stan G, Ciobanu CV, Parthangal PM, Cook RF (2007) Nano Lett 7:3691
57. Oulevey F, Gremaud G, Mari D, Kulik AJ, Burnham NA, Benoit W (2000) Scripta mater 42:31
58. Crozier KB, Yaralioglu GG, Degertkin FL, Adams JD, Minne SC, Quate CF (2000) Appl Phys Lett 76:1950
59. Cuenot S, Frétigny C, Demoustier-Champagne S, Nysten B (2004) Phys Rev B 69:165410
60. Mangamma G, Mohan Kant K, Rao MSR, Kalavathy S, Kamruddin M, Dash S, Tyagi AK (2007) J Nanosci Nanotechnol 7:2176
61. Reinstädtler M, Rabe U, Scherer V, Hartmann U, Goldade A, Bhushan B, Arnold W (2003) J Appl Phys 82:2604
62. Drobek T, Stark RW, Gräber M, Heckl WM (1999) New J Phys 1:15
63. Kawagishi T, Kato A, Hoshi Y, Kawakatsu H (2002) Ultramicroscopy 91:37
64. Caron A, Rabe U, Reinstädtler M, Turner JA, Arnold W (2004) Appl Phys Lett 85:6398
65. Song Y, Bhushan B (2005) J Appl Phys 97:083533
66. Scherer V, Reinstädtler M, Arnold W (2004) Atomic force microscopy with lateral modulation. In: Bhushan B, Fuchs H, Hosaka S (eds) Applied scanning probe methods, vol I. Springer, Berlin Heidelberg New York, p 75
67. Reinstädtler M, Kasai T, Rabe U, Bhushan B, Arnold W (2005) J Phys D: Appl Phys 38:R269
68. Mazeran PE, Loubet JL (1997) Trib Lett 3:125
69. Rabe U, Amelio S, Kester E, Scherer V, Hirsekorn S, Arnold W (2000) Ultrasonics 38:430
70. Turner JA, Wiehn JS (2001) Nanotechnology 12:322
71. Chang WJ (2002) Nanotechnology 13:510

72. Wu TS, Chang WJ, Hsu JC (2004) Microelectronic Engineering 71:15
73. Sthal F, Cretin B (1995) In: Jones JP (ed) Acoustical imaging, vol 21. Plenum Press, New York, p 305
74. Yaralioglu GG, Degertekin FL, Crozier KB, Quate CF (2000) J Appl Phys 87:7491
75. Yamanaka K, Nakano S (1998) Appl Phys A 66:S313
76. Olson S, Sankaran B, Altemus B, Geer R, Castracane J, Xu B (2006) J Microlith Microfab Microsyst 5:021197
77. Burnham NA, Gremaud G, Kulik AJ, Gallo PJ, Oulevey F (1996) J Vac Sci Technol B 14:1308
78. Stark RW (2004) Rev Sci Instr 75:5053
79. Schäffer TE, Fuchs H (1995) J Appl Phys 97:083524
80. Yamanaka K, Tsuji T, Noguchi A, Koike T, Mihara T (2000) Rev Sci Instr 71:2403
81. Rabe U (2007) private communication
82. Rabe U, Scherer V, Hirsekorn S, Arnold W (1997) J Vac Sci Technol B 15:1506
83. Yamanaka K, Maruyama Y, Tsuji T, Nakamoto K (2001) Appl Phys Lett 78:1939
84. Kobayashi K, Yamada H, Matsushige K (2002) Surf Interf Anal 33:89
85. Efimov E, Saunin SA (2002) In: Proc of the scanning probe microscopy conference 2002, p 79. Available at http://ntmdt.com/publications? year = 2002 (accessed August 2008)
86. Hurley DC, Kos AB, Rice P (2005) In: Kalinin SV, Goldberg B, Eng LM, Huey BD (eds) Proc of the MRS 838E. Mater Res Soc, Warrendale PA, p O8.2.1
87. Kos AB and Hurley DC (2008) Meas Sci Technol 19:015504
88. Tsuji T, Yamanaka K (2001) Nanotechnology 12:301
89. Tsuji T, Irihama H, Yamanaka K (2002) Jpn J Appl Phys 41:832
90. Striegler A, Pathuri N, Köhler B, Bendjus B (2007) In: Thompson DO, Chimenti DE (eds) AIP Conference Proceedings 894, Rev Prog QNDE 2006. AIP Publishing, Melville NY, p 1572
91. McGuigan AP, Huey BD, Briggs GAD, Kolosov OV, Tsukahara Y, Yanaka M (2002) Appl Phys Lett 80:1180
92. Hurley DC, Kopycinska-Müller M, Langlois ED, Kos AB, Barbosa N, (2006) Appl Phys Lett 89:021911
93. Sarioglu AF, Atalar A, Degertekin FL (2004) Appl Phys Lett 84:5368
94. Reinstädtler M, Rabe U, Scherer V, Turner JA, Arnold W (2003) Surf Sci 532–535:1152
95. Adams JD, York D, Whisman N (2004) Rev Sci Instr 75.2903
96. Sahin O, Yaralioglu G, Grow R, Zappe SF, Atalar A, Quate C, Solgaard O (2004) Sens Act A 114:183
97. Sadewasser S, Villanueva G, Plaza JA (2006) Rev Sci Instr 77:073703
98. Mendels DA, Lowe M, Cuenat A, Cain MG, Vallejo E, Ellis D, Mendels F (2006) J Micromech Microeng 16:1720
99. Arinero R, Lévêque G (2003) Rev Sci Instr 74:104
100. Shen K, Hurley DC, Turner JA (2004) Nanotechnology 15:1582
101. Espinoza Beltrán FJ, Scholz T, Schneider GA, Muñoz-Saldaña J, Rabe U, Arnold W (2007) In: Meyer E, Hegner M, Gerber C, Güntherodt H-J (eds) J Phys Conference Series 61, Proc ICN& T 2006. IOP Publishing, Bristol UK, p 293
102. Villarrubia JS (1996) J Vac Sci Technol B 14:1518
103. Villarrubia JS (1997) J Res Natl Inst Stand Technol 102:425
104. Itoh H, Fujimoto T, Ichimura S (2006) Rev Sci Instr 77:103704
105. Muraoka M (2005), Nanotechnology 16:542
106. Schwarz UD, Zwörner O, Köster P, Wiesendanger R (1997) J Vac Sci Technol B 15:1527
107. Jesse S, Kalinin SV, Proksch R, Baddorf AP, Rodriguez BJ (2007) Nanotechnology 18:435503
108. Humphris ADL, Miles MJ, Hobbs JK (2005) Appl Phys Lett 86:034106

109. Hansma PK, Schitter G, Fantner GE, Prater C (2006) Science 314:601
110. Batog GS, Baturin AS, Bormashov VS, Sheshin EP (2006) Tech Phys 51:1084
111. Schwarz UD (2003) J Coll Interf Sci 261:99
112. Ebert A, Tittmann BR, Du J, Scheuchenzuber W (2006) Ultrasound Med Biol 32:1687

6 AFM Nanoindentation Method: Geometrical Effects of the Indenter Tip

Lorenzo Calabri · Nicola Pugno · Sergio Valeri

Abstract. Atomic force microscopy (AFM) nanoindentation is presently not that widespread for the study of mechanical properties of materials at the nanoscale. "Nanoindenter" machines have greater accuracy and are presently more standardized. However, AFM could provide interesting features such as imaging the indentation impression right after the application of load. The AFM has, in fact, become an increasingly popular tool for characterizing surfaces and thin films of many different types of materials and recent developments have led to the utilization of the AFM as a nanoindentation device, increasing the accuracy of this machine.

In this work a new method for nanoindentation via AFM is proposed. It allows hardness measurement with standard sharp AFM probes and a simultaneous high-resolution imaging (which is not achievable with standard indenters – *Cube Corner* and *Berkovich*). How the shape of the indenter and the tip radius of curvature affect the hardness measurement at the nanoscale is herein analyzed with three different approaches: experiments, numerical simulations, and theoretical models. In particular the effect of the tip radius of curvature, which is not negligible for the real indenters, has been considered both in the nature of the indentation process, and in the practice of imaging via AFM.

A complete theoretical model has been developed and it includes the effect of the tip radius of curvature as well as the variable corner angle. Through this model we have been able to define a correction factor C that permits us to evaluate the actual hardness of the material, once measured the actual geometry of the tip.

Key words: Nanoindentation, Atomic Force Microscopy, Finite element method, Indentation shape effect, Tip radius of curvature effect

6.1
Introduction

As defined by Fisher-Cripps indentation testing is "a simple method that consists essentially of touching the material whose mechanical properties such as elastic modulus and hardness are unknown with another material whose properties are known" [1]. Nanoindentation differs from conventional macro-indentation in the length scale of the penetration, which is of the order of nanometers rather than microns or millimeters. Over the last few years, interest in nanomechanics has grown exceptionally. In particular the mechanical properties of materials at the nanoscale have been carefully investigated from a theoretical and experimental point of view, but theoretical work strongly depends upon accurate experimental results. The nanoindentation technique, in particular, has been extensively exploited by many researchers all over the world in order to study hardness, Young's modulus, and other mechanical properties of thin films and coatings such as the strain-hardening exponent, fracture toughness (for brittle materials), and viscoelastic properties [2–5]. The technique has also been thoroughly investigated in order to understand its main features at the nanoscale [6].

The idea of nanoindentation, in fact, arose from the necessity to measure the mechanical properties of very small volumes of materials. In principle, if a very sharp tip is used, the volume of material that is tested can be made arbitrarily small but in this case it is very difficult to determine the indentation area. In a conventional hardness test, in fact, the characteristic contact area of the indentation is calculated from direct measurements of the residual impression left in the specimen surface. In a nanoindentation test, the size of the residual impression is on the sub-micrometer range and too small to be conventionally measured (optical microscopy). The hardness is in fact defined as the ratio between the maximum applied load ($Pmax$ – easily measurable during the indentation) and the projected area of the indentation impression (A_p – not measurable directly by conventional methods). This area can be evaluated measuring the depth of indentation into the surface, which provides an indirect measurement of the contact area, knowing the actual geometry of the indenter. For this reason nanoindentation can be considered a special case of the more general Depth Sensing Indentation (*DSI*) methods [7–10]. In particular most of the recent studies concerning material nanohardness, are based on the analysis of the load-displacement curves resulting from the nanoindentation test using the Oliver and Pharr (O-P) method [8, 9, 11]. The O-P method allows hardness measurement without imaging the indentation impression, since it establishes a relationship between the projected area of the indentation impression (A_p), the maximum depth of indentation (h_{max}), and the initial unloading stiffness (S), where h_{max} and S are both measurable from the load-displacement curve (Fig. 6.1 [9]).

The atomic force microscopy (AFM) approach to nanoindentation [5, 12, 13] on the contrary permits a direct measurement of the projected area of the indentation

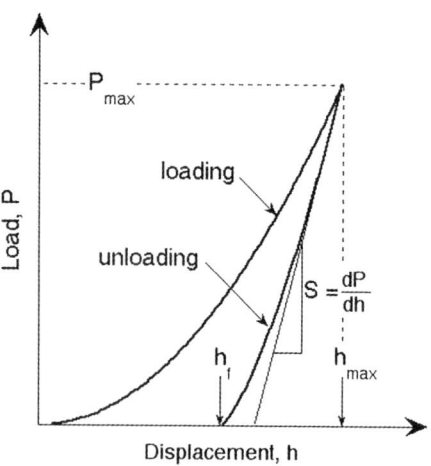

Fig. 6.1. Schematic illustration of indentation load-displacement data [9] showing important measured parameters. [Reprinted from [9], Copyright (2004) Materials Research Society. Reproduced with permission.]

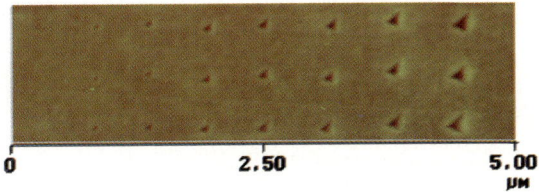

0   2.50   5.00
                                            µM

Fig. 6.2. AFM image of the plastic impressions [13] remaining in the BCB material after indentation (height scale is 0 to 20 nm from *black* to *white*). [Reprinted from [13], Copyright (2001) reproduced with permission.]

impression. As a matter of fact, AFM could provide interesting features such as imaging the indentation impression right after the application of load (Fig. 6.2, [13]).

The AFM has become an increasingly popular tool for characterizing surfaces and thin films of many different types of materials and recent developments have led to the utilization of the AFM as a nanoindentation device. During operation of the AFM in indentation mode, the probe tip is first lowered into contact with the sample, then indented into the surface, and finally lifted off the sample surface. AFM software has been modified and diamond-tipped probes have been developed (Fig. 6.3.) specifically for indenting and scratching materials with nanoscale spatial resolution. The software modification allows the surface to be imaged in tapping mode immediately before and after indentation.

With this approach it is thus possible to obtain the exact morphology of the indentation impression with high resolution at the nanoscale (Fig. 6.2) and to directly measure the actual value of the projected area A_p. This approach allows us also to consider the presence of piled-up material (Fig. 6.4 [14]), which is a major topic for indentation at the nanoscale [14, 15]. The pile-up effect is usually neglected with the O-P approach and leads to an overestimation of the hardness value [16–18]. Beegan et al. [14] used, in the case reported in Fig. 6.4, specific software (Matlab®) in order

Fig. 6.3. SEM image of an AFM diamond-tipped probe (Cube Corner indenter) customized specifically for indenting and scratching

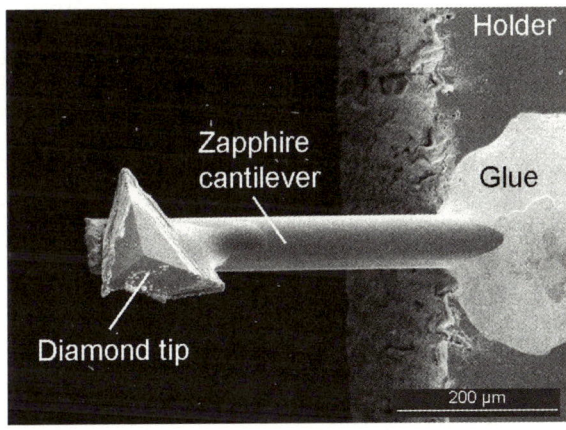

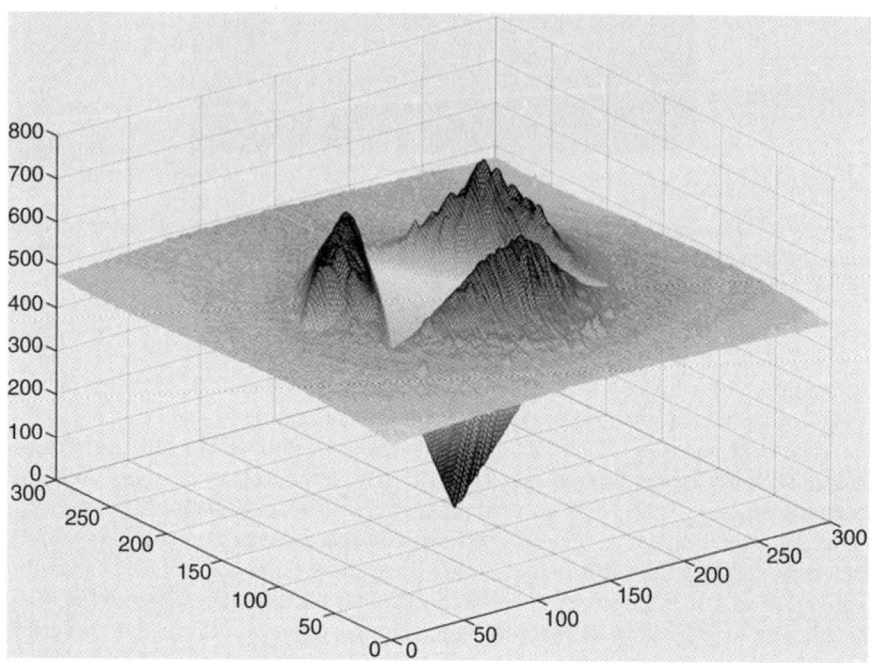

Fig. 6.4. Matlab® mesh surface plot of 40-mN indent in 750-nm Cu film. [Reprinted from [14], Copyright (2005), with permission from Elsevier.]

to characterize and precisely quantify the amount of piled-up material, exploiting 3D images obtained by AFM investigation.

Hardness, as previously mentioned, is the mean pressure that a material bears under load. This parameter is only "nominally" an intrinsic constant factor and it is experimentally affected by several geometrical uncertainties, such as penetration depth, size and shape of the indenter [19–22]. Nix et al. in an early work about nanoindentation [20], illustrate the size effects in crystalline materials, showing that the hardness of a material strongly depends upon depth of indentation, tending asymptotically to the macroscopic hardness of the material (Fig. 6.5, [20]).

Much of the early work on indentation was reviewed by Mott [23]. Afterwards Ashby [24] proposed that geometrically necessary dislocations [25] would lead to an increase in hardness measured by a flat punch. The problem of a conical indenter has been thoroughly investigated by Nix et al. [20], showing a consistent agreement with micro-indentation experiments. However, recent results that cover a greater range of depths show only partial [21, 26] or no agreement [27] with this model [20]. Thus, the model initially proposed by Nix et al. was extended by Swadener et al. [21] in order to cover a greater range of depth (Fig. 6.6, [21]) and also to treat indenters with different sizes and shapes. The results were compared with micro-indentation experiments, but limitations for small depths of indentation were observed, as pointed out by the same authors. In the last few years a new model capable of matching as limit cases all the discussed indentation laws, simultaneously capturing the deviation

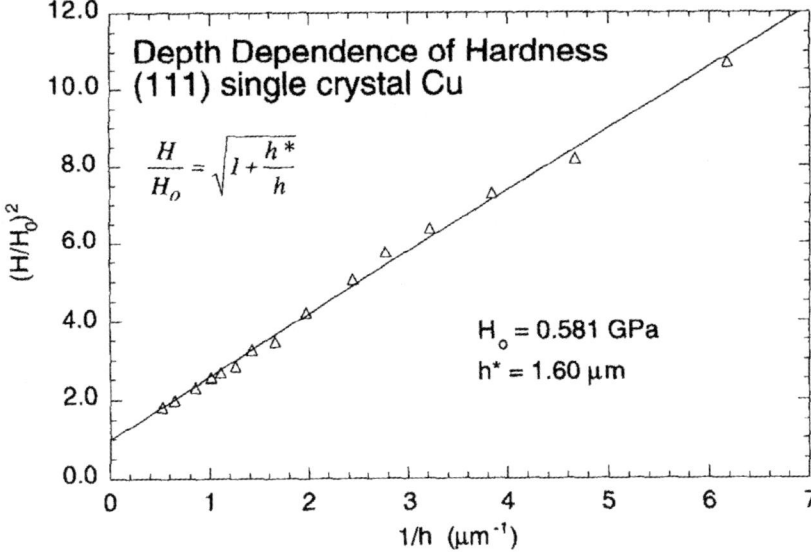

Fig. 6.5. Depth dependence of the hardness of (111) single crystal copper plotted according to the equation reported in the graph. [Reprinted from [20], Copyright (1998), with permission from Elsevier.]

observed towards the nanoscale, has been proposed by Pugno [28] and will be thoroughly discussed in Sect. 6.4.

In this chapter the geometrical uncertainties related to the indenter tip (shape and radius of curvature) are investigated in order to find a way to compensate for these effects. In particular, considerations about the tip shape, in terms of tip corner angle, have been deduced in order to understand what the effect of indenter geometry is on the material hardness evaluation. Thanks to this approach a new method for nanoindentation is proposed. It allows hardness measurement with standard sharp AFM probes (thus with variable corner angle); the use of these probes enables a simultaneous high-resolution imaging (which is not easily achievable with standard indenters – *Cube Corner* and *Berkovich*). How the shape of the indenter [19] and the tip radius of curvature [29] affect the hardness measurement is herein analyzed to find a relationship between the measured hardness of a material, the corner angle of the pyramidal indenter, and its tip radius of curvature. To experimentally understand this effect a photoresist material (*Microposit s1813 photo resists*) has been indented with Focused Ion Beam (FIB) nanofabricated probes with different corner angles [19]. We then compared the results obtained experimentally with those obtained by numerical simulations and by theoretical models [28].

The comparison between these three approaches reveals a good agreement in the hardness behavior even if an overestimation of the experimental results was noticed at small corner angles [19]. This is related to the tip radius of curvature of the real indenters, which is not negligible and affects the experimental results, both in the nature of the experimental procedure, and then in the process of imaging with AFM.

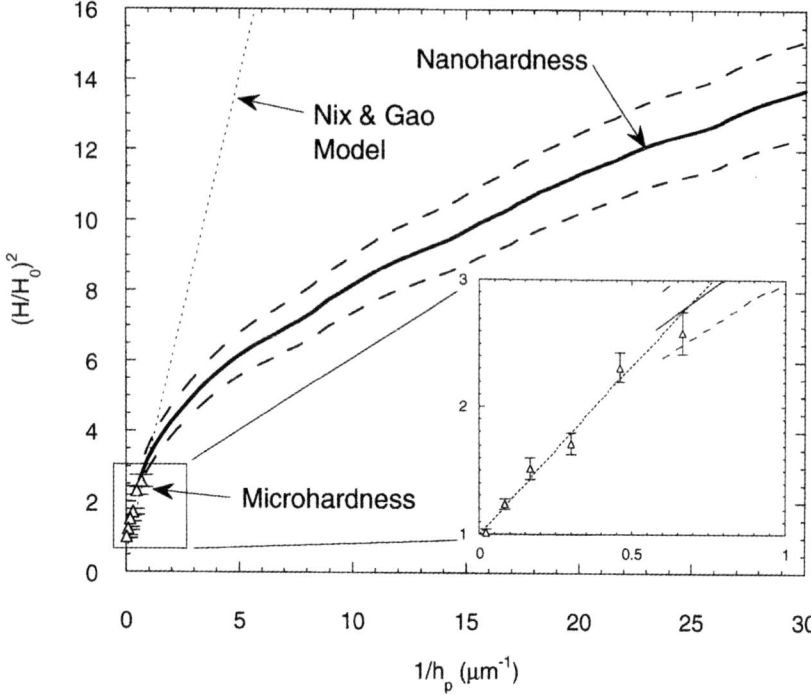

Fig. 6.6. Indentation size effect in annealed iridium measured with a Berkovich indenter (Δ and *solid line*) and comparison of experiments with the Nix et al. [20] model for $H_0 = 2.5$ GPa and $h_p = 2.6\,\mu$m (*dotted line*). The *dashed lines* represent $+$ and $-$ one standard deviation of the nanohardness data. [Reprinted from [21], Copyright (2002), with permission from Elsevier.]

The presence of a non-zero tip radius of curvature is ascribable to wear during the indent-imaging process or to manufacturing defects. It could affect the indentation process because the indenter, no longer ideal, will deform the specimen with a different geometry. However, it could also affect the process of imaging, as long as the non-ideal tip interacts with the morphology, convoluting the asperities depending on its actual shape. For this reason the theoretical and numerical models have been modified in order to compensate for these effects and to obtain a closer match between experimental and numerical approaches. In this way we were also able to define a correction factor C which permits us to evaluate the actual hardness of the material, filtering the experimental data.

6.2
Experimental Configuration

In this section the experimental approach proposed by the authors in a previous work [19] is reviewed and extended.

Table 6.1. Mechanical properties of the specimen material (photoresist). [Reprinted from [19], Copyright (2007), with permission from IoP.]

Mechanical property	Reference value
Microhardness	$\sim 200\,\mathrm{MPa}$
Ultimate tensile strength	$51.2\,\mathrm{MPa}$
Yield tensile strength	$43\,\mathrm{MPa}$
Elongation at break	0.6%
Young's modulus	$8\,\mathrm{GPa}$
Poisson's ratio	0.33

The set of indenters used for the nanoindentation experimental analysis are commercial silicon AFM tips, in particular we used *MPP-11100-Tap300 Metrology Probes* from *Veeco®*. The commercial silicon AFM tips are easy to find, cheap and reshapable with the FIB nanofabrication process. These kinds of tips are silicon made and consequently they could not provide a high mechanical profile in terms of hardness and non-deformability. For this reason we decided to use them to indent a soft substrate in order to keep a high ratio between the hardness of the indenter and the hardness of the sample. In this way the silicon tips, even if they provide a poor hardness value, will be basically non-deformable when pressed on the selected soft material. The substrate we used is a photoresist material, namely a *Microposit s1813 photo resists* by *Shipley®*. It is a *positive* photoresist based on a *NOVOLAC* polymer. Its mechanical properties are reported in Table 6.1 [19]. This material is ideal for this kind of study, because it is very soft, thus easy to indent with a silicon tip, and at the same time it is very flat, allowing an accurate measurement of the indentation projected area.

6.2.1
FIB Nanofabrication

The pristine geometry of the probe tip is a quadratic pyramid (Fig. 6.7a,b). As reported in the Veeco® probes catalogue the characteristic geometry of the tip is listed in the table inset in Fig. 6.7b.

In this work, in order to obtain a set of indenters with a variable corner angle, we functionalized the pristine probes with a FIB apparatus in order to transform them into a triangular pyramid (as nanoindenters usually are). The FIB system is a Dual Beam machine (*FEI StrataTM DB 235*) combining a high-resolution FIB column equipped with a Ga Liquid Metal Ion Source (LMIS) and a SEM column equipped with a Schottky Field Emission Gun (SFEG) electron source. FIB offers the ability to design, sculpt or pattern nano- and micro-structures on different materials with spatial resolution down to 20 nm.

By means of the FIB machine we proceeded in cutting the tip with a plane positioned with a proper different orientation. In this way we utilized two pristine faces of

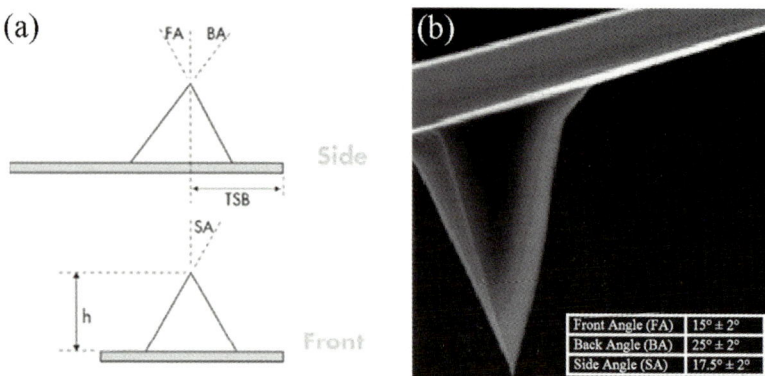

Fig. 6.7. (**a**) Schematic of the tip geometry (from Veeco® probes catalogue); (**b**) image of the pristine probe (from Veeco® probes catalogue)

the original tip and we just created the third face. In Fig. 6.8 is reported the reshaping procedure step by step [19].

With this procedure we realized three different indenters. In Fig. 6.9 the SEM images of the three probes obtained by FIB nanofabrication are reported [19]. The orientation of the cutting planes is an important feature in order to fabricate a tip that approaches the sample perpendicularly to the surface. To obtain these geometries we always consider the 12° angle of the AFM probe holder (Fig. 6.8b). The second probe is obtained cutting the tip with three different planes (there was no chance in fact to utilize any pristine face of the original probe) and the shape was completely recreated (Fig. 6.9b).

In Fig. 6.10 one can see the final shape of customized probe n°1 [19], observed with the SEM microscope (Fig. 6.10a,b) and with AFM (Fig. 6.10c) using a calibration grid composed of an array of sharp tips (test grating *TGT1 – NT-MDT®*).

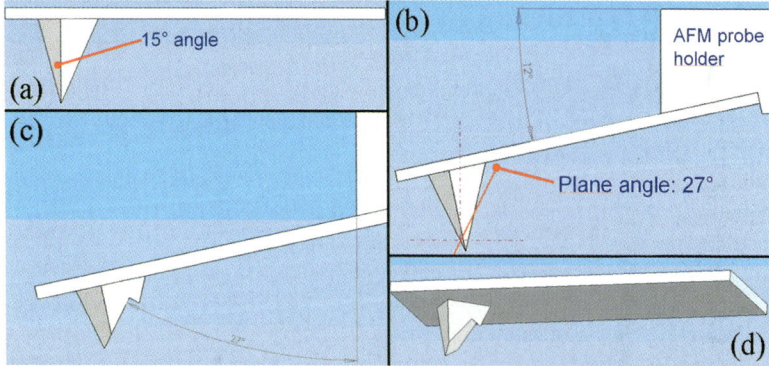

Fig. 6.8. (**a**) Probe pristine geometry; (**b**) position of the cutting plane; (**c**) tip profile after the reshaping phase; (**d**) axonometric projection of the customized probe. [Reprinted from [19], Copyright (2007), with permission from IoP.]

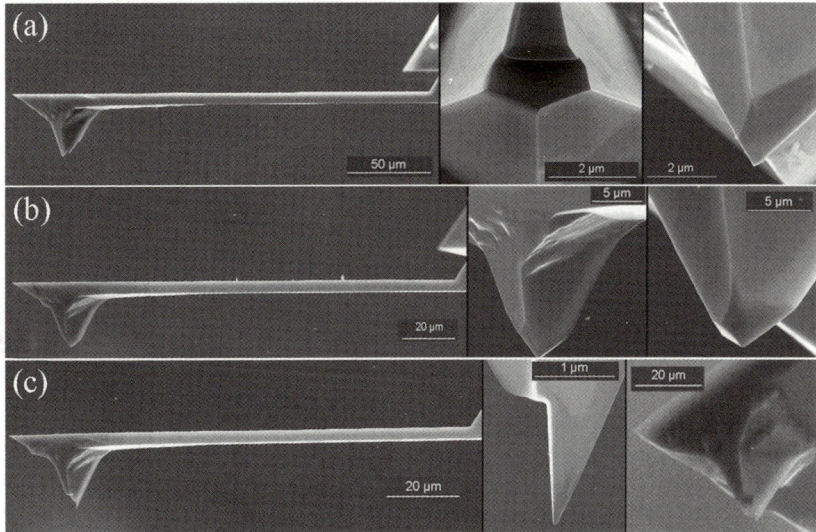

Fig. 6.9. SEM images of the customized probes [19]: (**a**) indenter n°1 – equivalent corner angle of 62°; (**b**) indenter n°2 – equivalent corner angle of 97°; (**c**) indenter n°3 – equivalent corner angle of 25. [Reprinted from [19], Copyright (2007), with permission from IoP.]

The final shape of the indenters is a triangular pyramid with a customized geometry. To codify the new geometry of the nanofabricated probes the characteristic parameter which has been used is the *equivalent corner angle*. The equivalent corner angle of a triangular pyramid is defined as the corner angle of a conical indenter with the same area function. Using this kind of codification we obtained for the three functionalized indenters the angles listed in Table 6.2 [19].

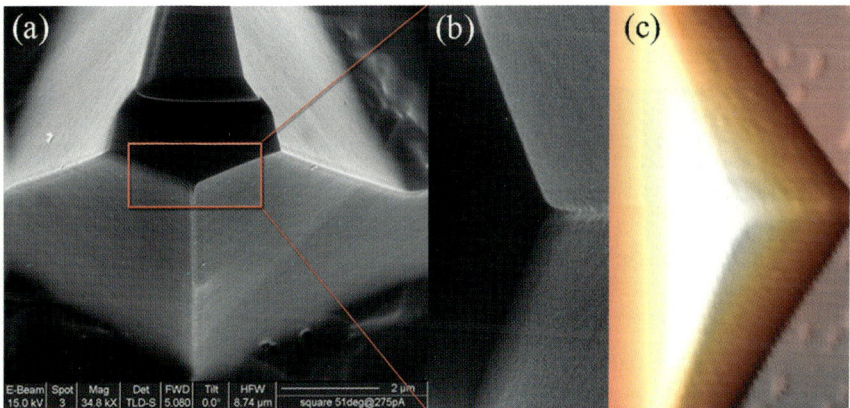

Fig. 6.10. (**a**,b) SEM images of the customized probe n°1; (**c**) AFM image on the calibration grid of the customized probe n°1 [19]. [Reprinted from [19], Copyright (2007), with permission from IoP.]

Table 6.2. Equivalent corner angles of the nanofabricated probes. [Reprinted from [19], Copyright (2007),
with permission from IoP.]

Indenter	Equivalent corner angle [°]
Probe n°1	62
Probe n°2	97
Probe n°3	25

6.2.2
Tip Radius of Curvature Characterization

In the AFM indentation procedure, the radius of curvature at the tip affects the hardness measurement in two different ways: (1) it has an influence on the nature of the penetration process, as long as the indenter, no more ideal, will deform the specimen with a different geometry; (2) it affects the process of AFM imaging, as long as the non-ideal tip will interact with the morphology, convoluting the asperities depending on its actual shape. For this reason it is necessary to determine the real shape of the indenter (in terms of tip radius of curvature) in order to modify and develop the theoretical and numerical models.

This characterizing procedure, yet introduced in a previous work [29], is herein reviewed. The topography of the customized indentation tips has been obtained by a SEM microscope equipped with a SFEG electron source and also by an AFM working in tapping mode. Using the SEM we are able to directly obtain the geometry of the tip (Fig. 6.11a, [29]), although the image obtained is a 2D projection of the tip. The result achieved in this way does not concern the actual three-dimensional structure of the system, but just a planar view of it.

Using the AFM, it is in addition possible to obtain a 3D topography of the tip (Fig. 6.11b, [29]), scanning the probe on a calibration grid (test grating *TGT1 – NT-MDT®*). The image obtained in this way is three-dimensional and represents the actual shape of the indenter. Using the "Tip characterization" tool equipped with the

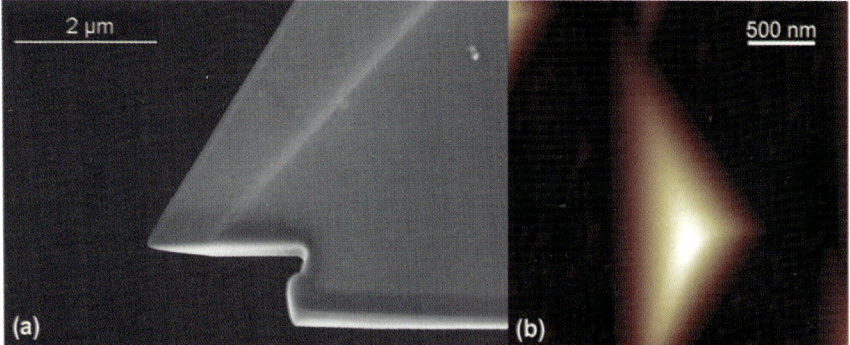

Fig. 6.11. (a) SEM image of the customized probe n°1; (b) AFM image on the calibration grid of the customized probe n°1 [29]

Table 6.3. Tip radius of curvature of the nanofab-ricated probes [29]

Indenter	Tip radius of curvature [nm]
Probe n°1	21
Probe n°2	26
Probe n°3	25

SPIP^{TM} software that we used for the AFM image analysis, we are able to precisely detect the tip radius of curvature of the three customized probes used for the nanoindentation procedure (Table 6.3, [29]).

6.2.3
Nanoindentation Experimental Setup

The whole experimental analysis has been carried out using AFM nanoindentation. The instrument that we used was a *Digital Instruments EnviroScope Atomic Force Microscope* by *Veeco*®. This instrument allowed us to indent the sample and image it right after the indentation. The experiments consisted of a matrix of 25 indentations (Fig. 6.12) performed for each probe under exactly the same conditions. The indentations reported in Fig. 6.12 reveal a clear pile-up. This phenomenon is related to the material, which is very soft, and also to the geometry of the tip (decreasing the corner angle of the tip increases the plastic deformation of the material and thus its pile-up).

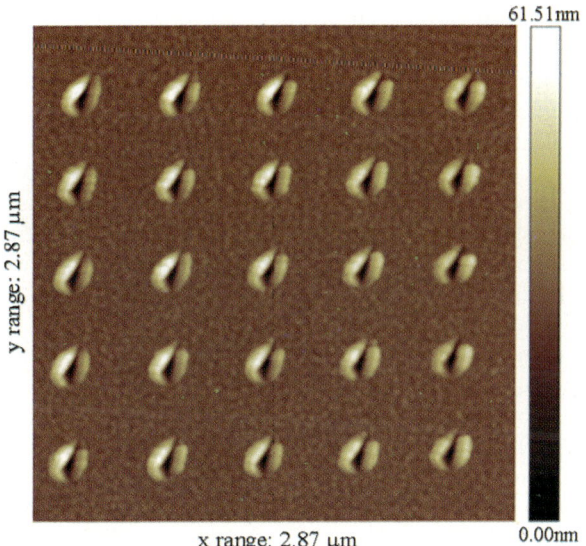

Fig. 6.12. AFM image of the indentation matrix, composed of 25 indentations, performed for each probe under exactly the same conditions

Table 6.4. Elastic spring constant and deflection sensitivity of the probes.
[Reprinted from [19], Copyright (2007), with permission from IoP.]

Indenter	Spring constant [N/m]	Deflection sensitivity [nm/V]
Probe n°1	49.0	34.2
Probe n°2	35.2	34.5
Probe n°3	38.4	34.8

The load applied to the sample is 1.6 V in terms of photodetector voltage.
Thus, considering the cantilever spring constant (obtained using the *Sader* approach
[30, 31]) and the deflection sensitivity, we obtain a maximum load for the three indenters of about 2,000 nN.

The deflection sensitivity has been calibrated using the load-displacement curves
and it corresponds to the slope of the force plot in the contact region. The deflection
sensitivity is the factor that allows us to convert the cantilever deflection from volts to
nanometers. The results in terms of elastic spring constant and deflection sensitivity
for the three probes are reported in Table 6.4 [19].

6.3
Numerical Model

The Finite Element Method (FEM) is herein used to simulate the indentation process
in order to find a numerical correlation between the hardness value and the shape of
the indenter, considering also the effect of the tip radius of curvature.

In this analysis we approach the numerical model as a non-linear contact problem. Both the indenter and the specimen are considered bodies of revolution and
the pyramid indenter is approximated by an axisymmetric cone (Fig. 6.13 [29]) with
the same equivalent corner angle. In this way it is possible to avoid a high com-
puting time connected with the three-dimensional nature of the problem, with no
introduction of considerable errors. Using a 3-D pyramid indenter, in fact, there
will be an elastic singularity at its edges, influencing the stress-strain response of
only a tiny area close to these edges. On the contrary this will not affect the continuum plastic behavior of the material, with no interference in its load-deflection
response [32].

In the present model it is assumed that the indenter is perfectly rigid and the test
material is isotropic homogeneous, elasto-plastic with isotropic hardening behavior,
obeying von Mises' yield criterion; the material was assumed to be elastic-fully plastic, thus with no strain hardening [19, 29]. This is an acceptable hypothesis, since the
material is a polymer-based photoresist which presents a perfectly plastic regime
characterized by a constant yield stress [33].

The indentation process is simulated moving the indenter with a downward-
upward displacement (100 nm). This causes the indenter to push into the surface
and then release, until it is free of contact with the specimen.

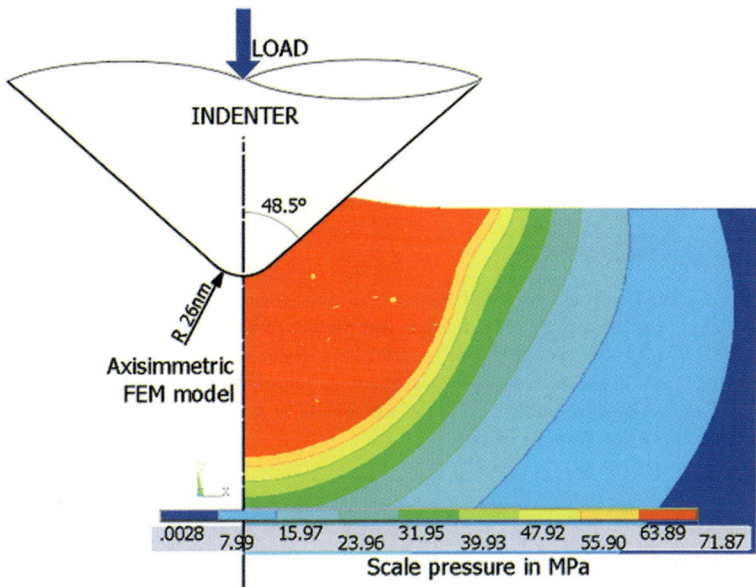

Fig. 6.13. Stress distribution in the indentation model for the probe n°2 (corner angle $= 97°$); the *vertical blue arrow* represents the applied load direction during the indentation process [29]

The indenter is modeled using the equivalent corner angles designed for the customized probes (62, 97, and $25°$ – Table 6.2) and using the tip radius of curvature obtained from the tip characterization (21, 26, and 25 nm – Table 6.3).

6.4
Theoretical Model: A Shape/Size-Effect Law for Nanoindentation

In this section the approach proposed by Pugno [28] is reviewed.

Consider an indenter with a given geometry $h = h(r, \vartheta)$, with r and ϑ polar coordinates. The previous models [20, 21] assume that plastic deformation of the surface is accompanied by the generation of sub-surface geometrically necessary dislocation loops (supposed to be of length $l(h)$ in this treatment). The deformation volume (V) is assumed to be a hemispherical zone below the projected area (A) of the indentation impression, with radius $a = \sqrt{A/\pi}$ (Fig. 6.14). Its value can be obtained by:

$$V = 2\pi/3(A/\pi)^{3/2}$$

(6.1)

Thus, the total length L of the geometrically necessary dislocation loops can be evaluated by summating the number of steps on the staircase-like indented surface (Fig. 6.14):

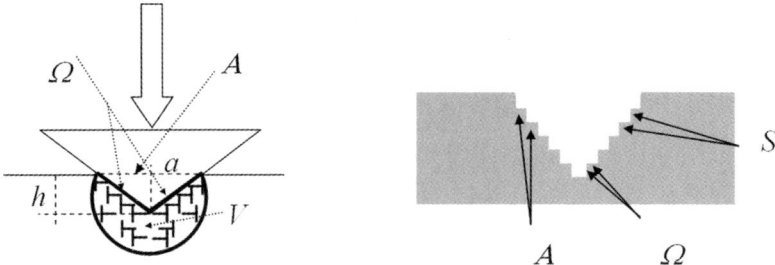

Fig. 6.14. Geometrically necessary dislocations during indentation: h is the indentation depth; a is the radius of the projected area A; Ω is the contact area and V is the dissipation domain (proportional to a^3). Note that the indented surface at the nanoscale appears in discrete steps due to the formation of dislocation loops, i.e., of quantized plasticity. In our model, the scaling law is predicted to be a function only of S/V, where $S = \Omega - A$. [Reprinted from [28], Copyright (2006), with permission from Elsevier.]

$$L = \sum_{steps} l_i = \frac{\Omega - A}{b} = \frac{S}{b} \approx \frac{1}{b}\int_0^h l(h)\mathrm{d}h \qquad (6.2)$$

where Ω is the contact area of the indented zone and b is (modulus of the) Burger's vector. Considering that the indented surface at the nanoscale appears in discrete steps due to the formation of dislocation loops, Ω can be defined as the sum of the "vertical" surfaces (S) and of the "horizontal" ones (A – Fig. 6.14). Thus, the surface S can be interpreted as the total surface through which the energy flux arises, positive if outgoing (Ω) and negative if incoming (A) in the indenter. Note the generality of the result in Eq. (6.2), which does not need any specification of the form of h, as usually required.

According to Eq. (6.2), the average geometrically necessary dislocation density is:

$$\rho_G = \frac{L}{V} = \frac{S}{bV} \qquad (6.3)$$

while the total dislocation density is usually assumed to be:

$$\rho_T = \bar{r}\rho_G + \rho_S \qquad (6.4)$$

Where $\bar{r}\rho_G$ can be considered as the actual number of dislocations that must be generated to accommodate plastic deformation, that we could call geometrically "sufficient" dislocations, and it is greater than the number of geometrically necessary dislocations ρ_G [34] by the so-called Nye factor $\bar{r}$ ($\sim$ 2, [21]). ρ_S is the statistically stored dislocation density [20]. However, we note here that, according to Eq. (6.4), $\rho_T(V/(bS) \to 0) \to \infty$, i.e., the total dislocation density at the nanoscale diverges, whereas it must physically present a finite upper-bound, which we call $\rho_T^{(nano)}$. The existence of such an upper-bound has recently been confirmed [28, 35].

Note that ρ_T is related to the shear strength τ_p by the Taylor's hardening model [36], i.e.:

$$\tau_p = \alpha \mu b \sqrt{\rho_T} \tag{6.5}$$

where μ is the shear modulus and α is a geometrical constant usually in the range 0.3–0.6 for FCC metals [37]. Thus, the upper-bound $\rho_T^{(nano)}$ is related to the existence of a finite nanoscale material strength $\tau_p^{(nano)}$, that only at the true atomic scale is expected to be of the order of magnitude of the theoretical material strength.

Accordingly, the upper-bound $\rho_T^{(nano)}$ is straightforwardly introduced in our model through the following asymptotic matching:

$$\frac{1}{\rho_T} = \frac{1}{\bar{r}\rho_G + \rho_S} + \frac{1}{\rho_T^{(nano)}} \tag{6.6}$$

Note that at the atomic scale, as a consequence of the quantized nature of matter, $\rho_T^{(nano)}$ must be (at least theoretically) of the order of b^{-2} as for a pure single dislocation. This is also reflected in the fact that $\beta = 1/(b^2 \rho_T^{(nano)}) = (\alpha \mu / \tau_p^{(nano)})^2$ is of the order of the unity, since $\alpha \mu$ is of the same order of magnitude as the theoretical material strength. Note the analogy with Quantized Fracture Mechanics (QFM) [38] that, quantizing the crack advancement as must (particularly) be at the nanoscale, predicts a finite theoretical material strength, in contrast to the results of the continuum-based linear elastic fracture mechanics [39].

The flow stress is related to the shear strength by von Mises' rule, i.e. $\sigma_p = \sqrt{3}\tau_p$, and the hardness is related to the flow stress by Tabor's factor [40], i.e., $H = 3\sigma_p$ [20, 21]; thus $H = 3\sqrt{3}\tau_p$. Introducing in this equation the shear strength given by Eq. (6.5), after having substituted the total and geometrically necessary dislocation densities according to Eq. (6.6) and Eq. (6.3) respectively, we derive $H = 3\sqrt{3}\alpha\mu b \left\{ (\bar{r}S/(bV) + \rho_S)^{-1} + \left(\rho_T^{(nano)} \right)^{-1} \right\}^{-1/2}$. Finally, rearranging this relation and introducing dimensionless parameters, we deduce the following hardness scaling law [28]:

$$\frac{H(S/V)}{H_{nano}} = \left(\frac{\delta^2 - 1}{\ell S/V + 1} + 1 \right)^{-1/2}, \quad \frac{H(S/V)}{H_{macro}} = \left(\frac{\delta^2 - 1}{\delta^2 V/(\ell S) + 1} + 1 \right)^{1/2},$$

$$\delta = \frac{H_{nano}}{H_{macro}} \tag{6.7}$$

where

$$H_{nano} \equiv H(\ell S/V \to \infty) = 3\sqrt{3/\beta}\,\alpha\mu$$

$$H_{macro} \equiv H(\ell S/V \to 0) = \frac{3\sqrt{3}\alpha\mu b}{\sqrt{\rho_S^{-1} + \beta b^2}}$$

$$\ell = \frac{\bar{r}}{\rho_S b}$$

H_{macro} represents the macro-hardness, H_{nano} the nano-hardness, and ℓ a characteristic length, governing the transition from nano- to macro-scale. From a physical point of view note that $\ell S/V = \bar{r}\rho_G/\rho_S$, i.e., it is equal to the ratio of the geometrically "sufficient" and statistically stored dislocation densities, whereas $\delta = \sqrt{1 + \rho_T^{(\text{nano})}/\rho_S}$. The two equivalent expressions in Eq. (6.7) correspond respectively to a bottom-up view or to a top-down view. Equation (6.7) is a general shape/size-effect law for nanoindentation that provides the hardness as a function of the ratio between the net surface through which the energy flux propagates and the volume where the energy is dissipated; or, simply stated, as a function of the surface over volume ratio of the domain where the energy dissipation occurs.

The law of Eq. (6.7) can be applied in a very simple way to treat any interesting indenter geometry. However, to make a comparison, it is useful to focus on the axisymmetric profiles (i.e., $h = h(r)$), yet to be investigated by other researchers [21].

Conical indenter. Considering a conical indenter with corner angle φ, its geometry will be defined by: $h(r) = \tan((\pi - \varphi)/2)r$; thus by integration with Eq. (6.2) we found $S/V = \frac{3\tan^2((\pi-\varphi)/2)}{2h}$, which can be introduced into Eq. (6.7), giving the following expression:

$$H_{\text{cone}}(h, \varphi) = H_{\text{macro}}\sqrt{1 + \frac{\delta^2 - 1}{\delta^2 h/h^*(\varphi) + 1}} \qquad (6.8)$$

where

$$h^*(\varphi) = 3/2\ell \tan^2((\pi - \varphi)/2)$$

From physical considerations about Eq. (6.8), we can observe that, for $h/h^* \to 0$ or $\varphi \to 0$, $H_{\text{cone}} \to H_{\text{nano}}$, whereas for $h/h^* \to \infty$ or $\varphi \to \pi$, $H_{\text{cone}} \to H_{\text{macro}}$; only for the case of $\delta \to \infty$ (which means that $H_{\text{cone}} \to H_{\text{nano}} = \infty$ for $h \to 0$), $H_{\text{cone}} = H_{\text{macro}}\sqrt{1 + h^*/h}$ as derived by Nix et al. [20] (with the identical expression for $h^*(\varphi)$). Note that such a scaling law was previously proposed by Carpinteri [41] for material strength (with h structural size).

We have here derived S by integration of $l(h)$, according to Eq. (6.2) and for consistency with Swadener et al. [21]. A more direct calculation considers the difference between the lateral (Ω) and base (A) surface areas of Eq. (6.2), leading to a slightly different value of h^* $[h^*(\varphi) = 3/2\ell \tan^2((\pi - \varphi)/2)(1/\sin((\pi - \varphi)/2)) - 1/\tan((\pi - \varphi)/2)]$; with respect to the calculated previous one $[h^*(\varphi) = 3/2\ell \tan^2((\pi - \varphi)/2)]$. The ratios $h^*(\varphi)/\ell$ evaluated with the two different procedures were compared in [28] for a conical indenter: the related difference was moderate and unessential in our context. Thus, we conclude that both the methodologies can be applied to fit experiments.

Parabolic (spherical) indenter. Consider the case of a parabolic indenter with radius at tip R, i.e., $h = r^2/(2R)$, that for not too large an indentation depth corresponds also to the case of a spherical indenter. By integration we found $S/V = 1/R$, that introduced into Eq. (6.7) gives:

$$H_{parabola}(R) = H_{macro}\sqrt{1 + \frac{\delta^2 - 1}{\delta^2 R/R^* + 1}} \qquad (6.9)$$

where

$$R^* = \ell$$

Thus, the hardness is here not a function of the indentation depth h. We can observe that for $R/R^* \rightarrow 0$, $H_{parabola} = H_{nano}$, whereas for $R/R^* \rightarrow \infty$, $H_{parabola} = H_{macro}$; only for the case of $\delta \rightarrow \infty$, $H_{parabola} = H_{macro}\sqrt{1 + R^*/R}$, as derived in [21] (with the identical expression for R^*). This law describes a true size-effect and agrees with the Carpinteri's law [41].

Conical indenter with a rounded tip. Assuming the presence of a non-vanishing tip radius of curvature (R) in a conical indenter, the tip geometry that we consider in order to find a theory which models this kind of problem is the one reported in Fig. 6.15 [29]. Note that geometrically $h_S = R(1 - \sin \varphi)$, $h_C = R(1 - \sin \varphi)/\sin \varphi$ and r depends on the depth of indentation (h) and it is: $r = \sqrt{2Rh - h^2}$ for $h \leq h_S$ or $r = (h + h_C)\tan \varphi$ for $h > h_S$.

Thus, the term $\frac{S}{V}(h, \varphi, R)$ can be deduced as a function of h, φ, and also R, by geometrical consideration as [29]:

$$\frac{S}{V}(h, \varphi, R) = \begin{cases} \dfrac{3h^2}{2(2Rh - h^2)^{3/2}} & h \leq h_S \\[2mm] \dfrac{[(h + h_C)^2 - (h_S + h_C)^2](1/\sin \varphi - 1)\tan^2 \varphi + h_S^2}{2/3 \cdot (h + h_C)^3 \tan^3 \varphi} & h > h_S, \end{cases}$$

$$(6.10)$$

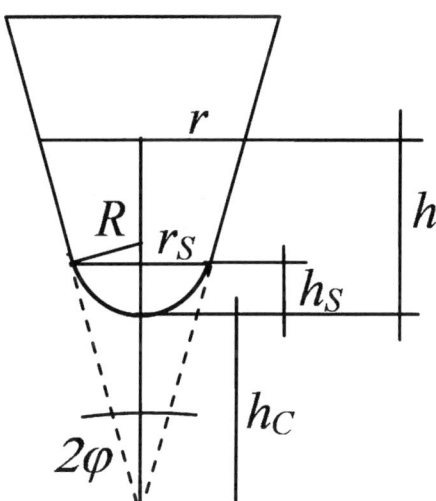

Fig. 6.15. Conical indenter with worn spherical tip [29]

Introducing Eq. (6.10) into Eq. (6.7) it is possible to have an expression for the material hardness as a function of: (1) depth of indentation, (2) tip corner angle, (3) tip radius of curvature ($H(h, \varphi, R)$).

Note the continuity of the function $S/V(h, \varphi, R)$ around h_S and note that $S/V(h \gg R, \varphi, R) \equiv S/V(h, \varphi, R = 0)$. This means that the role of the tip radius of curvature becomes negligible for large indentation depths (in agreement with the macro-scale results). Obviously $S/V (h \leq h_S, \varphi, R)$ is not a function of φ, just representing a pure spherical indentation.

Imagine performing experiments with an ideal tip (thus with $R \to 0$), in order to measure the ideal material hardness $H_{ideal} \equiv H(h, \varphi, R = 0)$. Unfortunately, we would experimentally measure $H_{measured} \equiv H(h, \varphi, R_{exp})$ and thus the ideal material hardness will be $H_{ideal} = CH_{measured}$, thus the correction factor C will be defined as [29]:

$$C = \frac{H(h, \varphi, R = 0)}{H(h, \varphi, R_{exp})} \tag{6.11}$$

Since $S/V(h, \varphi, R)$ Eq. (6.10) increases by decreasing R, correction factors C larger than one have to be expected ($C \geq 1$). For this reason, wear would imply hardness underestimations, in agreement with the finite element results (see Fig. 6.20a).

In particular introducing Eq. (6.7) and Eq. (6.10) into Eq. (6.11) gives:

$$C(h, \varphi, R) = \sqrt{\frac{1 + \dfrac{\delta^2 - 1}{\delta^2 \ell^{-1} V/S(h, \varphi, 0) + 1}}{1 + \dfrac{\delta^2 - 1}{\delta^2 \ell^{-1} V/S(h, \varphi, R_{exp}) + 1}}} \tag{6.12}$$

which allows us to deduce the correction factors C for the nanofabricated probes used for the experimental part of this work (Sects. 6.2.1 and 6.2.2). The values of the correction factors are reported in Table 6.5 [29].

Flat indenter. Consider the case of a flat indenter of radius a, geometrically we found $S/V = \frac{2\pi a h}{2/3\pi a^3}$, that introduced into Eq. (6.7) gives:

$$H_{flat}(a, h) = H_{macro} \sqrt{1 + \frac{\delta^2 - 1}{\delta^2 \, a^2/(3h\ell) + 1}} \tag{6.13}$$

For $a/\ell \to 0$, $H_{flat} \to H_{nano}$, whereas for $a/\ell \to \infty$, $H_{flat} \to H_{macro}$; interestingly, for $h/\ell \to 0$, $H_{flat} \to H_{macro}$, whereas for $h/\ell \to \infty$, $H_{flat} \to H_{nano}$, showing an inverse h-size-effect, in agreement with the discussion by Swadener

Table 6.5. Tip radius of curvature correction factors for the nanofabricated probes [29]

Indenter	Correction factor (C)
Probe n°1 ($R = 21$, nm, $\varphi = 62°$)	1.156
Probe n°2 ($R = 24$, nm, $\varphi = 97°$)	1.840
Probe n°3 ($R = 26$, nm, $\varphi = 25°$)	1.036

et al. [21] and with intuition (the contact area does not change when the penetration load or depth increase) [24]. This suggests a new intriguing methodology to derive the nanoscale hardness of materials by a macroscopic experiment, using large flat punches, even if the finite curvature at the corners is expected to affect the results. This case was only discussed in [21], because of the complexity in their formalism to treat such a cuspidal geometry. Note that for $h \propto a$ and $\delta \to \infty$ the size-effect law again coincides with that introduced by Carpinteri [41].

6.5
Deconvolution of the Indentation Impressions

The presence of a radius of curvature at the tip in an AFM indenter, affects not only the process of indentation, but also the process of AFM imaging, as long as the non-ideal tip interacts with the morphology, convoluting the asperities depending on its actual shape. For this reason, in order to deal with this problem, which could dramatically influence the AFM hardness results, the AFM images have been geometrically deconvoluted [29], considering the tip radius of curvature that we measured during the topography characterization (Sect. 6.2.2).

The software that we used to measure the indentation impression area (SPIP^{TM} software), considers a "tangent height" of the indentation (which is the depth of indentation) reduced by 10%, in order to avoid any roughness influence (see Fig. 6.16a where h is 10% of the whole depth of indentation H). In Fig. 6.16a the red dashed line is the artifact image obtained by AFM, assuming the presence of a tip radius of curvature R, while the black continuous line is the ideal profile. Thus, considering the "tangent height," the difference between the measured indentation impression and the ideal one is only related to the length x (Fig. 6.16a,b), [29]), which could be obtained by geometrical considerations as $x = R \cdot \cos \alpha$, with $\alpha = \arcsen \left(1 - \frac{h}{R} \right)$.

The actual projected area (A^I – hatched area in Fig. 6.16b) could be thus obtained from the measured one (A_p – pink area in Fig. 6.16b) with the following relation:

$$A^I = \frac{\sqrt{3}}{4} \cdot \left(\sqrt{\frac{4 \cdot A_p}{\sqrt{3}}} + 2 \cdot \sqrt{3} \cdot x \right) \tag{6.14}$$

6.6
Results

The nanohardness was first calculated following the Oliver–Pharr approach [8, 9, 11, 19], analyzing the load-unload curves performed during the experimental campaign. In this case the material is highly deformable in the plastic regime and a huge pile-up occurs. For this reason the hardness value will be extremely overestimated using the O-P method. In some cases the material that piles up aside the indentation, almost doubles the indentation depth. This means that the projected area

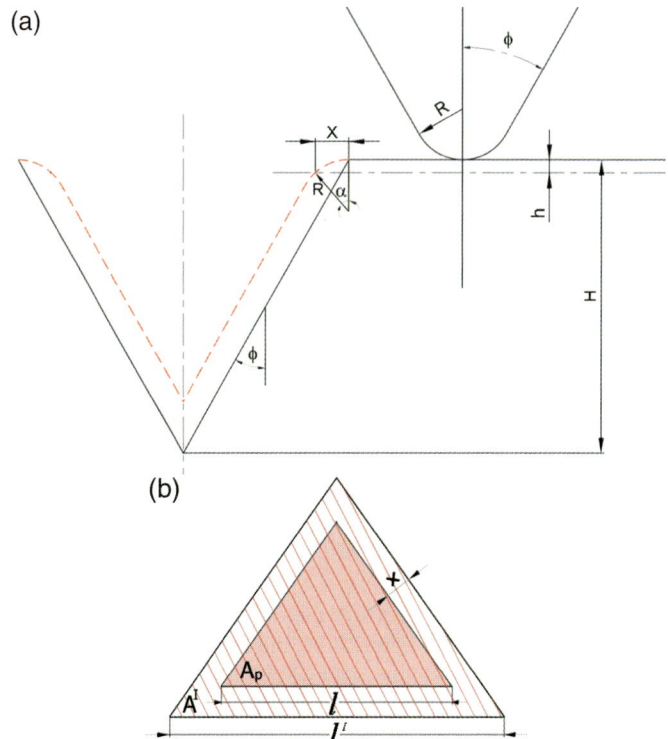

Fig. 6.16. Schematic of the deconvolution process; **a** interaction during the AFM imaging phase between the tip and the indentation impression; **b** projected area of the indentation impression [29]

(which is proportional to h^2) should be approximately four times larger and the hardness value four times smaller than the O-P values [19].

For this reason we adopted a direct method to measure the projected area of the indentation impression (Sect. 6.2.3). The hardness results obtained in this way are reported in Fig. 6.17 (blue dots) where a higher statistic has been obtained in comparison with the results shown in [19], increasing the number of indentations on the same material [29]. With a direct measurement of the projected area it is possible to take into account the pile-up effect. In the legend the best-fit parameters obtained with the theoretical model described in Sect. 6.4 for a conical indenter, are reported. The macro-hardness (H_{macro}) appears quite similar to the actual value of the polymer material and also the other best-fit parameters are plausible and this confirms that the theoretical model is self-consistent.

As introduced in Sect. 6.3, a FEM simulation has been performed. This numerical approach allowed us to verify the experimental results and to understand better how the indenter shape affects the hardness measurement. As a matter of fact, the study of the distribution of pressure in the contact area between the specimen and the indenter, could provide several useful pieces of information [19].

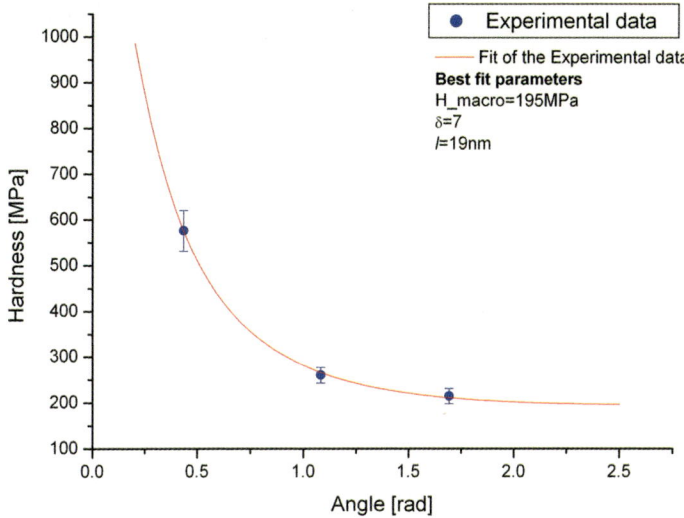

Fig. 6.17. Experimental results (*blue dots*) for the three customized indentation probes obtained with a direct measurement of the projected area and theoretical interpolation (*red line*)

In Fig. 6.18 the numerical hardness results are reported for the range of angles under study. The corner angle 2_φ assumes the following values for the three probes investigated (Table 6.2): 25° (0.436 rad), 62° (1.082 rad), 97° (1.693 rad).

In Fig. 6.19 [29] a direct comparison between the experimental and numerical results is reported. This comparison reveals a good agreement in the hardness

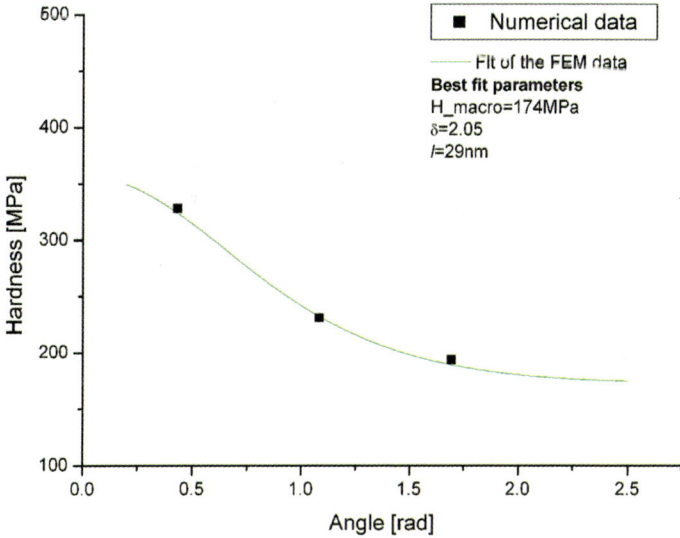

Fig. 6.18. Numerical results (*black squares*) for the indentation model over the whole range of the indenter corner angles and theoretical interpolation (*green line*)

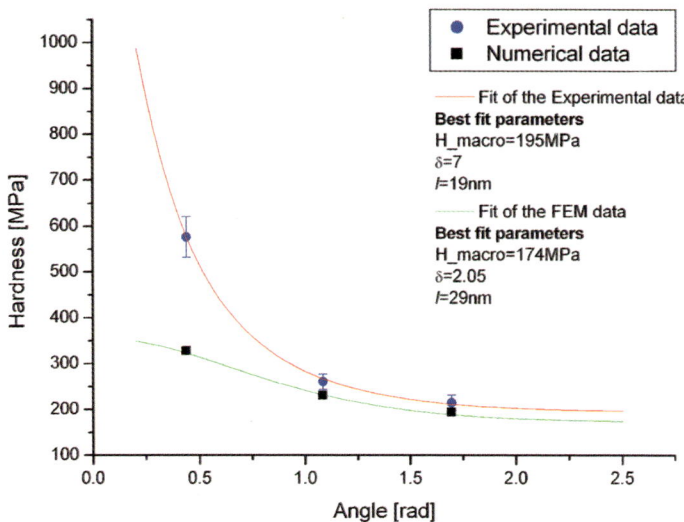

Fig. 6.19. Comparison between the experimental and numerical results, both interpolated using the "general size/shape-effect law for nanoindentation" [29]

behavior although an overestimation of the experimental results is evident at small corner angles as a consequence of the tip radius of curvature effect. We should thus expect that the presence of a rounded tip on the indenter gives an overestimation of the measured hardness, justifying the gap between experimental and numerical results, observable in the graph. On the contrary, as highlighted from numerical simulations (Fig. 6.20a) and confirmed from theoretical models (Sect. 6.4 – *Conical indenter with a rounded tip*), the effect of the tip radius of curvature on the hardness measurement in terms of penetration process is opposite. The hardness value, numerically evaluated modeling the indenter as ideal (Fig. 6.20a – red dots), is in fact higher than that evaluated using a worn tip (Fig. 6.20a – blue squares). These numerical data have also been fitted with the theoretical Eqs. (6.7, 6.10), considering in the first case (red curve) a vanishing tip radius ($R = 0$) while in the second case (blue curve) the actual tip radii of the three customized probes is considered (Table 6.3). The best fit parameters reported in the inset of Fig. 6.20a have exactly the same values for the two interpolations, confirming that the theoretical approach perfectly agrees with the numerical one.

The mismatch observed in Fig. 6.19 is therefore not ascribable to the tip radius of curvature effect on the indentation process, but it is ascribable to this effect on the AFM imaging process. As reported in Sect. 6.5, in fact, the measured projected area of an indentation impression is slightly smaller than the real one, because of the convolution effect of the AFM tip. Thus, considering this effect and correcting the measured areas with the procedure described in Sect. 6.5, we are able to estimate the actual value of the material hardness. In Fig. 6.20b [29] raw experimental data (green triangles) vs. deconvoluted experimental data (magenta triangles) are reported. It is possible to observe the difference, especially for small corner angles, in the hardness values. In the same

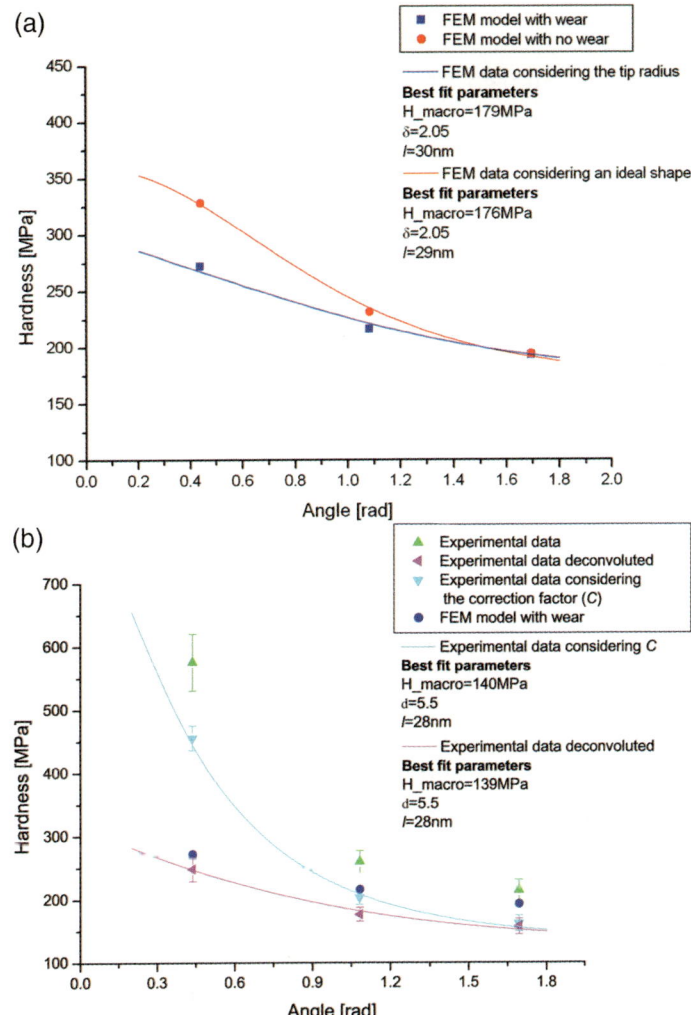

Fig. 6.20. (**a**) Numerical hardness results considering in the FEM model an indenter with an ideal shape (*red dots*) and with a worn shape (*blue squares*). The numerical data are theoretically interpolated (*red and blue line*, respectively); (**b**) comparison between the experimental results deconvoluted (*magenta triangles*) and numerical data (*blue dots*). In the same graph are also reported the raw experimental results (*green triangles*) and the experimental results considering the correction factor C (*cyan triangles*) [29]

graph the numerical data of the modeled worn indenter are also reported, show-ing now an excellent agreement along the whole range of the corner angles chosen. Lastly, we also filtered the deconvoluted experimental results (magenta triangles) by the correction factor C, theoretically evaluated for the three cus-tomized probes (Table 6.5). These corrected data are also reported in Fig. 6.20b (cyan triangles). The results filtered by the deconvolution process (magenta triangles) and those corrected by the correction factor C (cyan triangles) have both been theoretically interpolated (magenta and cyan curve, respectively). The result of this process reveals a complete agreement in terms of best-fit parameters (inset of Fig. 6.20b), which are exactly the same, confirming that the theoretical model is self-consistent [29].

6.7
Conclusion

An AFM-based nanoindentation study on the effect of geometrical uncertainties of the indenter tip on the hardness measurement is proposed herein. In particular three different-shaped indentation probes have been designed and realized with a FIB machine. A whole experimental analysis has been performed with these indenters in order to quantify how the hardness measurement is affected by two important parameters: (1) the tip corner angle [19] and (2) the tip radius of curvature of the nanoindenters [29]. A FEM model has also been designed in order to better understand the process of indentation and it has been further developed in order to take into account the tip radius of curvature effect. In parallel, a theoretical approach, based on a recent theory on nanoindentation [28], has been optimized for a worn indenter. These two approaches allow us to interpret the experimental results, showing that the differences between the experimental, the numerical, and the theoretical data are related mainly to the tip radius of curvature effect, which affects not only the penetration process during indentation, but also, and in a significant way, the AFM imaging process. The hardness has been in fact obtained in a direct way, measuring the projected area of the indentation impression by the AFM high-resolution images. A geometrical deconvolution process has been utilized in order to correct the systematic error related to the tip radius of curvature effect.

In this way it is possible to deduce a theoretical relation that links the measured hardness value with the shape of the indenter and with its tip radius of curvature. In particular a correction factor C has been defined and it allows us to correct the experimental data, obtained by AFM nanoindentation, from the geometrical effects of the indenter tip.

Acknowledgments. The authors would like to acknowledge the FIB laboratory (a CNR-INFM-S3 Lab).

This work has been supported by Regione Emilia Romagna (LR n.7/2002, PRRIITT misura 3.1A) and Net-Lab SUP&RMAN (Hi-Mech district for Advanced Mechanics Regione Emilia Romagna).

The author Nicola Pugno was supported by the "Bando Ricerca Scientifica Piemonte 2006" – BIADS: Novel biomaterials for intraoperative adjustable devices for fine tuning of prostheses shape and performance in surgery.

References

1. Fischer-Cripps AC (2004) Nanoindentation. Springer, New York
2. Li X, Bhushan B (2002) A review of nanoindentation continuous stiffness measurement technique and its applications. Mater Charact 48(1):11–36
3. Saha R, Nix WD (2002) Effects of the substrate on the determination of thin film mechanical properties by nanoindentation. Acta Mater 50(1):23–38
4. Zhang F, Saha R, Huang Y, Nix WD, Hwang KC, Qu S, Li M (2007) Indentation of a hard film on a soft substrate: strain gradient hardening effects. Int J Plasticity 23:25–43
5. Bhushan B, Koinkar VN (1994) Nanoindentation hardness measurements using atomic force microscopy. Appt Phys Lett 64(13):1653–1655
6. Bhushan B (2004) Handbook of nanotechnology. Springer, Berlin
7. Doerner MF, Nix WD (1986) A method for interpreting the data from depth-sensing indentation instruments. J Mater Res 1(4):601–109
8. Oliver WC, Pharr GM (1992) An improved technique for determining hardness and elastic modulus using load and displacement sensing indentation experiments. J Mater Res 7(6):1564
9. Oliver WC, Pharr GM (2004) Measurement of hardness and elastic modulus by instrumented indentation: advances in understanding and refinements to methodology. J Mater Res 19(1):3–20
10. VanLandingham MR (2003) Review of Instrumented Indentation. J Res Natl Inst Stand Technol 108:249–265
11. Pharr GM, Oliver WC, Brotzen FR (1992) On the generality of the relationship among contact stiffness, contact area, and elastic modulus during indentation. J Mater Res 7(3):613–617
12. Withers JR, Aston DE (2006) Nanomechanical measurements with AFM in the elastic limit. Adv Colloid Interface 120(1–3):57–67
13. VanLandingham MR, Villarrubia JS, Guthrie WF, Meyers GF (2001) Nanoindentation of polymers: An Overview, Macromol. Symp. 167. In: Tsukruk VV, Spencer ND (eds) Advances in scanning probe microscopy of polymers. Wiley-VCH Verlag GmbH, Weinheim, Germany, pp 15–43
14. Beegan D, Chowdhury S, Laugier MT (2005) Work of indentation methods for determining copper film hardness. Surf Coat Technol 192:57–63
15. Miyake K, Fujisawa S, Korenaga A, Ishida T, Sasaki S (2004) The effect of pile-up and contact area on hardness test by nanoindentation. Jpn J Appl Phys 43(7B):4602–4605
16. Poole WJ, Ashby MF, Fleck NA (1996) Micro-hardness of annealed and work-hardened copper polycrystals. Scr Mater 34(4):559–564
17. Pharr GM (1998) Measurement of mechanical properties by ultra-low load indentation. Mater Sci Eng A 253(1–2):151–159
18. Saha R, Xue Z, Huang Y, Nix WD (2001) Indentation of a soft metal film on a hard substrate: strain gradient hardening effects. J Mech Phys Solids 49(9):1997–2014
19. Calabri L, Pugno N, Rota A, Marchetto D, Valeri S (2007) Nanoindentation shape-effect: experiments, simulations and modelling. J Phys: Condens Matter 19:395002–395013
20. Nix WD, Gao H (1998) Indentation size effects in crystalline materials: a law for strain gradient plasticity. J Mech Phys Solids 46:411–425
21. Swadener JG, George EP, Pharr GM (2002) The correlation of the indentation size effect measured with indenters of various shapes. J Mech Phys Solids 50:681–694
22. VanLandingham MR (1997) The effect of instrumental uncertainties on AFM indentation measurements. Microscopy Today 97:12–15
23. Mott BW (1956) Micro-indentation hardness testing. Butterworths, London

24. Ashby MF (1970) The deformation of plastically non-homogenous materials. Philos Mag 21:399–424
25. Nye JF (1953) Some geometric relations in dislocated crystals. Acta Metall 1:153–162
26. Poole WJ, Ashby MF, Fleck NA (1996) Micro-hardness tests on annealed and work-hardened copper polycrystals. Scripta Mater 34:559–564
27. Lim YY, Chaudhri MM (1999) The effect of the indenter load on the nanohardness of ductile metals: an experimental study on polycrystalline work-hardened and annealed oxygen-free copper. Philos Mag A 79:2879–3000
28. Pugno N (2006) A general shape/size-effect law for nanoindentation. Acta Materialia 55:1947–1953
29. Calabri L, Pugno N, Menozzi C, Valeri S (2008) AFM nanoindentation: tip shape and tip radius of curvature effect on the hardness measurement. J Phys: Condens Matter (unpublished)
30. Sader JE, Chon JWM, Mulvaney P (1999) Calibration of rectangular atomic force microscope cantilevers. Rev Sci Instrum 70:3967–3969
31. Sader JE, Larson I, Mulvaney P, White LR (1995) Method for the calibration of atomic force microscope cantilevers. Rev Sci Instrum 66:3789–3798
32. Bhattacharya AK, Nix WD (1988) Finite element simulation of indentation experiments. Int J Solids Struct 24:881–891
33. Yoshimoto K, Stoykovich MP, Cao HB, de Pablo JJ, Nealey PF, Drugan WJ (2004) A two-dimensional model of the deformation of photoresist structures using elastoplastic polymer properties. J Appl Phys 96:1857–1865
34. Arsenlis A, Parks DM (1999) Crystallographic aspects of geometrically-necessary and statistically-stored dislocation density. Acta Mater 47:1597–1611
35. Huang Y, Zhang F, Hwang KC, Nix WD, Pharr GM, Feng G (2006) A model of size effects in nano-indentation. J Mech Phys Solids 54:1668–1686
36. Taylor GI (1938) Plastic strain in metals. J Inst Met 13:307–324
37. Wiedersich H (1964) Hardening mechanisms and the theory of deformation. J Met 16:425–430
38. Pugno N, Ruoff R (2004) Quantized fracture mechanics. Philos Mag 84/27:2829–2845
39. Griffith AA (1920) The phenomena of rupture and flow in solids. Phil Trans Roy Soc A 221:163–199
40. Tabor D (1951) The hardness of metals. Clarendon Press, Oxford
41. Carpinteri A (1994) Scaling laws and renormalization groups for strength and toughness of disordered materials. Int J Solids Struct 31:291–302

7 Local Mechanical Properties by Atomic Force Microscopy Nanoindentations

Davide Tranchida · Stefano Piccarolo

Abstract. The analysis of mechanical properties on a nanometer scale is a useful tool for combining information concerning texture organization obtained by microscopy with the properties of individual components. Moreover, this technique promotes the understanding of the hierarchical arrangement in complex natural materials as well in the case of simpler morphologies arising from industrial processes. Atomic Force Microscopy (AFM) can bridge morphological information, obtained with outstanding resolution, to local mechanical properties. When performing an AFM nanoindentation, the rough force curve, i.e., the plot of the voltage output from the photodiode vs. the voltage applied to the piezo-scanner, can be translated into a curve of the applied load vs. the penetration depth after a series of preliminary determinations and calibrations. However, the analysis of the unloading portion of the force curves collected for polymers does not lead to a correct evaluation of Young's modulus. The high slope of the unloading curves is not linked to an elastic behavior, as would be expected, but rather to a viscoelastic effect. This can be argued on the basis that the unloading curves are superimposed on the loading curves in the case of an ideal elastic behavior, as for rubbers, or generally in the case of materials with very short relaxation times. In contrast, when the relaxation time of the sample is close to or even much larger than the indentation time scale, very high slopes are recorded.

Where AFM nanoindentations are concerned, one observes a dependence of the penetration, i.e., the relative motion between the sample and the tip (indenter), on the elastic properties of a material when using equivalent loads. This relationship becomes visible on samples that are homogeneous down to the scale of nanoindentation. The elastic modulus can be obtained by applying Sneddon's elastic contact mechanics approach, since the contact between the tip and the sample is dominated by an elastic behavior with negligible plastic deformation. Under such circumstances, the dependence of the penetration on the load follows an exponent of 1.5, consistent with elastic contact mechanics and justified on the basis of the large elastic range exhibited by polymers, on the constraints due to the geometry of the deformation during indentation and to the critical yielding volume needed in order to induce plasticity. As a result, elastic moduli taken from AFM force curves show a very good agreement with bulk values obtained by macroscopic tensile testing. This is true for a broad range of polymers, from materials with rubbery to semi-crystalline, or even glassy behaviors. This result confirms that AFM nanoindentations in polymers take place mostly in the elastic range and opens the possibility of characterizing the mechanical behavior of polymers on an unparalleled small scale as compared to commercial depth-sensing instruments (DSIs), which use much blunter indenters.

A further application is discussed where, upon decreasing the load, and consequently the penetration depth, the scale becomes comparable to that of the underlying texture which is probed as opposed to the bulk material. Although this apparently presents a limitation on the resolution of the scale that can be mapped, this feature is discussed and shown to open the possibility of identifying properties of individual phases with their surroundings as well as the role of the connectivity among the phases.

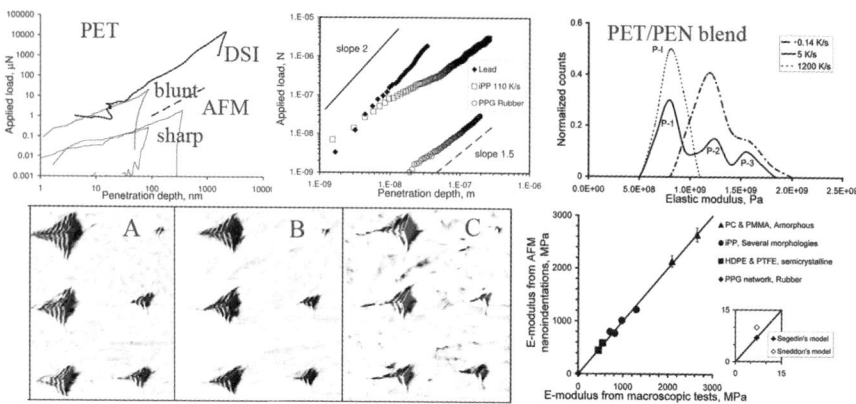

Key words: Atomic force microscopy, Nanoindentation, Soft materials, Polymers, Elastic young's modulus, Nanoscale mapping, Mechanical properties

7.1
Introduction

The measurement of mechanical properties on the nanometer scale is a fundamental tool that can find applications in all fields of materials science bordering with other disciplines, such as biology, tissue engineering, electronics, structure distribution in complex polymeric articles, and so forth.

Nature provides us with a variety of what can be referred to as composite structures with remarkable mechanical properties [1, 2]. The variables that decide the mechanical properties of these materials can be divided into two categories: compositional and structural. Compositional variables are determined by the mechanical properties of the individual components as well as the strength of the interactions between these components. The structural variables, on the other hand, are governed by the mutual arrangement of the components in the composite material. A hierarchy of arrangements of component phases is often recognized in nature and in certain cases industrial processes can mimic such situations although the optimization that can be reached is always rather poor [3–5].

It has been made possible to measure the mechanical behavior of very small volumes of material by the development of instruments that can sense and apply extremely small forces and displacements. Such instruments include AFMs and nanoindenters [6]. These devices are often capable of producing contact areas and penetration depths characterized by nanometer dimensions while also providing lateral motion capabilities for studying tribological behaviors [7]. One objective when using such devices is to provide methods for characterizing the mechanical response of material systems with a nanoscale spatial resolution. Such measurements are key when investigating both compositional as well as structural features of materials, and open the possibility of disclosing the hierarchy of arrangements of individual components in complex materials [8].

Among the mechanical properties of interest, one or more of which can be obtained using commercial and specialized indentation testers, can be mentioned elastic–plastic deformation behavior, hardness, Young's modulus of elasticity, scratch resistance, substrate adhesion, residual stresses, time-dependent creep and relaxation properties, fracture toughness, and fatigue. Indentation measurements can assess structural heterogeneities on and underneath the surface, such as diffusion gradients, precipitates, presence and properties of buried layers, grain boundaries, and modification of surface composition [9–11].

When dealing with polymers, indentation tests on the nanometer scale require the use of very small applied loads [12], in the range of 0.5–$5\,\mu N$. AFM is thus a useful tool, since the upper limit for this instrument is a few microNewton [13]. With respect to nanoindenters, AFM renders it possible to collect images of the sample morphology and to indent specific areas as a result of the sharp AFM tip allowing an outstanding lateral resolution. These possibilities have turned out to be extremely important when studying inhomogeneous samples (such as blends or rubber/matrix systems) or samples with a structure distribution. The possibility of visualizing the morphology and measuring the Young's modulus in-situ, represents a fundamental tool for linking the structure distribution with the final physical properties [14].

Nanoindentation via AFM is performed in a nonscanning mode, termed force mode, in which the AFM probe is moved toward the sample surface, pushed into the surface, and then lifted back off. A deflection-displacement curve is produced from which a force-penetration curve related to the nanoindentation process can be determined. The applied force is related to the deflection (bending) of the cantilever beam through the cantilever spring constant, and the penetration into the material is the difference between the overall vertical displacement of the cantilever and that of the probe tip related to cantilever bending, which must be calibrated [15–17].

Limitations related to this approach have already been partly discussed in the literature for a variety of polymeric materials ranging from polyurethanes to polystyrene, HDPE-LLDPE mixtures, and polyethylene terephthalate (PET) [13, 18–23]. An evaluation of certain parameters characterizing the two interacting bodies is required. The elastic constant of the cantilever allows a remarkable change of the indentation conditions: in fact its reduction permits an enhanced force resolution and offers the opportunity to investigate adhesion behavior by studying pull-off and jump-to-contact forces [24–26]. On the other hand, increasing the elastic constant makes it possible to obtain a harder indentation of the samples although this technique cannot be employed on soft materials [18]. The elastic constant of the cantilever thus has to be carefully chosen, depending on the latitude of the physical properties of the materials under study [27]. However, determining the elastic constant of the cantilever is not an easy subject as was demonstrated by Burnham et al. [28] who recently compared several methods. Other studies have been carried out by Holbery and Eden [29], who compared the performances and qualities of numerous commercially available cantilever brands, or by Matei et al. [30] who determined the accuracy of the thermal calibration method. Recent approaches include the use of cantilever arrays [31] or piezolevers in order to know, a priori, the applied deformation [32, 33], and also laser Doppler vibrometry [34].

As far as knowledge of the tip geometry is concerned, manufacturer data are often insufficient as a result of production scatter. Moreover, when carrying out force curve analysis with large tip penetrations, the tip geometry can be significantly modified due to residues of the material under study being removed and deposited on the tip and, under severe conditions of use, due to tip damage and wear [35]. The tip condition can be characterized in terms of an apex curvature radius, which can, although a tip radius is typically on the order of 10 nm, be difficult to measure even by electron microscopy. Despite the crucial role of the tip radius in characterizing mechanical properties through indentation tests, its value is seldom reported. However, methods such as blind estimation [36] can be used for routinely verifying changes in tip geometry, tuning the operating conditions for a more reliable determination of the tip radius [37] and choosing appropriate samples for characterizing the tip [38]. Because of a tip convolution effect, the wear of an AFM tip, even under extremely low (normal) loads, is one of the most crucial issues in AFM as well as in other probe-based applications [39].

Nanoindentation measurements by AFM require an accurate knowledge of the vertical distance traveled by the piezo, however, this distance often suffers from errors of creep and hysteresis of the piezo elements. The calibration of the response of piezoelements has been carried out by means of laser interferometry [40, 41], capacitance dilatometer [42], or by an inductive linear gauge [43], but the handling of these methods is definitely complex. A more convenient and direct way to calibrate the AFM in the large scan regime, is to scan a grid with an appropriate lattice constant. Such a grid can be manufactured by means of instance e-beam lithography. An internal lookup table is built, which, depending on the scan rate, applies the necessary calibration in the three principal directions. More recently, closed-loop instruments capable of overcoming such problems have been produced [44, 45].

Finally, sample roughness and thickness must respectively be kept as small and as large as possible, in order to model the sample with an elastic half-space [17]. All things considered, contact models can only be applied if these parameters are known.

Contact mechanics models have been developed in the case of elastic, viscoelastic and, although with a heuristic approach, elastic-plastic contacts [46–49]. A sample is usually modeled as a smooth half space, wherefore the effect of its roughness and thickness can be neglected, whereas a vast variety of geometries are available for the indenter shape. These models are reviewed in a large number of recent publications, and constant reference is made in the next paragraphs concerning the analysis of the nanoindentation force curves, i.e. the plot of the applied load vs. the penetration depth. The three most frequently used models in the literature are those of Oliver–Pharr (O/P) [50], Hertz [46], and Johnson–Kendall–Roberts/Derjaguin–Muller–Toporov (JKR/DMT) [51, 52]. The Oliver–Pharr model [50] requires a double calibration by indentations performed on a reference material, usually aluminum, tungsten, or silicon, and is based on the hypothesis of a purely elastic unloading recovery. The Hertz model [46] relates the mechanical properties of the sample to the penetration depth of the tip into the material and assumes a purely elastic contact. The DMT [52] and JKR [51] models are based on the Hertz approach but introduce corrections due to adhesion forces. There are also simpler approaches based on energetic considerations in which elastic and plastic work estimates provide the elastic modulus in addition to other interesting parameters linked to the influence of plasticity on indentation [53]. Further, the Sneddon [47] model is based on the hypothesis that the contact takes place within the elastic behavior of the substrate, and that the indenter is a rigid body. This last requirement is easy to fulfill in the case of AFM nanoindentations of soft matter, since AFM tips are normally fabricated of silicon or silicon nitride. Finally, there is the theory of Ting [48] that analyzes the contact problem in the case of a viscoelastic response. A more systematic analysis of the limitations of these various approaches is described in one of the following sections that discusses their application to the experimental data collected from force curves on several homogeneous polymer samples.

This chapter presents investigations of the contact mechanics of AFM nanoindentation in order to demonstrate that, under appropriate experimental conditions, the contact is dominated by an elastic response and that the evaluation of the Young's modulus is accurate [54]. On the other hand, it is also shown that the unloading curve of the nanoindentation cannot be used, due to the onset of relaxation phenomena rendering approaches such as that of Oliver and Pharr not useful [55]. The technique is validated by analyzing a set of polymeric samples with a vast range of moduli [54,56]. These materials run from rubbery samples with moduli of a few tens of MPa, up to hard glassy samples of a few GPa, and even include semi-crystalline materials of varying morphologies that are intrinsically inhomogeneous on a local level. A validation carried out through the comparison of moduli, as determined by AFM nanoindentation with bulk tensile moduli, can be afforded only if samples are homogeneous from the scale of the penetration induced by the indentation, i.e., a few tens of nm, up to the scale of the macroscopic samples for tensile testing; in other words, if the mechanical behavior on both scales can be assumed to represent a continuum. The techniques used for sample preparation are therefore described in some detail in order to give an account of the bulk homogeneity [57].

Finally, an example where the continuum behavior breaks down is presented in order to display the possibility of discerning the mechanical properties of single components in the complex texture of semi-crystalline materials [58]. This is afforded by the use of a statistical approach to probe local elastic moduli: for larger penetrations, the range of values determined is described by a monomodal distribution around the average value detected. On the other hand, upon decreasing the penetration, the distribution broadens and eventually breaks into multiple peaks representative of single component mechanical properties. This opens the possibility of linking the morphology with the mechanical properties of component phases, as well as of disclosing the hierarchical arrangement in complex materials.

7.2
The Scale of AFM Nanoindentations

AFM nanoindentations can test mechanical properties on an extremely small length scale. This is a clear advantage and opens a wealth of applications in the case of heterogeneous materials. Moreover, the small detection scale should not be considered a limitation since the meaning of the measurement can be grasped upon comparing its values to the bulk mechanical property obtained by macroscopic tests. In other words, AFM can give a hint as to the contribution of various micro- or nano-phases to the macroscopic behavior.

Fig. 7.1. A nanoscale mechanical test: crosshatched lamellae in an iPP sample compared with a typical indentation (pointed out by the *arrow*). The load applied to the tip was $3.02\,\mu\text{N}$. [56] Copyright Elsevier reproduced with permission

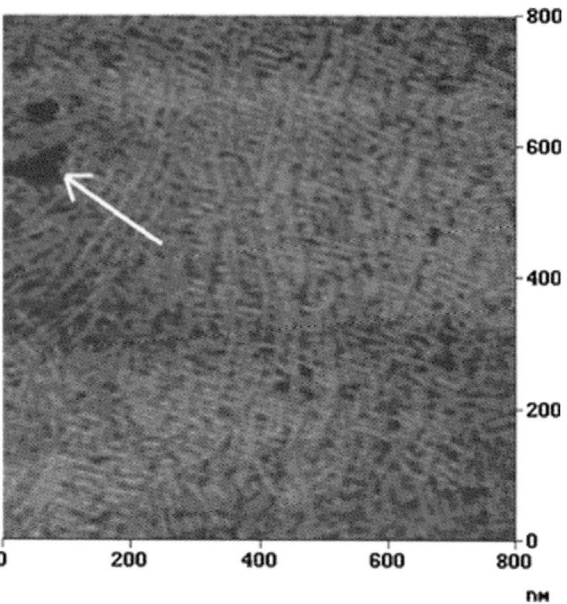

This observation is clearly perceivable from Fig. 7.1 in which the scale of indentations is compared to the underlying morphology of a semi-crystalline polymer, i.e., an iPP sample [56]. In this case, the typical cross-hatched semi-crystalline morphology is shown together with a residual indentation imprint on the top left of the image. Although the overall stress field, and thus the volume tested, extends over a larger volume as compared to the residual imprint, the scale of the nanoindentations that determines the mechanical response clearly involves the contribution of a few crystalline lamellae, i.e., of the material behaving as a continuum.

Therefore, a fine-tuning of the load applied to the tip permits an investigation of the mechanical properties of materials with an underlying nanostructure on a variety of length scales, e.g., semi-crystalline polymers or even materials with a hierarchical structure. It is easy to conceive that nanoindentations at relatively high penetration depths will reveal the continuum mechanical property behavior of the sample whereas, on the other side, the use of low penetration depths can give insights to the mechanical properties of the single components.

This is even better demonstrated in Fig. 7.2, where deformed lamellae of polyethylene are shown within the residual imprints at six loads [59]. Depending on the load applied to the tip, it is possible to test a single, see upper right corner of Fig. 7.2A, up to a bunch of lamellae, upper left corner.

Moreover, Fig. 7.2 shows the evolution with time of the indentation shape and underlying morphology, and highlights a fundamental piece of information: the residual depth cannot be considered as the result of a purely dissipative process, but rather, a long-term energy release of a viscoelastic nature must be taken into account. This means that energy is stored in the material during the loading process with three competing and simultaneous processes: (1) elastic energy that is released during the unloading, (2) viscoelastic energy that is partially released during the unloading,

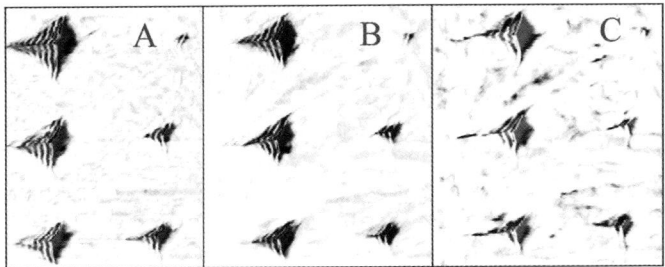

Fig. 7.2. AFM phase images of the residual imprints left behind from nanoindentations on polyethylene, clearly displaying the complex morphology evolution for indents performed at 0.7, 1.4, 2.1, 2.8, 3.5, 4. 1 µN (from top right clockwise) after 6 **A**, 400 **B**, and 1,400 **C** min. [59] Copyright American Chemical Society. Reproduced with permission

depending on the respective time scales between the unloading time and the relevant time scale for the material relaxation, and partially released in long-term relaxation of the imprint, and (3) dissipation processes, i.e., plasticity that causes a permanent imprint.

In the case of polyethylene lamellae, Fig. 7.2 shows that crystalline lamellae rearrange their shape and geometry during the long-term recovery processes. The evolution with time of the corresponding profile of a nanoindentation performed at a relatively low applied load is shown in Fig. 7.3 [59]. The vertical recovery can take place to a large extent, and is accompanied by morphological rearrangements of lamellae as shown in Fig. 7.2. The residual depth is for instance halved after 24 h. More insights into the importance of this observation is given in the following sections.

As a first step, one can start by analyzing the elastic recovery that takes place during the unloading of the AFM nanoindentation. These data are collected in a force curve. The high three-dimensionality of the stress field, especially when a sharp AFM tip is used as the indenter, makes it very difficult to properly analyze the strain inside the sample [60]. However, a phenomenological analysis can be based only on the comparison of the penetration depth under full load and the residual indentation depth [54].

Fig. 7.3. The evolution of the indent profile on a HDPE sample; a few minutes after the indentation (*black circles*), 400 min after the indentation (*light gray circles*), and 24 h after the indentation (*dark gray circles*), with an applied load of 0. 7 µN. [59] Copyright American Chemical Society. Reproduced with permission

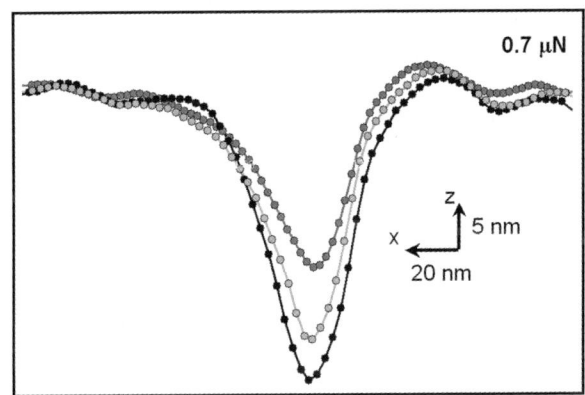

0.7 µN

z

5 nm

x

20 nm

High stresses are, due to the sharpness of the tip itself [61], expected during an AFM indentation, and such stresses point towards a significant contribution of plastic deformation dissipative processes in the sample. However, despite these stresses, the recovery during unloading is substantial. Figure 7.4 shows the variation of the elastic recovery parameter, ξ, with changing indentation rate, measured as a function of the frequency of the voltage applied to the piezo in order to bring the sample into contact with the tip. This ξ parameter is bound between 0 and 1 according to the equation

$$\xi = 1 - \frac{i}{p_{max}} \qquad (7.1)$$

Here, p_{max} is the penetration depth under full load and i is the indentation depth as imaged immediately after the nanoindentation. Therefore, ξ vanishes in the case of plastic indentation as no recovery takes place and $p_{max} = i$, while ξ is equal to 1 in the case of a fully elastic contact when no residual imprint is left behind after load removal. Since long-term viscoelastic effects are not taken into account by this parameter, its use is not perfectly rigorous but it makes it possible to capture several features of the nanoindentation process. In particular, a nanoindentation is considered to be mainly plastic in the case of a low recovery parameter as measured immediately after the nanoindentation. On the contrary, a high value of the ξ parameter, close to 1, obviously points to a contact dominated by an elastic response since most of the deformation is immediately recovered.

Nevertheless, Figs. 7.2 and 7.3 seem to contradict this observation. Apparently, a large part of the residual imprint is recovered with time, thus casting doubt on the plastic contribution that arises in a nanoindentation even in the case of a low ξ value. However, placing a dichotomy that is based on the use of this parameter is of value at least in order to discriminate the elastic behavior from the others. It will thus be used

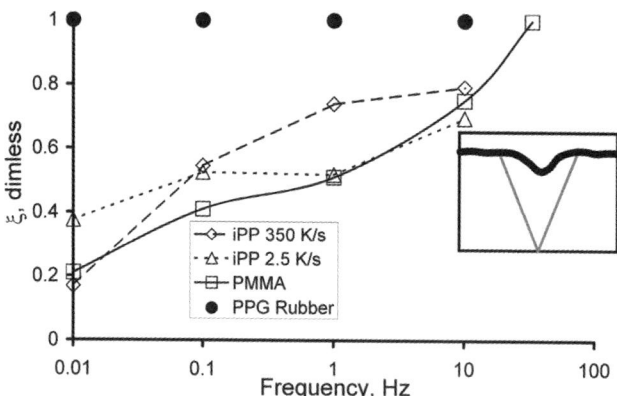

Fig. 7.4. The dependence of the ξ parameter on the loading rate for a wide range of materials: from a rubbery network, to a glassy PMMA, as well as two iPP samples solidified with semi-crystalline and mesomorphic morphologies. The *inset* shows the residual penetration (thick profile) as imaged by AFM as well as the penetration depth under full load (thin profile) in the case of an indentation performed at 10 Hz. [54] Copyright American Chemical Society. Reproduced with permission

in the following to identify those situations that clearly point to an elastic-dominated contact behavior.

In particular, Fig. 7.4 shows that while, for a rubber, the contact is elastic at all loading rates, for a wide variety of mechanical behaviors (e.g. amorphous, semi-crystalline, and mesomorphic polymers), the residual depth, as imaged immediately after the nanoindentation, amounts to only one tenth of the penetration under full load when high loading rates are used, see the points at 10 Hz. For instance, the profiles in the inset of Fig. 7.4 display a comparison between penetration at full load, as obtained from a force curve collected at a high loading rate, and the residual depth, imaged as soon as an image collection is possible, after the indentation. Therefore, in the case of high loading rate, the contribution of viscoelastic and plastic phenomena can be reasonably neglected and the nanoindentation can be thought to occur as dominated by elastic contact such that it can also be modeled with elastic contact models.

Low loading rates, on the other hand, show a more limited elastic recovery, e.g., a residual depth that is half as deep as the maximum depth under maximum load. In the former case, i.e., at high loading rates, very little viscoelastic energy can be assumed to be stored in the sample, whereas in the latter case, the sample is considered to store a larger amount of viscoelastic energy. This viscoelastic energy is subsequently released in long-term recovery phenomena as shown in Figs. 7.1, 7.2, 7.3 and can be enhanced by a low loading rate.

7.3
The Relationship with Microhardness and Analysis of the Unloading Curve

The concept that the residual imprint recovers with time raises a fundamental question. The typical approach of microhardness [62, 63] consists in measuring the residual imprint left by the indentation and relating the ratio of the applied load, L, as well as the imprint area, to mechanical properties of the sample, in particular to the yield stress, σ_Y, according to

$$MH = 3\sigma_Y = C\frac{L}{d^2} \tag{7.2}$$

Here, d is the diagonal of the residual imprint, C is a geometrical factor depending on the indent geometry, and MH is the microhardness value, which is obviously a characteristic of the material.

With the AFM one can perform the nanoindentation and image the resulting imprint with the same tip, as shown in Figs. 7.1 and 7.2. Seeing as the indentation is performed with the same tip as that used for imaging, one can expect the residual imprint to be very sharp and its imaging to be of limited use by convolution with the tip shape [64]. The process of deconvolving the tip shape from the image would be very demanding in the case of a routine use of AFM nanoindentations. However, the elastic recovery of the imprint takes place mostly in the vertical direction, similarly to what happens in the case of viscoelastic recovery as shown in Fig. 7.3. Moreover, elastic recovery amounts to at least one half of the maximum penetration depth under load, as displayed in Fig. 7.4, whereas horizontal recovery is much smaller [59]. This means that the residual imprint is much blunter than the profile of the AFM tip and, consequently, the imaging process does not actually suffer from tip convolution effects. The measurement of the residual imprint can thus be considered accurate.

However, the residual imprint keeps on recovering and therefore the estimate of the nanohardness is subject to severe errors. For example, Fig. 7.5 shows the evolution of the applied load as a function of the squared measured residual depth, at three different times. The abscissa, scaled by a geometrical factor dependent only on the geometry of the indenter used, essentially represents the residual area of the imprint. The slope of these plots can thus be considered as a measure of the nanohardness, see Eq. (7.2), which clearly changes with time as the images are collected.

For this fundamental reason, but also due to the intrinsic difficulty and time demanding approach of performing nanoindentations followed by imaging of the imprints, one usually prefers to register penetration depth and applied load on-line, during the nanoindentation. The plot of these two quantities is called a force curve.

Normally, the relative motion between a tip and the material being indented is obtained by either moving the cantilever or the tip at a constant rate. After the tip–sample contact, the cantilever starts bending and a load that is proportional to the cantilever deflection through the elastic constant of the cantilever is applied. The careful choice of the cantilever's elastic constant is essential for the measurements. Indeed, a stiff cantilever may cause very large indentations with small deflections, and, oppositely, a compliant cantilever might bend without indenting. Although this issue may seem a limitation, in reality it increases the flexibility of the technique.

A rule of thumb, as suggested by Tsukruk et al. [13] and shown in Fig. 7.6, is that one can choose the elastic constant of the cantilever with respect to the expected

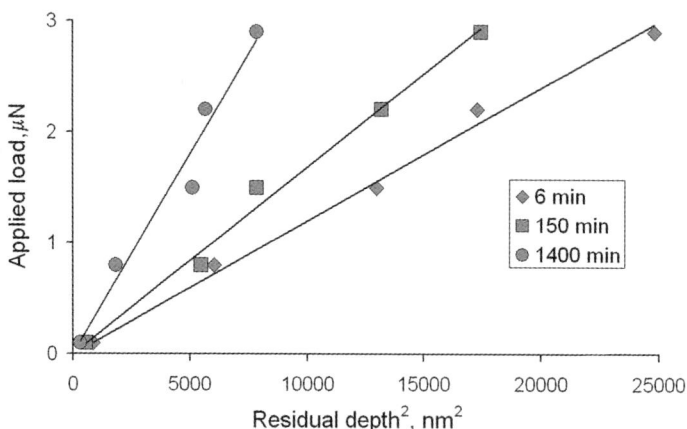

Fig. 7.5. The evolution of the applied load as a function of the measured residual depth squared, at 6, 150 and 1,400 min. The slope of these plots is proportional to the nanohardness, showing that this quantity apparently changes with respect to the time required for collecting the residual imprint after indentation. [59] Copyright American Chemical Society. Reproduced with permission

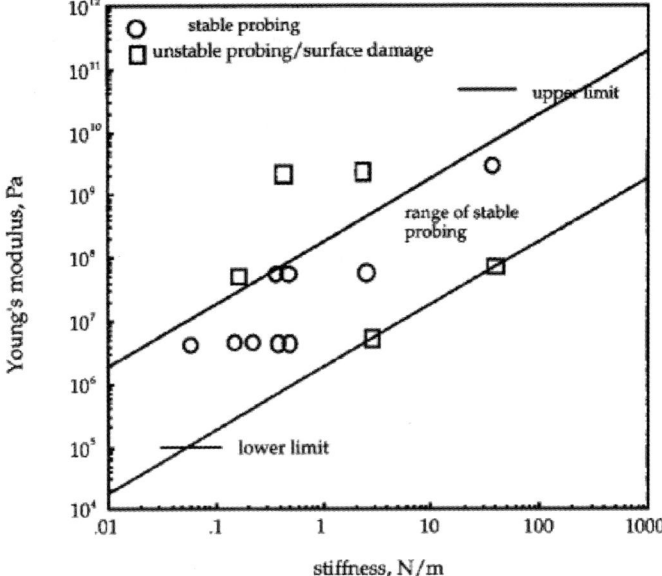

Fig. 7.6. The range of Young's moduli that can be studied by nanoindentation performed with AFM cantilevers of given stiffnesses. [27] Copyright Society of Chemical Industry. Reproduced with permission

range of the sample's Young's modulus. It is worth noticing that hard materials cannot be studied with standard commercial cantilevers, and the technique thus seems to be more suitable for the analysis of soft matter.

Depending on the mechanical properties of the sample, a larger or smaller penetration depth is obtained, and this depth cannot be controlled a priori. This means that a trigger can obviously be set to a certain load, but the loading history cannot be controlled a priori. Further insight into this matter will however be given at the end of the section, after introduction of certain concepts required for a proper analysis.

The penetration depth, p, is evaluated from the vertical movement and the deflection of the cantilever (or sample) according to

$$p = z - \delta \tag{7.3}$$

where z is the piezo displacement and δ is the cantilever deflection. Since both these quantities can be measured with high accuracy by AFM, penetration depths in the nanometer range can be easily monitored. However, although the penetration depth can be evaluated in real time, the penetration rate cannot be controlled for technical constraints since it obviously depends on the material being indented. This issue is further discussed during the introduction of contact models for interpreting force curves.

Mechanical properties from nanoindentations are commonly obtained by analysis of the unloading part of the force curve. The Oliver–Pharr approach [50, 65], based on the estimate of the contact stiffness upon unloading where the contact is presumed to be elastic, is often applied for this analysis, and although this approach can be successful when analyzing force curves collected with commercial depth-sensing instruments on hard samples, the application to AFM nanoindentation of soft materials is rarely successful [66]. Values of Young's modulus are indeed often overestimated, up to 300%. Although the cause is still under debate [67, 68], some sources of error clearly stem from the fact that the AFM nanoindentation process is not and cannot be easily controlled, as well as from the onset of physical phenomena not taken into account in the O/P approach. This is indeed based on the model of Sneddon [47], assuming an elastic contact and a conical indenter. Since viscoelastic phenomena probably play a major role in the unloading curve of soft materials, and the AFM tip cannot be assumed to be an ideally pointed cone, deviations can be severe. In particular, the basic assumption behind the O/P approach is a quadratic relationship of the applied load with regard to the penetration depth, at least in the upper part of the unloading curve. However, experiments show that this exponent is often much higher [55, 66]. AFM nanoindentations on isotactic polypropylene (iPP) samples with various morphologies, from mesomorphic to semi-crystalline, have demonstrated that the unloading curve did not bear any information regarding the mechanical properties of the sample [56, 66].

Moreover, nanoindentations on PET have confirmed this finding over a wide range of indentation depths and applied loads, as well as for tips with different geometries. This can be seen in Fig. 7.7 where four force curves obtained with different instruments and indenter geometries for a broad range of applied loads and penetration depths are reported. Two force curves were obtained with sharp AFM tips at two different loads, one was obtained with a blunt AFM tip especially suitable for nanoindentations; and the last curve was obtained with an MTS instrument at a higher load level and using a Berkovich indenter with a loading rate of $100 \mu N/s$.

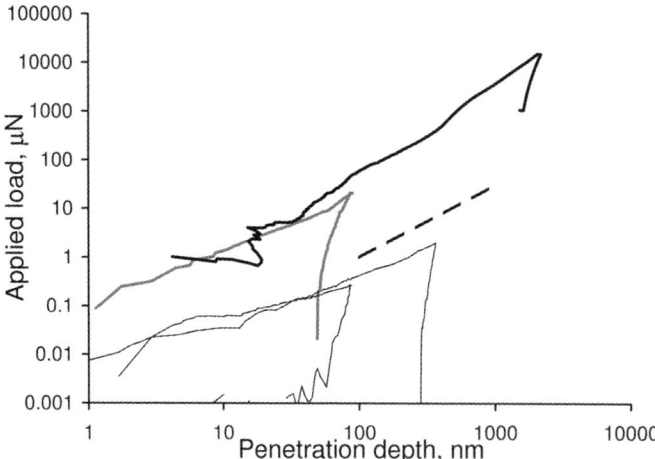

Fig. 7.7. Force curves collected by AFM and DSI nanoindentations. The two *thin-lined curves* both refer to a sharp AFM tip with a low elastic constant, on the order of 30 N/m. The *thick gray* line refers to a blunt AFM tip with a larger elastic constant, ca. 150 N/m and the *thick black* line refers to a commercial MTS nanoindenter, equipped with a Berkovich indenter. The material substrate was in all cases a PET sample solidified at high cooling rates so as to be homogeneously amorphous. [55] Copyright Wiley-VCH Verlag GmbH & Co. KGaA. Reproduced with permission

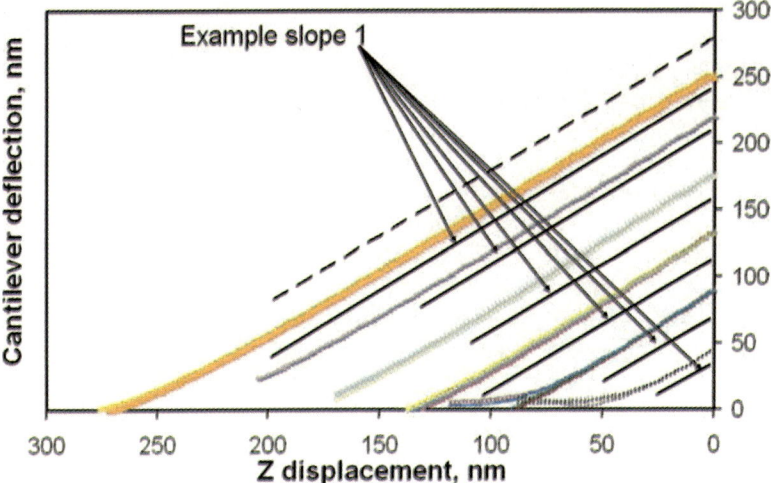

Fig. 7.8. The scaling of the cantilever deflection vs. the piezo displacement upon loading the same substrate at different loads. The slope of the curves is unaffected by the material recovery which will eventually take place at much longer times than the piezo retraction. In other words, the slope is unaffected by the mechanical behavior of the material. [57] Copyright Formatex. Reproduced with permission

In the case of AFM nanoindentation, although the loading was neither load- nor deformation-controlled, the average loading rates were ca. $100\,\mu N/s$ for the thick curve and ca. $10\,\mu N/s$ for the thin one. These values were estimated on the basis of total load and total indentation time.

Figure 7.7 points out the very high slopes of the unloading curves; the dashed line represents a curve relating the applied load and the penetration depth as a power law with an exponent of two. In all cases, the experimental unloading slopes were much larger than two. This condition is reported in the literature as characteristic of plastic materials with a rather limited reversible deformation range: i.e., when the load is removed, the elastic recovery is very poor and, as a result, the penetration depth remains almost unaltered thus causing the slope of the unloading portion of the applied load vs. penetration depth curve to become very large.

However, images of the imprints left behind by the nanoindentations show that the residual indentation is often much smaller than the penetration depth under maximum load, as also illustrated in Fig. 7.4. Therefore, the high slope of the unloading curves is not linked to the plastic behavior, as would be expected, but rather to a viscoelastic effect. This can be argued on the basis that the unloading curves are superimposed on the loading curves in the case of an ideal elastic behavior, as for rubbers, or, in general, in the case of materials with very short relaxation times. In contrast, when the relaxation time of the sample is close to or even much larger than the indentation time scale, very high slopes are immediately recorded. It is noteworthy that this feature seems to be related also to the fact that an AFM nanoindentation is run without load or displacements control. This observation casts serious doubts on the O/P analysis as applied to polymeric materials and usually carried out to extract the Young's modulus from the contact stiffness value without ever checking the physical meaning of the unloading exponent.

A further and final example, useful for stressing the limits of the unloading portions of force curves collected by AFM, is given in Fig. 7.8 [57]. In this case, only the unloading curves for a HDPE sample are shown, as the cantilever deflection (δ) vs. the piezo displacement, z. This can be translated into applied load vs. penetration depth with Eq. (7.3). It can be noted that upon unloading, δ was approximately equal to $z-z_0$ in a very broad range of applied loads. Here, z_0 was a constant roughly equal to the value assumed by the piezo displacement during unloading when the cantilever deflection vanished. Recalling Eq. (7.3), this means that the penetration depth remains almost unaltered when the load is removed. The elastic recovery is thus very poor, and as a result, the slope of the unloading portion of the curve of F vs. p is very large and, in particular, does not scale with a power law with an exponent equal to two. This thus confirms the results shown in Fig. 7.7.

7.4
Models to Describe the Force Curves

The force curve, i.e., the relationship between the applied load and the penetration depth, is commonly analyzed by contact mechanics models. Unfortunately, there is no valid generalized method, but rather, different approaches have been proposed in the literature depending on the local mechanical behavior: plastic, elastic, or viscoelastic, determining the contact between the indenter and the material. A brief account is outlined in the following to point out the difficulties associated with describing a material behavior where very large deformations, with a three-dimensional strain field, are often involved and where neither finite elements approaches nor material constitutive equations are able to present the necessary accuracy.

The loading curve for nanoindentations of plastic materials is often expressed by Kick's law [69], which is a parabolic relation of the applied load and the penetration depth, i.e.

$$F = Kp^2 \tag{7.4}$$

Here, F is the applied load and K is a constant that depends on the geometry of the contact and on the material properties. This equation has been used in the literature for sharp indenters, for example, conical [70, 71], Vickers [72, 73], and Berkovich [74–77]. Since it relates the ratio of the applied load and the squared penetration depth to a certain measure of average contact pressure for a plastic contact, the constant K can be considered as a hardness parameter [78, 79]. The value of this constant depends on the shape of the indenter tip and on the material properties of the indented material. Hainsworth et al. [80] suggested the following relation for K:

$$K = E \left(\Phi\sqrt{\frac{E}{H}} + \Psi\sqrt{\frac{H}{E}} \right)^{-2} \tag{7.5}$$

In this equation, E and H are the Young's modulus and hardness, respectively, of the indented material, Φ and Ψ are two empirical constants, for which the values $\Phi = 0.194$ and $\Psi = 0.903$ have been obtained [80] after analyzing indentation load-displacement curves measured on a vast variety of materials. The loading curves have, however, been observed to display deviations from Eq. (7.4) [81–83]. The explanation behind this finding can be obtained when taking into consideration that Eq. (7.4) implies a purely plastic deformation, which is not correctly fulfilled for nanoindentations of real materials.

In order to correct the quadratic Kick's law by means of an additional linear characteristic, Bernhardt [84] introduced the following formula,

$$F = K_1 p + K_2 p^2 \tag{7.6}$$

Obviously, the use of two fitting constants gives rise to more accuracy in approaching the experimental force curve although deviations have been observed. The physical meaning of the two constants has been suggested by Bernhardt [84] and Fröhlich et al. [78], who considered these constants as surface and volume contributions to microhardness. An alternative explanation was reported by Li and Bradt [85], suggesting that a quantity called the "proportional specimen resistance" ($PSR = K_1 p$) represents the elastic resistance of the test specimen and the friction between the indenter facets and the specimen, while the constant K_1 is a load-independent coefficient.

An alternative formula was also proposed by Bückle [86, 87] to identify the loading function with a limited sum of integer powers of the penetration depth. In comparison to previous relations, this method should formally allow a better approach of the experimental curve thanks to a greater number of fitting coefficients ($m + 1$ in this case):

$$F = K_0 + K_1 p + \cdots + K_m p^m \tag{7.7}$$

It has often been suggested that the K_0 term corresponds to a load threshold for an indenter to make a permanent indentation but it has such a low magnitude that it can be ignored in most situations [88]. To overcome the difficulty arising from the fact that the exponent in Eq. (7.4) usually has a value of less than 2, Hays and Kendall [89, 90] adopted a slightly different approach with an additional purely Newtonian constant term, denoted W. This term constitutes a small fraction of the total load that does not contribute to the indentation but rather acts as a reaction force of the specimen. The relation introduced is

$$F = W + K_{HK} p^2 \tag{7.8}$$

All the abovementioned approaches can be unified, as shown by Attaf [76], by the Meyer equation [91]. This equation describes the load-displacement relationship with a power law in the form

$$F = C p^m \tag{7.9}$$

where m is known as Meyer's index and C is a constant. It is worth mentioning that the original Meyer law was, to begin with, specified for a spherical indenter. Nonetheless, several authors have confirmed that it can be used even in the case of pointed indenters such as the Berkovich, Vickers, or conical indenters [92–94]. A comparison between Eqs. (7.9) and (7.4) shows that Kick's law is nothing but a particular case of Meyer's law in which the constant C and the exponent m are two fitting parameters that ensure a better modeling. Meyer's law is often used in the analysis of materials' behavior, but due to the fact that m is generally a noninteger, it has been hard to find a convincing interpretation of how to connect this law to depth-sensing experiments. Fröhlich et al. [78] clearly stated that this power law is often used from an exclusively empirical point of view.

All the equations referred to above are obviously empirical, and more or less far from an exact model that would provide an accurate fitting of the nanoindentation loading curve in a plastic material. Concerning elastic models and before going into details, it is worth mentioning a relationship obtained by Cheng and Cheng [95] through dimensional analysis. Taking into account a frictionless contact between a rigid conical indenter and the sample, the dimensional analysis of the physical quantities involved gives

$$\frac{F}{Ep^2} = \Pi\left[\frac{l}{p}, \frac{\sigma_Y}{E}, n, \theta\right] \tag{7.10}$$

Here, n is the strain hardening exponent and θ is the semi-opening angle of the cone, while l is introduced in order to better model the geometry of the contact. Indeed, when dealing with nanometer scale contact and keeping in mind obvious technological constraints, the tip cannot be properly modeled as an ideal pointed cone. Therefore, the quantity l represents a typical length scale taking into account the tip shape deviations from an ideal conical geometry.

Equation (7.10) implies that a power law relationship between the applied load and the penetration depth needs to be found, with an exponent equal to two in the case of an ideal cone, but differing from two when taking into account the tip roundness at its apex.

In the case where the contact is elastic, the models of Hertz and Sneddon can be used. The one suggested by Hertz [46] simulates the contact of two spheres, valid only at vanishing penetration depths or when the contact radius is much smaller than the penetration depth. Setting one of the radii to infinity, the contact between a sphere and a flat half-space is obtained. It is worth mentioning that the assumption of a vanishing contact radius, a, in comparison with the sphere radius, R, is particularly restrictive. Taking into account a typical 5–10-nm radius of an AFM tip, a penetration depth of only 1 nm would indeed imply a contact radius of a few nm, which is comparable to the radius of the tip, see Fig. 7.9.

Other models should thus be used in order to properly study AFM indentations [54]. Sneddon [47] modeled the contact of a rigid body, eventually characterized by several geometries, with a flat surface and this model has no limitations with regard to the penetration depth range as long as the material behavior can safely be assumed to be elastic. As pointed out in Sect. 7.3, this assumption generally requires that specific experimental conditions be verified. In particular, Sneddon's model distinguishes among five possible geometries for the indenter, i.e., flat punch, sphere, cone, paraboloid, and ellipsoid. The choice of the proper geometry must then be dictated by the judicious analysis of the experimental conditions and the length-scales involved, but the exponent that relates the applied load and the penetration depth can provide a useful hint. For example, with the assumption of an elastic contact, this exponent is 1 in the case of a flat punch, 2 in the case of an ideally pointed cone, and 1.5 in the case of a paraboloid. More complicated relationships are given for other geometries [47].

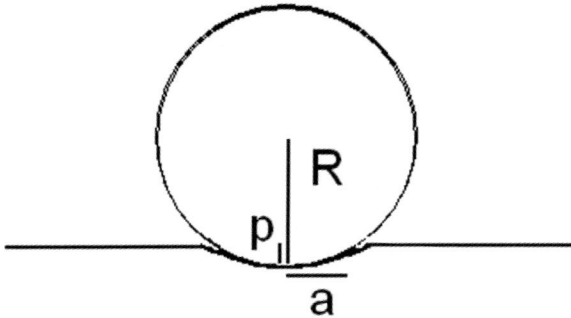

Fig. 7.9. The limitation of the Hertz approach, which consists of the assumption of a vanishing contact radius, a, in comparison to the sphere radius, R, is particularly restrictive for AFM since typical tip radii range from 5–10 nm. A penetration depth of only 1 nm, implying a contact radius of a few nm, is thus comparable to the tip radius

Chizhik et al. [13] compared the use of various models in order to achieve a reliable nanomechanical characterization of several polymers, i.e., polystyrene (PS), polyvinyl chloride (PVC), polyurethane (PU), and an isoprene rubber. Figure 7.10 shows the change in Young's modulus evaluated at various penetration depth levels in the case of the isoprene rubber. It can be seen, despite a certain amount of scattering which was probably related to the instability of the contact at the low penetration depth, that the Sneddon model, applied by describing the AFM tip as a paraboloid, provided a substantially depth-independent measurement. On the other hand, it is interesting to note that by assuming the AFM tip to be a cone, or pyramid, a noticeable change in the results was obtained. Indeed, this points to the fact that the power

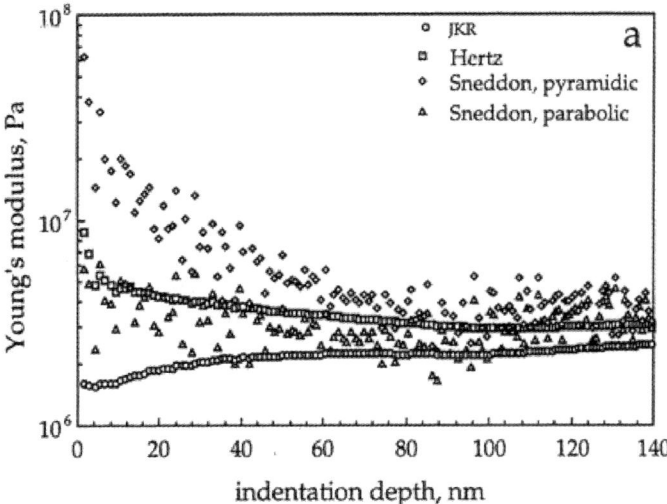

Fig. 7.10. The application of various models in order to extract the Young's modulus from nanoindentations of an isoprene rubber. The graph shows the dependence of the modulus on the penetration depth. [13] Copyright American Chemical Society. Reproduced with permission

law relationship between the load and the penetration depth was correct in the case of a paraboloid, i.e., giving no dependence in modulus vs. penetration, whereas this was not the case for the pyramid, as will be discussed thoroughly later on. The application of the JKR model, also shown in Fig. 7.10, which also accounts for adhesion effects, will be discussed in the next section.

In order to strengthen this information, it can be useful to compare force curves obtained from materials with very different behaviors. Figure 7.11, for example, shows force curves collected for two model materials: lead as a model plastic material and a rubber as a model elastic material. First of all, it can be noticed that the curves are linear, thus confirming a power law relationship. However, the slope of the double logarithmic plot, corresponding to the exponent that scales the applied load and the penetration depth, was equal to two in the case of a plastic contact, in agreement with the previously introduced Kick's law. On the other hand, the exponent in the case of the elastic material was close to 1.5, which corresponds to the contact of a paraboloid with an elastic surface.

The occurrence of this exponent is not surprising, following the dimensional analysis previously introduced in Eq. (7.10), since the AFM tip cannot reasonably be considered as an ideal pointed cone as it typically displays a radius. Indeed, the tip radius is a characteristic length scale of the paraboloid assumed to describe tip shape, which can be modeled in cylindrical coordinates as

$$\rho^2 = 4qz \qquad\qquad\qquad (7.11)$$

where q is a length scale, characterizing the paraboloid. It is then clear that a parabolic geometry properly accounts of the AFM tip shape in order to describe the force curves collected for a rubber and a sample of iPP in Fig. 7.11. The elastic

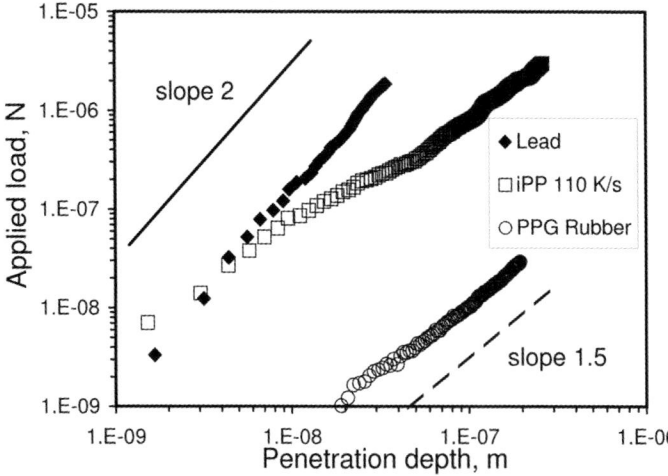

Fig. 7.11. Logarithmic plots of the applied load vs. the penetration depth collected on two polymer samples: a PPG rubber and an iPP sample. The typical slope of 1.5 is compared to a metal sample, lead, for which a slope of 2 is observed in agreement with an elasto-plastic behavior, see Kick's law [Eq. (7.4)]. [54] Copyright American Chemical Society. Reproduced with permission

constant of the cantilever as well as the tip radius should however be determined with good accuracy.

While different procedures have been introduced and refined, as described in the introduction, in order to calibrate the elastic constant, evaluating the tip radii still represents an issue. Manufacturer data for tip radii may be inaccurate and lead to errors, depending on production scattering and tip wear or modifications. In order to achieve a reliable estimate, transmission electron microscopy (TEM) imaging of tip apexes is possible [96] although it cannot be considered a routine technique in the case of high-throughput analysis. The estimation of tip radii from images collected on calibration samples with well-known shapes, and incidentally very sharp features, is also possible [97, 98]. Most often, blind estimations of tip shapes from images collected on unknown, so-called characterizer, samples are used. Samples displaying a broad range of asperities are for this purpose accurately chosen [37].

In the case of a tip whose shape can be described as a paraboloid in contact with an elastic half-space, the equation that relates the most significant physical quantities can be written as

$$F = \frac{4E}{3(1 - v^2)}(2qp^3)^{1/2} \tag{7.12}$$

Figure 7.12 shows a plot of the maximum applied load normalized by the independently measured bulk elastic modulus vs. the maximum penetration depth to the power of 1.5, for several polymers with quite different mechanical properties and also clearly belonging to different mechanical behaviors. In this case, each point represents one force curve. According to Eq. (7.12), a straight line satisfactorily fits the data, and a certain amount of scattering to the main trend should be attributed to the change in tip radius as different cantilevers were used for the different materials.

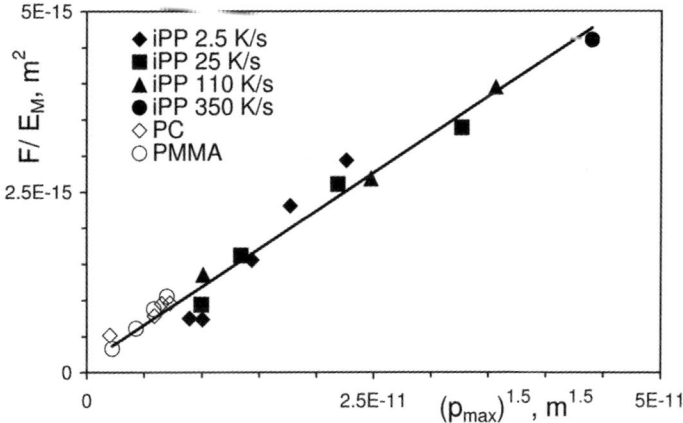

Fig. 7.12. The dependence of the applied load, normalized by the elastic modulus measured through bulk tensile tests, on the maximum penetration depth to the power of 1.5 for several samples covering a vast range of elastic properties. For iPP, the solidification conditions are reported in terms of cooling rate, K/s. [54] Copyright American Chemical Society. Reproduced with permission

This plot suggests that an accurate estimate of a sample's Young's modulus can be obtained. Examples in the literature range from glassy polymers, e.g., polycarbonate (PC) and polymethylmethacrylate (PMMA), to the rather different semicrystalline morphologies of iPP obtained by rapid cooling from the melt, and finally to rubbery polymers. The agreement obtained is outstanding, as demonstrated in Fig. 7.13, with respect to the error bars confined within the symbols. The accuracy of the evaluations rely mostly on accurate calibrations of all the necessary parameters that determine the fitting of Eq. (7.12): i.e., the radius of curvature of the tip, the elastic constant of the cantilever, and the piezo linearity.

In the case of very compliant materials, however, the penetration depth can be very large and the assumption that the tip can be described as a paraboloid can lead to errors of up to 50%. Cappella et al. [99] recently showed that deviations from the 1.5 exponent were indeed observed at large enough applied loads. In particular, two distinctive trends were noticed at small and large penetration depths and the authors argued that the intersection between the straight lines fitting the behaviors at the two extremes could be identified as a yielding load, see Fig. 7.14. It is worth mentioning that the 250-nm cantilever deflection, as shown in Fig. 7.14, corresponds to the relatively large load of approximately 11 μN applied to the cantilever used in that study [99].

The force curve in Fig. 7.14 was also fitted with the hyperbolic relationship

$$D^{3/2} = f(\delta_c) = (\beta\delta_c - \varepsilon) + \sqrt{\alpha^2\delta_c^2 - 2\varepsilon(\beta - \gamma)\delta_c + \varepsilon^2} \qquad (7.13)$$

where D, in this case, is the penetration depth, δ_c is the cantilever deflection, and where there are certain constraints on the fitting parameters (i.e., α, β, γ, $\varepsilon > 0$ and $\beta - \alpha < \gamma < \beta + \alpha$). It was thus shown that the Young's modulus could be calculated with good accuracy starting from the fitting parameters α and β.

An alternative approach in the case of a large penetration depth, and before the yielding point is reached, is given by the model proposed by Segedin [100]. In this case, a more realistic shape of the tip is obtained, taking into account both the conical

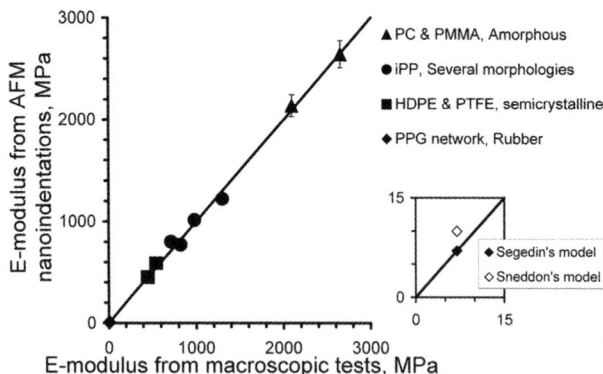

Fig. 7.13. A comparison of the elastic modulus evaluated by Sneddon's model, with the macroscopic ones obtained from bulk tensile testing. [54] Copyright American Chemical Society. Reproduced with permission

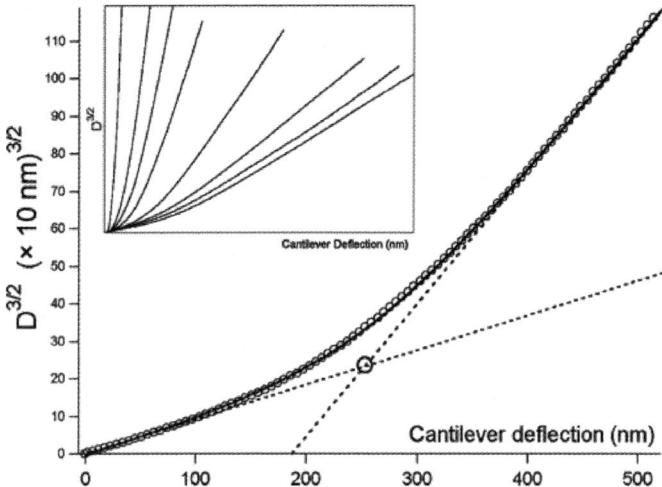

Fig. 7.14. A force curve obtained at a high maximum load, displaying two trends at their extremes. Instead of the applied load vs. the penetration depth, the force curve is in this case plotted as the penetration depth to the power of 1.5 vs. the cantilever deflection. [99] Copyright American Chemical Society. Reproduced with permission

shape of the tip, given by technological constraints of its manufacturing, and its roundness at the apex as measured by the abovementioned methods. For example, both geometries, conical ending with a parabolic tip, can be superimposed with the constraint of continuity of the derivative for the two profiles as the condition for the transition, as shown in Fig. 7.15.

In this case, the profile can be fitted with a power law relationship through a set of parameters c_n as

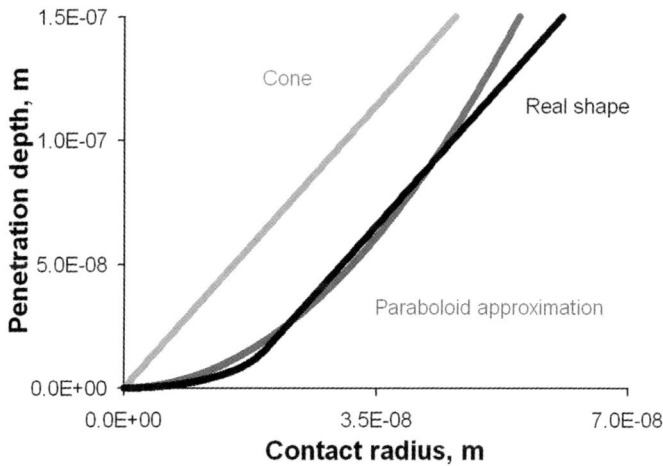

Fig. 7.15. The penetration depth as a function of the contact radius

$$z = \sum_n c_n x^n \tag{7.14}$$

The Segedin model [100], also confirmed independently by Sneddon [47], allows one to estimate both the penetration depth and the applied load from certain values of the contact radius, a.

$$p = \sqrt{\pi} \sum_{n=1}^{+\infty} \frac{\Gamma(1/2n + 1)}{\Gamma(1/2n + 1/2)} c_n a^n \tag{7.15a}$$

$$F = \frac{2\sqrt{\pi} \mu a}{1 - \nu} \sum_{n=1}^{+\infty} \frac{\Gamma(1/2n + 1)}{\Gamma(1/2n + 1/2)} c_n a^n \tag{7.15b}$$

A force curve relating the penetration depth and the applied load is then derived by coupling these two values, obtained at the same contact radius. If the contact is elastic, the accuracy with which this model can predict a force curve depends only on the accuracy with which the tip shape is modeled through the parameters c_n. The Young's modulus of the sample is at this point the only fitting parameter capable of forcing the experimental curve to fit the theoretical force curve. An elastic modulus obtained by this procedure is very accurate, even in the case of a compliant material, as shown in the insert of Fig. 7.13. However, this analysis is not required if the penetration depth is kept smaller than 100–120 nm as shown in Fig. 7.15, where the profile of the paraboloid tip is sufficiently close to a more realistic modeling.

Force curves obtained at very different penetration depths are shown in Fig. 7.16 together with their fitting according to Segedin [54, 100]. The Young's modulus, used as a fitting parameter, was found to be 7 MPa for the polypropyleneglycol (PPG)-based rubber and 1.99 GPa for PC. The latter is very similar to the value calculated by Sneddon's approach since in this case the penetration depth is so small that a paraboloid modeling of the shape of the AFM tip is adequate.

In a case where the tip–substrate contact is dominated by a viscoelastic behavior, the force curve should be analyzed by the models of Ting [48] or Lee and Radok

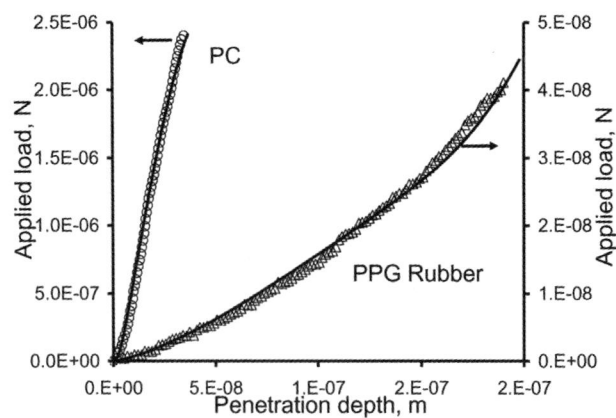

Fig. 7.16. The fitting of two force curves, obtained by the Segedin contact model, see Eq. (7.15), as compared to experimental data referring to two extreme cases: a rubbery PPG network and a PC sample. [54] Copyright American Chemical Society. Reproduced with permission

[101]. These models give rise to equations for viscoelastic contact but are however very difficult to apply. This has the effect that solving the problem in the case of an unknown loading history is an overwhelmingly difficult task. If a series of load-controlled and displacement-controlled experiments can be run, the equations can be highly simplified. In particular, registering the load evolution after the application of a step loading followed by a constant penetration depth, thus mimicking a relaxation experiment, should be performed. In a second series of experiments, one should carry out a penetration depth evolution after step loading on the substrate and while keeping the load constant. Such an analysis is called a creep experiment. The relaxation and creep curves should then constitute the experimental input to Ting's analysis, which, in turn, should be able to provide the time dependence of the elastic modulus and Poisson's ratio. This, if carried out within a broader time range or at different temperatures, could provide the complete viscoelastic characterization of a material. Such an analysis has, to the best of our knowledge, never been performed and rather simplified models have been used in order to extract only the relaxation modulus. At this point, it is worth stressing that such an approach implies the use of a constant Poisson ratio which is an extremely restrictive assumption for a viscoelastic contact [102, 103].

However, the application of viscoelastic models is often very difficult, and, if proper experimental conditions can be identified, the best option is represented by running experiments that maximize the elastic contribution and utilize Sneddon's model to interpret the force curves, thereby obtaining a meaningful elastic modulus.

7.5
Final Remarks on Loading History, Adhesion, Roughness Effects

It is wise to make a further comment regarding the loading history applied during an AFM nanoindentation, as exemplified in Fig. 7.17 in the case of three very different morphologies and presumably quite different viscoelastic material behaviors. Upon using Eq. (7.12), i.e., the elastic contact model of Sneddon, it is possible to observe that the applied load and penetration depth scale with an exponent equal to 1.5 thus confirming that the contact is elastic. Keeping in mind that the cantilever deflection is proportional to the applied load through the elastic constant of the cantilever, and that time is proportional to the piezo displacement through the voltage saw tooth applied to the piezo, one can write

$$t \propto z \propto (p + \delta) \propto (p + p^{1.5}) \tag{7.16}$$

From this follows

$$\frac{dt}{dF} \propto \frac{d(p + p^{1.5})}{dF} \propto \frac{dp}{dF} + \frac{dp^{1.5}}{dF} \tag{7.17}$$

The last term, inversely proportional to the elastic modulus and therefore a constant, is responsible for a linear increase of the load with time. The term dp/dF, on the other hand, is responsible for a nonlinear loading history that becomes more relevant if the penetration increases to a greater extent with increasing applied load, i.e., when the material is compliant. Thus, a possible interpretation of the traces

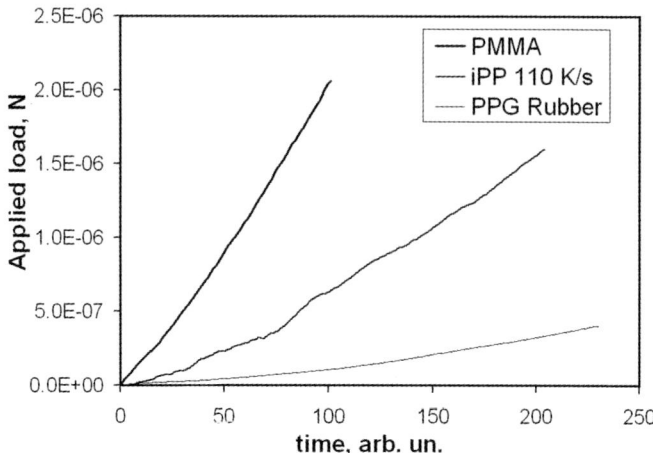

Fig. 7.17. Load histories obtained during AFM nanoindentations of three samples with dramatically different mechanical properties. [57] Copyright Formatex. Reproduced with permission

reported in Fig. 7.17 is that while the time dependence of the load for PMMA is almost linear, the one for the PPG rubber shows significant deviations from linearity.

In the case of adhesive contacts, the JKR [51] and DMT [53] models are often used to analyze the force curves. However, these models represent a correction to the Hertz model taking into account the adhesion forces, and are therefore bound by the same limitation as the Hertz model, as discussed in the previous section. In other words, the contact radius is seldom much smaller than the tip radius. As an alternative, Sun et al. [104] recently applied the general equations proposed by Sneddon in the case of the contact of an arbitrary indenter profile to the adhesive contact. In particular, the tip was there modeled as a hyperboloid according to Fig. 7.18.

Sneddon's equations relate the applied load and the penetration depth to the contact radius, and by coupling these relationships with a criterion for the breakage of the adhesive contact, Griffith's criterion in that case [105], the following equations were derived [104]

Fig. 7.18. A SEM image of an AFM contact tip characterized by a higher tip radius as compared to standard noncontact tips. The indenter apex, modeled as a hyperboloid, is highlighted by the *black lines*. [104] Copyright American Chemical Society. Reproduced with permission

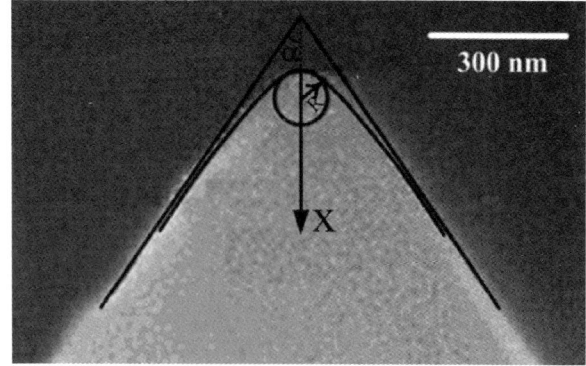

$$p = \frac{aA}{2R}\left[\frac{\pi}{2} + \arcsin\left(\frac{(a/A)^2 - 1}{(a/A)^2 + 1}\right)\right] - \left(\frac{2a\pi(1 - v^2)W_{12}}{E}\right)^{1/2} \quad (7.18a)$$

$$F = \frac{2E}{1 - v^2}\left[\frac{A}{2R}\left[aA + \frac{a^2 - A^2}{2}\left(\frac{\pi}{2} + \arcsin\frac{(a/A)^2 - 1}{(a/A)^2 + 1}\right)\right]\right.$$

$$\left. - a\left(\frac{2a\pi(1 - v^2)W_{12}}{E}\right)^{1/2}\right] \quad (7.18b)$$

Here, W_{12} is the interfacial energy of the tip and the sample, and $A = R \cot\alpha$, where α is twice the tip angle as shown in Fig. 7.18. The application of this approach requires a two-step calculation that is far from straightforward and whose details can be found in ref. [104]. In most of the cases considered in the following, the adhesion between the tip and the sample was negligible with respect to the applied forces as it eventually played a role only during the unloading portion of the force curve, and thus was not used for the elastic modulus calculations.

A final remark concerns the influence of the surface roughness; this parameter is altogether neglected once the sample for nanoindentations is modeled as a half-space. However, surfaces are always rough and the size of their asperities can vary from the length of the sample down to the atomic scale [106]. By convention, surface irregularities are classified according to errors in form, waviness, roughness, and subroughness [107]. The term subroughness is used in reference to roughness of very smooth surfaces whose asperities are on the order of nanometers [107]. The subroughness describes the fine structure of a real surface and is closely related to the so-called physical relief. It involves an accidental and imperfect location of crystallographic planes, chaotically distributed grains and islet-type films including oxide as well as adsorbed ones. For semi-crystalline polymers, the subroughness can be connected with alternating crystalline and amorphous regions of dozens of nanometers in size. Because of the surface roughness and also the short range character of inter-atomic forces, contact between nominally flat surfaces is limited to a small fraction of the apparent contact area, to a multitude of discontinuous junctions randomly distributed on the surface.

These spots, called single asperities, are responsible for the tribological properties of the two solid bodies that are in contact and can provide new insight into the theory of friction, lubrication, and wear since this level of roughness is associated with corresponding types of contact areas (i.e., nominal, contour, real, and physical areas of contact) [106]. Therefore, a great amount of research has been devoted to the investigation of contact mechanics and frictional properties of rough surfaces [108–110]. On the theoretical side, models have been developed for surfaces with roughnesses comprised within a narrow bandwidth of length scales. This is the case for Gaussian distributed spherical asperities that have become the prototype for successive developments and refined calculations [110, 111]. Besides ideal surfaces, the contact between a flat elastic solid moving on a slightly wavy surface, i.e., contact with a simple one-length scale roughness of sinusoidal profile, was studied by Carbone and Mangialardi [112]. Finally, Buzio et al. [113] investigated the contact mechanics and friction forces between AFM probes and self-affined fractal carbon films, and demonstrated that, for a multi-asperity contact [114], the penetration depth depends strongly on the fractal dimension and surface roughness whereas the friction coefficient is related only to the bulk parameters. In conclusion, the surface roughness constitutes an important parameter creating interference with the measurement of mechanical properties through nanoindentations and should be determined with accuracy in addition to the instrumental calibrations already mentioned.

A final comment can however be advanced concerning the role of the roughness when measuring the mechanical properties by AFM nanoindentations of compliant materials. The large elastic range of polymers and the use of very sharp tips render the effect of the roughness less crucial than in the case of hard materials. This is due to the contact surface being small in comparison with the lower limit of the size of the asperities that can be observed in polymers. In fact, this size is only one order of magnitude smaller than that of the tip. One could thus debate whether, in the case of polymer nanoindentations, roughness is a fractal quantity or not; a concept that certainly deserves further quantitative investigation.

7.6
Selected Applications

Discerning the bulk mechanical properties of samples characterized by a fine-tuning of the morphology, or measuring the mechanical behavior of single component phases thus distinguishing their contribution in a complex texture, are two very challenging problems. They are discussed in this last section.

In the first case, by exploiting the use of different cooling rates with a continuous cooling approach, polypropylene samples were prepared so as to present varying morphologies. Differently from metallurgy, the adoption of a low sample thickness allows a constant thermal gradient such that homogeneous morphologies can be obtained [115]. This renders it possible to compare mechanical properties determined by bulk methods with those measured by AFM nanoindentations, thus reliably comparing macro- and nano-mechanical properties. It is worth mentioning that the thickness is always tenfold, or even up to several hundred times, larger than the maximum penetration depth. Thereby, substrate effects can be neglected with certainty.

Isotactic polypropylene samples were solidified from the melt with five different cooling rates. Their structures are described by the so-called "Solidification Curve" reported in Fig. 7.19, which quantitatively describes the density dependence on the cooling rate. The figure also presents corresponding wide-angle X-ray diffraction, WAXD, patterns measured at specific cooling rates, showing the decrease of the stable α-monoclinic phases with increasing cooling rates up to a lower density plateau. Here, the stable phases disappear leaving only the mesomorphic metastable iPP phase [116]. Thus, density values and WAXD patterns provide similar albeit mutual information on the dependence of the morphology on the cooling rate. In association with this continuous change of the morphology, the mechanical properties of the samples were expected to change accordingly, with a modulus of 1.2 GPa in the case of the semi-crystalline morphology, as measured on macroscopic samples, to a

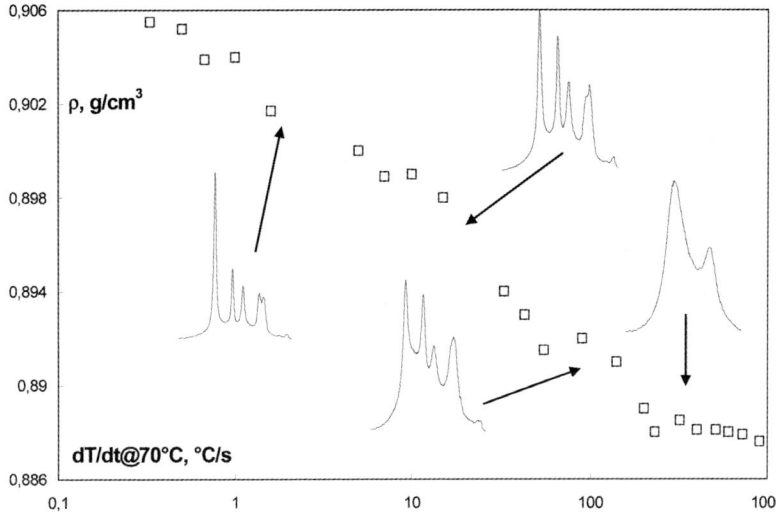

Fig. 7.19. The range of iPP morphologies upon solidification from the melt. Their dependence on cooling rate is quantitatively described by both density values and WAXD patterns. The mechanical properties of five morphologies were investigated by AFM nanoindentations. As the samples were homogeneous, also the bulk mechanical properties could be measured for comparison. [56] Copyright Elsevier. Reproduced with permission

lower bound of ca. 800 MPa in the case of the mesomorphic morphology. In this relatively small range, a high accuracy is required in order to determine the mechanical properties.

Figure 7.20 shows a plot of Young's modulus, as obtained by both macroscopic tensile tests and AFM nanoindentations, vs. the cooling rate [56]. The different series refer to measurements obtained at various levels of loading rate. As shown in Sect. 7.2, the use of low loading rates implies a mechanical contact with the ξ parameter of Eq. (7.1) far from the limit for elastic behavior, thus causing the elastic modulus determined by Sneddon's model to be unreliable. Therefore, at low frequencies, i.e., at low piezo velocities, low moduli are measured. This is in agreement with the observation that under such circumstances both plastic phenomena and viscoelastic relaxations cause a higher penetration depth at equivalent load levels. Elastic models would therefore anticipate that a low elastic modulus material is being tested.

However, as the loading rate increases, the contact becomes mostly elastic, see also Fig. 7.4. Thus, Sneddon's model gives rise to predictions for the Young's modulus of the samples that are in reasonable agreement with those of the bulk. It is worth mentioning that, although the estimation of the overall strain and the strain rate is a difficult task in the case of nanoindentations due to the high three-dimensionality of the stress field, the displacement rates in the case of nanoindentations performed at a frequency of 10 Hz are similar to those of the bulk tensile tests of Fig. 7.20, amounting to ca. 18 μm/s [56].

In the second case presented in this section, samples with a dispersed nanoscale phase were prepared starting from a polyethylene therephtalate/polyethylene naphtalate (PET/PEN) blend [58]. Macroscopic measurements, providing values of micro-

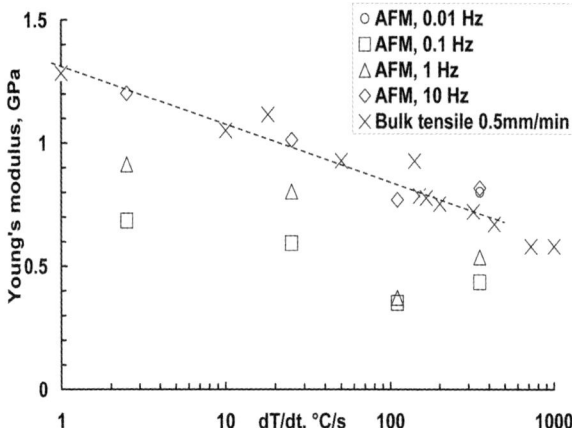

Fig. 7.20. The dependence of the elastic modulus on morphology, i.e., on the cooling rate for melt solidification, for an iPP resin. The *crosses* refer to values of the Young's modulus obtained from tensile tests on macroscopic samples. The *dotted line* is a guide for the eye. [56] Copyright Elsevier. Reproduced with permission

hardness and density, show complementary information, as can be seen in Fig. 7.21. In the case of density, its gradual change points out a variation of the morphology from semi-crystalline to amorphous. With an increasing cooling rate, distinct differences in the microhardness were measured among the different samples, and these differences were related to the onset of a state of an intermediate order, i.e., of crystalline clusters [58]. Although the density continues to decrease with increasing cooling rate, the presence of crystalline clusters, even at a rate above 2 K/s, is not immediately revealed from the density measurements, possibly due to the scarce

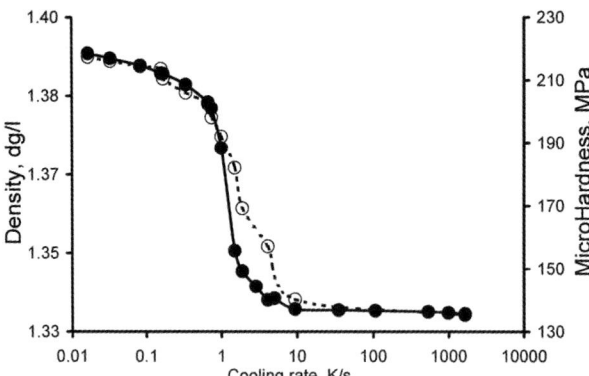

Fig. 7.21. The dependence of the density (*filled symbols*) and microhardness (*open symbols*) of PET on the cooling rate. An excess microhardness can be observed for cooling rates where the material is already amorphous. [58] Copyright Wiley-VCH Verlag GmbH & Co. KGaA. Reproduced with permission

density contrast between the clusters and the amorphous regions as also reported by Welsh et al. [117].

Microhardness, more sensitive to morphology than density, related to bulk crystalline content, does indeed display a different cooling rate dependence as opposed to the density, especially at cooling rates centered around 5 K/s. Here, a microhardness in "excess" to the amorphous expected value is observed, it confirms previous observations in other systems with semirigid chain polymers. [119, 120]

Amorphous semirigid chain polyesters, obtained from the melt at moderate cooling rates, are known to display a morphological order on the nanometer scale. The existence of a state of intermediate order was postulated also in PET, at a cooling rate above 2 K/s, with the absence of crystalline reflections in WAXD patterns along with the occurrence of SAXS (small-angle X-ray scattering) maxima [118] and with the onset of exothermic peak areas (from differential scanning calorimetry (DSC) analysis). A smectic structure, a precursor of crystalline structural order, was recognized to determine the higher mechanical performance upon careful annealing of the amorphous PET [119].

AFM nanoindentations were performed at varying penetration depths, i.e., different loads so as to probe the mechanical behavior of the material on varying scales. At relatively high penetration depths, for instance 100 nm with a 4-μN load, a cooling rate trend is observed that confirms the information obtained by the microhardness in Fig. 7.21. This trend involves the sample cooled at 5 K/s being stiffer than the amorphous one but more compliant than the semi-crystalline one confirming the microhardness data. One can even plot the 60 measurements in terms of frequency distribution, observing, in this case, a relatively low dispersion. This is put forward in Fig. 7.22 for both the amorphous (cooled at 1,200 K/s) as well as the 5 K/s sample, where the latter is clearly stiffer than the amorphous one.

However, Fig. 7.22 also shows a double-peaked distribution in the case of the semi-crystalline sample (solidified at 0.14 K/s); one can refer to Fig. 7.21 for its density and microhardness. The shape of the curve related to the semi-crystalline sample in Fig. 7.22 is due to the fact that a 92/8 PET/PEN blend cannot co-crystallize [28]

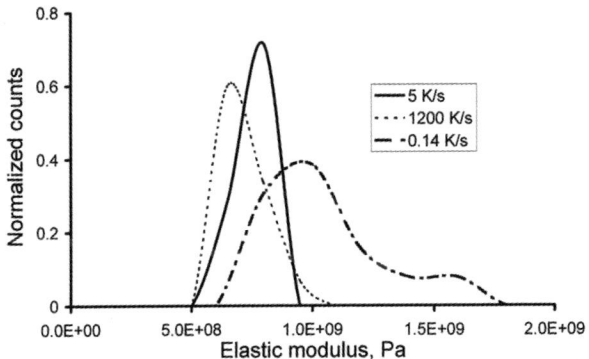

Fig. 7.22. The statistical distribution of the elastic modulus observed after 60 nanoindentations performed at 4 μN for samples solidified at three cooling rates. [58]Copyright Wiley-VCH Verlag GmbH & Co. KGaA. Reproduced with permission

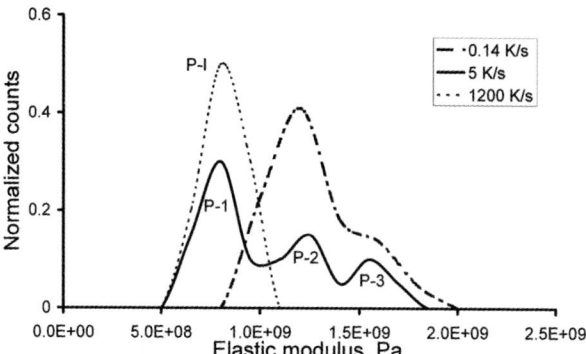

Fig. 7.23. The statistical distribution of the elastic modulus observed after 60 nanoindentations performed at 1 μN for samples solidified at three cooling rates. The onset of nanoclusters interferes with the scale of indentations, thus resulting in a perturbation of the modulus distribution observed for the 5-K/s sample. [58] Copyright Wiley-VCH Verlag GmbH & Co. KGaA. Reproduced with permission

therefore separate domains with long-range order of the two moieties are observed. Thus, a bimodal distribution, accounting for the separate contribution of PET and PEN is expected [120].

If lower loads are used, implying testing at lower penetration depths, novel results are obtained for the statistical distribution of elastic modulus as shown in Fig. 7.23. The samples cooled at low and high cooling rates do not display different behaviors with respect to the larger load; the distribution of the Young's modulus remained more or less unchanged between Figs. 7.22 and 7.23. On the other hand, the sample solidified at 5 K/s displays a distribution of the modulus split into three peaks, P-1, P-2, and P-3, corresponding to three phases of varying stiffness. The first peak is similar to the one obtained for the amorphous sample whose properties are found to occur with the larger probability. In this case, the nanoindentations probe the mechanical properties in areas where mostly the amorphous phase is present and where crystalline clusters are substantially absent. In the other two cases, i.e., for peaks P-2 and P-3, the two maxima can be seen to appear at the same values of the elastic modulus as those peaks characteristic of the semi-crystalline sample. These peaks correspond to the mechanical responses of the nanophases created by stacks of benzyl and napthyl rings of the blend. Indeed, due to the extremely small volume involved in the stress field at this load level, it is plausible that some nanoindentations are performed in cluster-free zones giving rise to the peak P-1, with the modulus correctly corresponding to that of the amorphous phase P-I, whereas other indentations fall in cluster-rich zones thus giving rise to the other two peaks, P-2 and P-3, centered at higher moduli. Interestingly, these latter two peaks can be related to the existence of nanoclusters arising from either PET (P-2), i.e., stacks of phenyl rings, as will be explained in the following, or from PEN (P-3), i.e., stacks of naphthyl rings.

The mechanical properties of each phase are also very well discerned in Fig. 7.23, reflected in the distinct, clear, and sharp peaks. This can be rationalized by noting that the material can be modeled as an amorphous matrix with a certain dispersion of harder particles and therefore the clusters situated directly below the depression are displaced downwards into the surrounding matrix. The local density of clusters thus increases, as compared to regions at a distance from the depression [121]. This results in the inter-particle spacing, especially along the vertical direction, being reduced in the region directly below the indentation. As the indenter moves downward, it encounters a resistance from a material with an increasingly greater concentration of hard clusters, to the extent that the mechanical properties become extremely well defined in Fig. 7.23.

On the basis of these results and observations, it is clear that the scaling down of penetration depths is not possible without loss of the meaning of the mechanical properties one intends to measure. As soon as the penetration depths are comparable to the scale of the texture, the material behavior is no longer representative of the continuum that is related to the bulk properties. Interference with texture implies an interaction of the tip with phases of individual components and their immediate surroundings, i.e., determined by the connectivity among the components. Therefore, the mapping of mechanical properties is limited, in the case of textured materials, to orders of magnitude involving dimensions larger than the individual components, as an example crystalline lamellae, in the case of crystalline polymers, a few nm thick.

This observation, even though it may seem a limitation, opens a wealth of possible scenarios for the characterization of the interaction of individual phases in nanocomposite materials, semi-crystalline polymers belonging to this class as a prerogative. A still unsolved problem that could be tackled by this technique concerns the modification that takes place in semi-crystalline metastable phases upon ageing. In this case, individual crystalline phase properties do not change whereas it is probable to ascribe most of the modifications to the intermediate constrained amorphous phases that are linked to the crystals. These phases are initially immobilized but slowly recover their mobility with time [122].

References

1. Fratzl P, Weinkamer R (2007) Prog Mat Sci 52:1263
2. Meyers MA, Chen P-Y, Lin AYM, Seki Y (2008) Prog Mat Sci 53:1
3. Stolz RRM, Daniels AU, VanLandingham MR, Baschong W, Aebi U (2004) Biophys J 86:3269
4. Jiang H, Liu X-Y, Lim CT, Hsu CY (2005) Appl Phys Lett 86:163901
5. Bhushan B, Chen N (2006) Ultramicroscopy 106:755
6. Garcia R, Magerle R, Perez R (2007) Nature Materials 6:405
7. Bhushan B (2001)Wear 250:1105
8. Kim TW, Bhushan B (2007) Ultramicroscopy 107:902
9. Mailhot B, Bussière P-O, Rivaton A, Morlat-Thérias S, Gardette J-L (2004) Macrom Rapid Comm 25:436
10. Sergei VVG, Chizhik A, Luzinov I, Fuchigami N, Tsukruk VV (2001) Macrom Symposia 167:167
11. Briscoe BJ, Fiori L, Pelillo E (1998) J Phys D Appl Phys 31:2395
12. Tsukruk VV, Huang Z, Chizhik SA, Gorbunov VV (1998) J Mater Sci 33:4905
13. Chizhik SA, Huang Z, Gorbunov VV, Myshkin NK, Tsukruk VV (1998) Langmuir 14:2606
14. Low IM, Che ZY, Latella BA (2006) J Mater Res 21:1969
15. Bhushan B, Koinkar VN (1994) Appl Phys Lett 64:1653

16. VanLandingham SHMMR, Palmese GR, Huang X, Bogetti TA, Eduljee RF, Gillespie JW (1997) J Adhesion 64:31
17. VanLandingham JSVMR, Guthrie WF, Meyers GF (2001) Macrom Symposia 167:15
18. Tsukruk VV, Huang Z, Chizhik SA, Gorbunov VV (1998) J Mater Sci 33:4905
19. Herrmann K, Jennett NM, Wegener W, Meneve J, Hasche K, Seemann R (2000) Thin Solid Films 377:400
20. Bischel MS, VanLandingham MR, Eduljee RF, Gillespie JW, Schultz JM (2000) J Mater Sci 35:221
21. Du B, Liu J, Zhang Q, He T (2001) Polymer 42:5901
22. Tsukruk VV, Sidorenko A, Yang H, (2002) Polymer 43:1695
23. Beake BD, Leggett GJ (2002) Polymer 43:319
24. Hao HW, Barò AM, Saenz M (1991) J Vac Sci Technol B 9:1323
25. Cappella B, Dietler G (1999) Surf Sci Rep 34:1
26. Butt HJ, Cappella B, Kappl M (2005) Surface Sci Reports 59:1
27. Tsukruk VV, Gorbunov VV, Huang Z, Chizhik SA (2000) Polym Int 49:441
28. Burnham NA, Chen X, Hodges CS, Matei GA, Thoreson EJ, Roberts CJ, Davies MC, Tendler SJB (2003) Nanotechnology 14: 1
29. Holbery JD, Eden VLJ. (2000) Micromech Microeng 10:85
30. Matei GA, Thoreson EJ, Pratt JR, Newell DB, Burnham NA (2006) Rev Sci Instrum 77:83703
31. Richard SG, Mark GR (2007) Rev Sci Instrum 78:86101
32. Saltuk BA, Joseph AT (2007) Rev Sci Instrum 78:43704
33. Langlois ED, Shaw GA, Kramar JA, Pratt JR, Hurley DC (2007) Rev Sci Instrum 78:93705
34. Benjamin O (2007) Rev Sci Instrum 78:63701
35. Khurshudov A, Kato K (1995) Ultramicroscopy 60:11
36. Villarrubia JS (1997) J Res NIST 102:425
37. Tranchida D, Piccarolo S, Deblieck RAC (2006) Meas Sci Technol 17:2630
38. Hiroshi I, Toshiyuki F, Shingo I (2006) Rev Sci Instrum 77:103704
39. Loubet JL, Belin M, Durand R, Pascal H (1994) Thin Solid Films 253:194
40. Albrektsen O, Madsen LL, Mygind J, March KA (1989) J Phys E 22:39
41. Riis E, Simonsen H, Worm T, Nielsen U, Besenbacher (1989) F Appl Phys Lett 54:2530
42. Vieira S (1986) IBM J Res Develop 30:553
43. Graffel B, Mueller F, Mueller A-D, Hietschold M (2007) Rev Sci Instr 78:053706
44. Chiara S, Arthur B, Stephen RB, Frederick S (2007) Rev Sci Instrum 78:36111
45. Younkoo J, Jayanth GR, Chia-Hsiang M (2007) Rev Sci Instrum 78:93706
46. Landau LD, Lifshitz EM (1986) Theory of Elasticity. Pergamon Press, Oxford
47. Sneddon IN (1965) Int J Eng Sci 3:47
48. Ting TCT (1966) J Appl Mech 33:845
49. Kramer D, Huang H, Kriese M, Robach J, Nelson J, Wright A, Bahr D, Gerberich WW (1998) Acta Materialia 47:333
50. Oliver WC, Pharr GM (1992) J Mater Res 7:1564
51. Johnson KL, Kendall K, Roberts AD (1971) Proc Roy Soc London A324:301
52. Derjaguin BV, Muller VM, Toporov YP (1975) J Coll Interf Sci 53:314
53. Arivuoli D, Lawson NS, Krier A, Attolini G, Pelosi C (2000) Mat Chem Phys 66:207
54. Tranchida D, Piccarolo S, Soliman M (2006) Macromolecules 39:4547
55. Tranchida D, Piccarolo S (2005) Macr Rap Comm 26:1800
56. Tranchida D, Piccarolo S (2005) Polymer 46:4032
57. Tranchida D, Kiflie Z, Piccarolo S (2007) Atomic Force Microscope Nanoindentations to Reliably Measure the Young's Modulus of Soft Matter. In: Méndez-Vilas A, Díaz J (eds) Modern Research and Educational Topics in Microscopy. Formatex, Badajoz, p 737
58. Tranchida D, Kiflie Z, Piccarolo S (2006) Macromolecular Rapid Commun 27:1584

59. Tranchida D, Kiflie Z, Piccarolo S (2007) Macromolecules 40:7366
60. Fischer-Cripps AC (2004) Nanoindentation, 2nd edn, Springer, Berlin
61. Wolf B (2000) Cryst Res Technol 35:377
62. Baltà-Calleja FJ (1985) Adv Pol Sci 66:117
63. Baltà-Calleja FJ (1994) Trends in Pol Sci 2:419
64. Goss CA, Blumfield CJ, Irene EA, Murray RW (1993) Langmuir 9:2986
65. Oliver WC, Pharr GM (2004) J Mater Res 19:3
66. Tranchida D, Piccarolo S, Loos J, Alexeev A (2007) Macromolecules 40:1259
67. Loubet JL, Georges JM, Meille J (1986) Nanoindentation Techniques in Materials Science and Engineering. ASTM, Philadelphia
68. Hochstetter G, Jimenez A, Loubet JL (1999) J Macromol Sc B-Phy 38:681
69. Kick F (1885) Das Gesetz der proportionalen Widerstunde und seine Anwendungen. Felix-Verlag, Leipzig
70. Cheng Y-T, Cheng CM (1998) J Appl Phys 84:1284
71. Lundberg B (1974) Int J Rock Mech Min Sci Geomech Abstr 11:209
72. Mata M, Alcalà JJ (2004) Mech Phys Solids 52:145
73. Malzbender J, de With GJ (2000) Mater Res 15:1209
74. Larsson P-L, Giannakopoulos AE, Söderlund E, Rowcliffe DJ, Vestergaard R (1996) Int J Solids Struct 33:221
75. Vaidyanathan R, Dao M, Ravichandran G, Suresh S (2001) Acta Mater 49:3781
76. Attaf MT (2004) Mater Lett 58:3491
77. Shan Z, Sitaraman SK (2003) Thin Solid Films 437:176
78. Fröhlich F, Grau P, Grellmann W (1977) Phys Status Solidi, A Appl Res 42:79
79. Suresh S, Nieh T-G, Choi BW (1999) Scr Mater 41:951
80. Hainsworth SV, Chandlera HW, Page TF (1996) J Mater Res 11:1987
81. Kim H, Kim TJ, (2002) Eur Ceram Soc 22:1437
82. Zeng K, Chiu C-H (2001) Acta Mater 49:3539
83. Andrews EW, Giannakopoulos AE, Plisson E, Suresh S (2002) Int J Solids Struct 39:281
84. Bernhardt EO (1941) Z Metöd 33:135
85. Li H, Bradt RC (1993) J Mater Sci 28:917
86. Bückle H (1959) Metall Rev 4;49
87. Bückle H (1965) Mikrohärterprüfung. Berliner Union Verlag, Stuttgart
88. Quinn JB, Quinn GD (1997) J Mater Sci 32:4331
89. Hays C, Kendall EG (1973) Metallurgica 6:275
90. Ghosh S, Das S, Bandyopadhyay TK, Bandhopadhyay PP, Chattopadhyay AB (2003) J Mater Sci 38:1565
91. Meyer E (1908) V.D.I. Zeitshrift 52:645
92. Dao M, Challacoop N, Van Vliet KJ, Venkatesh TA, Suresh S (2001) Acta Mater 49:3899
93. Rother B, Steiner A, Dietrich DA, Jehn HA, Haupt J, Giessler W (1998) J Mater Res 13:2071
94. Herrmann K, Hasche K, Pohlenz F, Seemann R (2001) Meas Sci Techn 29:201
95. Cheng Y-T, Cheng CM (1998) Appl Phys Lett 73:614
96. DeRose JA, Revel JP (1997) Micr Microan 3:203
97. Atamny F, Baiker A (1995) Surf Sci 323:L314
98. Ramirez-Aguilar KA, Rowlen KL (1998) Langmuir 14:2562
99. Cappella B, Kaliappan SK, Sturm H (2005) Macromolecules 38:1874
100. Segedin CM (1957) Mathematika 4:156
101. Lee EH, Radok JRM (1960) J Appl Mech 82:438
102. Hilton HH, Yi S (1998) Int J Solids and Struct 35:3081
103. Hilton HH (1996) Mech of Comp Mater Struct 3:97
104. Sun YJ, Akhremitchev B, Walker GC (2004) Langmuir 20:5837

105. Swedlow JL (1975) Int J Fract Mech 1:210
106. Bobji MS, Biswas SK (1998) J Mater Res 13:3227
107. Myshkin NK, Petrokovets MI, Chizhik SA (1999) Tribol Int 32:379
108. Bowden FP, Tabor D (1950) Friction and Lubrication of Solids. Clarendon Press, Oxford
109. Archard JF (1957) Proc R Soc Lond A 243:190
110. Greenwood JA, Williamson JBP (1966) Proc R Soc Lond A 295:300
111. McCool JI (1986) Wear 107:37
112. Carbone G, Mangialardi L (2004) J Mech Phys Sol 52:1267
113. Buzio R, Boragno C, Valbusa U (2003) Wear 254:917
114. Myshkin NK, Petrokovets MI, Chizhik SA (1998) Tribol Int 31:79
115. Brucato V, Piccarolo S, La Carrubba V (2002) Chem Eng Sci 57:4129
116. Piccarolo S (1992) J Macromol Sci – Phys B 31:501
117. Welsh GE, Blundell DJ, Windle AH (2000) J Mater Sci 35:5225
118. Baltà-Calleja FJ, Garcia Gutierrez MC, Rueda DR, Piccarolo S (2000) Polymer 41:4143
119. Flores A, Baltà-Calleja FJ, Asano T (2001) J Appl Phys 90:6006
120. Welsh GE, Windle AH (2001) Polymer 42:5727
121. Shen YL, Guo YL (2001) Modell Simul Mater Sci Eng 9:391
122. Piccarolo S. (2006) Polymer 47:5610

8 Thermal Activation Effects in Dynamic Force Spectroscopy and Atomic Friction

Mykhaylo Evstigneev

Abstract. Two experimental applications of an atomic force microscope (AFM) are considered: dynamic force spectroscopy and atomic friction. The former is aimed at determination of bond properties of biological complexes by means of subjecting them to a steadily increasing pulling force until the bonds break. On the other hand, in atomic friction experiments, one investigates the friction forces acting on the AFM tip brought into contact with the surface and pulled with respect to it; usually, the tip's motion proceeds via abrupt jumps from one lattice site of the surface to the next. Both forced rupture of chemical bonds and interstitial jumps are thermally activated events and are described within the same mathematical framework offered by Kramers' rate theory. Characterization of the force-dependent rate of bond rupture/interstitial jumps provides one with a valuable insight into the relevant energy scales of the system studied. The standard approach to data analysis is based on the single-step rate equation, from which the logarithmic relation between the pulling velocity and the most probable force of bond rupture/interstitial jump follows. An alternative method of analyzing the experimental data is discussed, which allows one to test the applicability of the single-step rate equation in a given experimental system, and to accurately deduce the transition rate from the experimental data. Application of this method to both dynamic force spectroscopy and atomic friction experiments indicated that, generically, the single-step rate equation cannot explain the experimentally observed statistics of rupture/jump events. In the former case, the discrepancy between theory and experiments is explained quantitatively in terms of heterogeneity of chemical bonds involved, while in the latter case, the discrepancy is attributed to the ageing of the contact between the AFM tip and the surface.

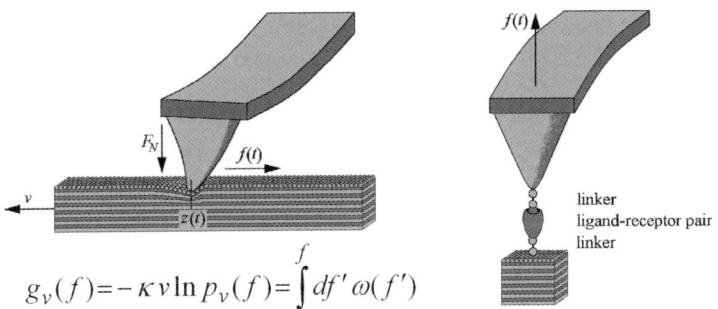

$$g_v(f) = -\kappa v \ln p_v(f) = \int^f df' \, \omega(f')$$

Key words: Atomic force microscopy, Dynamic force spectroscopy, Atomic friction, Thermal activation, Kramers' rate

8.1
Introduction

The invention [1] and continuous further development [2, 3] of atomic force microscopy (AFM) has led to a remarkable advancement of our abilities to experimentally investigate the properties of nanosized objects. In this chapter, two such research applications of AFM will be considered: one is aimed at determination of bond properties of biological complexes, and the other at the studies of microscopic mechanisms of friction.

Within the first research direction, dynamic force spectroscopy (DFS), a biological complex of interest is subjected to a constantly increasing force pulling it apart until the complex dissociates. On the basis of the observed distribution of rupture forces, one tries to gain insight into the properties of the chemical bonds involved. The interpretation of the experimental data represents a theoretical challenge of utmost importance not only for our understanding of the basic (bio)physical principles at work, but also for the purpose of controlling and manipulating the stability and decay of specific chemical bonds in biotechnological contexts.

With respect to the other research field, atomic friction, employing an AFM offers a unique opportunity to probe the frictional forces between a single asperity – the tip of an AFM cantilever – and an atomically flat surface. In this way, one can better understand the basic mechanisms of macroscopic friction, which involve interaction between multiple asperities of the two contacting surfaces. Experimentally, the AFM tip contacting the surface is set in motion with respect to it, and the lateral force due to the torsional deformation of the cantilever is measured. Typically, the tip performs the so-called stick-slip motion, consisting of sequential singular jumps between the adjacent surface sites.

Direct molecular dynamics simulations of both experimental situations are still very far from reaching experimentally realistic conditions [4–8]. The reason is the enormous time-scale separation between the microscopic atomic degrees of freedom and the AFM tip as a whole. In molecular dynamics simulations, one can study the evolution of the system during temporal intervals in the nanosecond range, whereas the typical experimental time-scale is about five orders of magnitude slower. This huge gap still cannot be bridged by today's computer facilities. Hence, nontrivial theoretical modeling steps, in particular, the concepts of nonlinear stochastic processes [9–13], are indispensable.

An exceptionally simple and successful approximate description of both experiments is offered by the rate theory. Within this approach, one employs the time-scale separation mentioned above to greatly simplify the analysis of the system. Namely, instead of working with a large number of equations of motion for all atoms involved, one deals with the probability to find the complex in the bound state (in the DFS pulling experiments), or to find the cantilever tip within the same lattice site on the surface (in nanofriction measurements). Both sets of experiments are then described within the same theoretical framework, as the temporal evolution of this occupation probability in both cases is governed by the same rate equation introduced in Sect. 8.2.

This chapter is organized as follows. In the next two sections, typical DFS and nanofriction experimental set-ups are described, and the standard approach to data analysis based on the single-step rate equation is discussed. Then, an alternative approach to the analysis of experimental data is introduced and applied to both experimental situations. It is shown that it is not only the rate equation that can provide one with information about the system at hand, but also that deviations of the experimental results from the predictions of the rate equation can give valuable insight into the investigated system.

8.1.1
Determination of Bond Strength of Biological Complexes

The specific binding of a ligand molecule to a receptor protein is an essential functional principle of molecular recognition processes and many biotechnological applications. The basic principle of dynamic AFM force spectroscopy is schematically sketched in Fig. 8.1. A single chemical bond of interest, for example in a ligand–receptor complex, is attached via two linker molecules to the tip of an AFM cantilever and a piezoelectric element. The latter is employed for pulling the attached linker molecule down at some constant velocity, which in turn increases the elastic reaction force of the cantilever. The magnitude of this force is determined by measuring the deformation of the cantilever with the help of a laser beam. The main quantity of interest is the value of the pulling elastic force at the moment when the bond breaks.

Important examples of such AFM-spectroscopic investigations of molecular compounds include the receptors avidin or streptavidin and the ligands biotin or biotin analoga [14–18], antigen–antibody complexes [19–22], separation of DNA strands [23], RNA dissociation [24–26], neural cell adhesion [27,28], protein unfolding [29], to name but a few.

Theoretical interpretation of the experimental data is a nontrivial task: upon repeating the same experiment with the same pulling velocity, the rupture forces are found to be distributed over a wide range, see Fig. 8.2. Furthermore, for different pulling velocities different such distributions are obtained. Hence, neither a single rupture event nor the average rupture force at any fixed pulling velocity can serve as a meaningful characteristic of a given chemical bond strength.

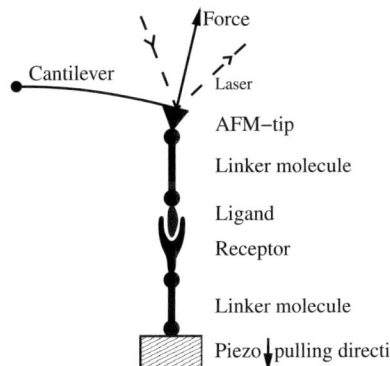

Fig. 8.1. Schematic representation of the experimental set-up in a typical pulling dynamic force spectroscopy experiment

Fig. 8.2.
Force-extension curves
for four different pulling
trials, two performed
with the pulling velocity
$v = 100\,\mathrm{nm/s}$ (*solid
lines*), and two at
$v = 5{,}000\,\mathrm{nm/s}$
(*dashed lines*). The bond
studied is formed by
expE1/E5 section of a
DNA and the regulatory
protein ExpG [30]

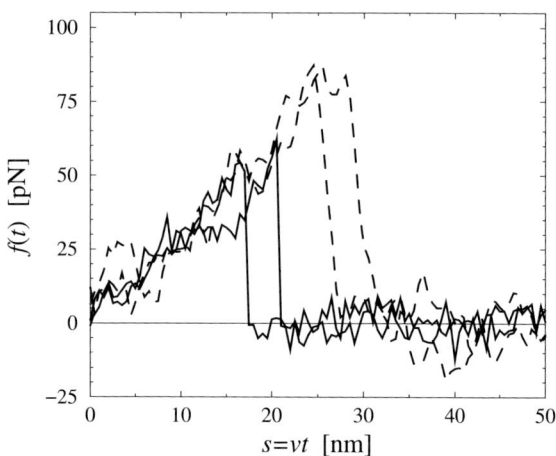

The main breakthrough in solving the puzzle came with the hallmark 1997 paper by Evans and Ritchie [31], whose precursor is due to Bell [32]. While Evans and Ritchie's original theoretical approach has been extended and refined in several important directions [31, 33–45], the essential physical picture has remained unchanged and has been the basis for evaluating the observed rupture data of all experimental investigations ever since [39, 46]. Namely, during each pulling trial, the elastic force developed within the cantilever increases approximately linearly, see Fig. 8.2. Superimposed on this linear increase are random fluctuations of the force due to thermal and instrumental noise. As the cantilever gets further away from the surface, the energy barrier separating the system from the dissociated state constantly reduces until the biological complex breaks due to thermal activation over this barrier. In other words, a forced bond rupture event is a thermally activated decay of a metastable state and is described within the general framework of Kramers' reaction rate theory [12, 47–49].

8.1.2
Atomic Friction

The usual macroscopic friction involves interaction of many asperities on two contacting surfaces and is of paramount practical importance and of notorious difficulty regarding its theoretical understanding [50–53]. In friction force microscopy (FFM) experiments, one investigates the friction forces acting on a single asperity – the tip of an AFM. Such studies of frictional forces between nanoscale objects are vital both for engineering of micromechanical devices and advancement of our understanding of the laws of nature acting in the nanoworld. Therefore, the research direction of friction force microscopy [54] had been initiated only a year after the invention of the AFM in 1986 [1] and became a subject of intensive studies since then (see the reviews [51, 55–58] and references therein).

In a typical FFM experiment [54], the tip of an AFM cantilever is brought in contact with an atomically clean surface by means of a normal load F_N, and the

Fig. 8.3. Schematic illustration of a nanofriction experiment. *Inset*: evolution of the lateral force $f(t)$

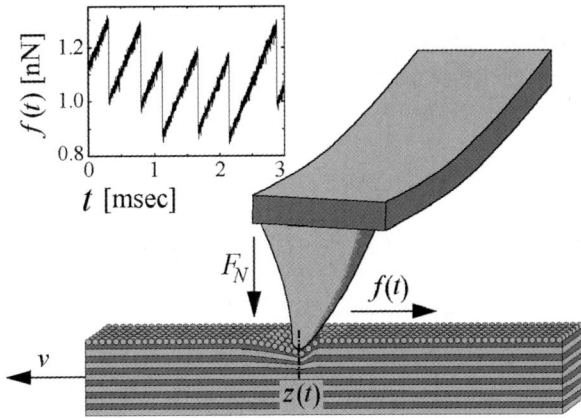

cantilever base is set in motion with respect to the surface (or vice versa) at a constant velocity v (see Fig. 8.3). The interaction between the tip and the surface leads to a torsional deformation of the cantilever. One can determine the magnitude of this deformation by optical means and thus deduce the resulting elastic force $f(t)$, which, by Newton's third law, equals the instantaneous force of friction. As a rule, the temporal evolution of the friction force proceeds in a sawtooth-like pattern [see the inset in Fig. 8.3 showing the results of numerical simulations; the experimentally observed force evolution is similar, see e.g. Fig. 8.6(b)]. This type of motion of the AFM cantilever is called stick-slip motion. It consists of the stick phases, in which the cantilever tip finds itself confined within one of the lattice sites of the surface and which are characterized by an approximately linear increase of the elastic force due to the accumulating deformation of the cantilever, and almost instantaneous slips into the next lattice site accompanied by sudden force drops. The central quantity of interest is the behavior of the time averaged friction force as a function of the pulling velocity, v. From the dependence of the friction force on v one tries to gain insight into the specific molecular properties of the probed surface and into the general microscopic principles of surface friction.

The laws of nanofriction differ drastically from those of macroscopic friction. In particular, it has been known from the time of Coulomb that the force of friction between two macroscopic bodies in contact is independent of their relative velocity. In contrast, friction force on the nanoscale exhibits nontrivial velocity dependence: in the stick-slip regime, it increases approximately logarithmically with v.

Though simpler than macroscopic friction, the adequate interpretation and modeling of nanofriction experiments still represents a formidable challenge. The above-mentioned time-scale separation between the atomic and cantilever degrees of freedom greatly facilitates the calculations within various nanofriction models, where the behavior of an AFM tip in contact with the surface is viewed as Brownian motion in a potential of the tip–surface interaction and of the elastic forces resulting from the deformations of the cantilever, the tip, and the surface in the contact region. The stick-slip motion of the cantilever tip is then pictured as a sequence of thermally activated interstitial transitions, which can be described, as in the case of pulling experiments, within the framework of Kramers' theory of thermally activated escape. The experimental evidence supporting this view is the observed fluctuations of the lateral force in each stick phase [59, 60], randomness of the slip events, the temperature dependence of the friction force [61,62], and the logarithmic force-velocity relation [63], which follows naturally from the thermally activated model of atomic friction.

8.2
Standard Method of Data Analysis

8.2.1
Rate Equation Approach

The systems depicted in Figs. 8.1 and 8.3, although small, possess a large number of microscopic degrees of freedom. Hence, their modeling by means of molecular dynamics simulations is feasible only for limited time intervals, typically in the range of tens of nanoseconds, whereas the characteristic experimental time-scale is of the order of milliseconds. A simple and efficient alternative approach is offered by the rate theory. It is valid when the system has several long-lived metastable states, whose lifetime is much longer than the characteristic time-scale of all other microscopic degrees of freedom. This time-scale separation condition allows one to use a simple yet very efficient description in terms of occupation probabilities of the metastable states and transition rates between them.

Referring to the system in Fig. 8.1, one can identify such a metastable state with the bound state of the complex. On the other hand, for atomic friction experiments, Fig. 8.3, the (equivalent) metastable states are associated with the location of the cantilever within specific lattice sites.

The main ingredients of the rate description are:

- the probability $P_v(t|t_0)$ to find the system in the metastable state at a moment of time t for pulling velocity v, provided that the system was in that state at the moment of time t_0, i.e. $P_v(t_0|t_0) = 1$;
- the off-rate $\omega[f(t)]$ to leave the metastable state under the action of the pulling force $f(t)$ due to the elastic deformation of the cantilever and, in the case of the atomic friction, Fig. 8.3, of the substrate within the deformed contact region.

This elastic force reduces the potential barrier, which needs to be surmounted in order for the cantilever to leave its metastable state. In the experiment, it increases approximately linearly in time as

$$f(t) = \kappa \, v(t - t_0) + f_0, \tag{8.1}$$

where κ is the elastic constant describing the cumulative effect of all system components in Figs. 8.1 and 8.3, and f_0 is the force value at $t = t_0$. Superimposed on this linear increase are random fluctuations due to the thermal and instrumental noise, see Fig. 8.2 and the inset in Fig. 8.3.

A rupture/slip event is viewed as a thermally activated transition out of the metastable state across the potential barrier. It is governed by the single-step reaction equation of the form

$$\dot{P}_v(t|t_0) = -\omega[f(t)]P_v(t|t_0), \tag{8.2}$$

where the overdot denotes time derivative. The only assumption implicit in this equation is that intramolecular thermal relaxation processes are much faster than the instantaneous lifetime $1/\omega[f(t)]$ of the metastable state and the time-scale on which the applied force $f(t)$ notably changes. For the experimentally feasible pulling speeds, this condition, and hence Eq. (8.2), is very well satisfied in most cases.

Under the very same conditions, reaction rate theory [12, 47–49] predicts an Arrhenius law for the off-rate of the form

$$\omega(f) = \Omega \exp\left(-\frac{\Delta E(f)}{k_B T}\right), \tag{8.3}$$

where the potential barrier against decay, $\Delta E(f)$, is large compared to the thermal energy $k_B T$, and the prefactor Ω representing the intramolecular attempt frequency depends subexponentially weakly on the force and is usually assumed to be constant; see, however, the works [64,65], where the force dependence of Ω is explicitly taken into account.

The concrete functional form of the force-dependent barrier height $\Delta E(f)$ can be established from the knowledge of the system's energy $V(x;f)$ as a function of the applied force and the reaction coordinate, x, see Fig. 8.4. The latter can be identified with the relative separation of the ligand-receptor pair in the case of the pulling experiments from Fig. 8.1. On the other hand, in the atomic friction measurements, Fig. 8.3, it can be associated with the location of the cantilever tip on the surface. If the cantilever spring is sufficiently soft, the energy can be decomposed as $V(x;f) = U(x) - fx$, where, $U(x)$ represents the energy of the ligand-receptor pair or the tip–surface interaction energy, and $-fx$ is the elastic energy; see [66,67] for a more precise evaluation of this function.

The barrier height as a function of the pulling force can be written as

$$\Delta E(f) = V(x_{\max}(f);f) - V(x_{\min}(f);f), \tag{8.4}$$

where $x_{\min,\max}(f)$ denote the force-dependent positions of the extrema of the potential $V(x;f)$. It decreases monotonically with its argument from the initial value $\Delta E_0 \equiv \Delta E(0)$ and assumes the value zero at some critical force f_c, at which the

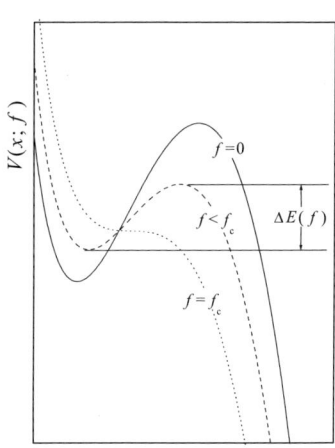

Fig. 8.4. A representative potential landscape $V(x; f)$ vs. the reaction coordinate x at different pulling forces f. *Solid line* depicts $V(x; f = 0)$ (no pulling force), *dashed line* corresponds to the subcritical pulling force $f < f_c$, and the *dotted line* to the critically tilted potential $f = f_c$

barrier disappears. For a large class of potentials $U(x)$, the force-dependent barrier height can be approximated with a power law

$$\Delta E(f) = \Delta E_0 \left(1 - \frac{f}{f_c}\right)^\gamma, \tag{8.5}$$

where we have introduced the exponent γ controlling the manner in which it decreases from ΔE_0 to zero. The Ansatz (8.5) can be regarded as the lowest-order term in the self-similar factor approximation [68, 69] to the true force-dependent barrier height.

Equations (8.3) and (8.5) imply that the off-rate

$$\omega(f) = \Omega \exp\left[-\frac{\Delta E_0}{k_B T}\left(1 - \frac{f}{f_c}\right)^\gamma\right] \tag{8.6}$$

is an increasing function of the applied force. Referring to the pulling experiments from Fig. 8.1, bonds whose lifetime ω^{-1} gets smaller with the pulling force, are called the slip bonds. They will be of primary interest within this review. We note that on the other side of the spectrum are the so-called catch bonds, whose lifetime increases with the applied force. Their existence was predicted by Dembo et al. in 1988 [70], and they were first detected with the help of an AFM some 15-years later [71].

As long as the energy barrier separating the unbound and bound states is sufficiently sharp and deep and/or the applied force is sufficiently weak, one can approximately linearize $\Delta E(f)$, resulting in the approximation for the rate

$$\omega^{\text{Bell}}(f) = \omega(0) \exp(\alpha f). \tag{8.7}$$

We will refer to this particular choice of the off-rate as Bell's Ansatz [32]. Here, $\omega(0)$ is the dissociation/jump rate in the absence of the applied force.

On the basis of the statistic of jump events, it is possible to reconstruct the force-dependent off-rate; the corresponding procedure is described in Sect. 8.3. Although unique reconstruction of the energy landscape $U(x)$ [or $V(x;f)$] from $\omega(f)$ is not possible [5], knowledge of this function can provide one with the information about the main features of the energy landscape. In particular, referring to the Bell's approximation, Eq. (8.7), the coefficient α multiplied by $k_B T$ can be identified with the spatial distance between potential minimum and maximum projected along the direction of the pulling force. In the more general expression (8.6), the exponent γ characterizes the geometry of the potential. For instance, if one approximates the function $U(x)$ with a cubic polynomial, then this exponent assumes the value 3/2 [72–75]; approximation of $U(x)$ with a parabolic function having a cusp at the transition point, or with two parabolas near the potential extrema, yields $\gamma = 2$ [76].

8.2.2
Most Probable Rupture Force

The currently predominant method for estimating the rate parameters from the experimentally observed rupture forces is based on the theoretical insight described above in conjunction with the following line of reasoning.

By making the change of variables according to Eq. (8.1) in the rate Eq. (8.2), one transforms it from the time domain into the force domain, where the derivative of the probability $P_v(f|f_0)$ to find the system in the same metastable state at the force value f for a given pulling velocity v is

$$P'_v(f|f_0) = -\frac{1}{\kappa v}\omega(f)P_v(f|f_0),$$ (8.8)

where the second argument indicates the initial condition $P_v(f_0|f_0) = 1$. The solution of this equation is written as

$$P_v(f|f_0) = \exp\left(-\frac{1}{\kappa v}\int_{f_0}^{f} df'\,\omega(f')\right),$$ (8.9)

as can be verified by differentiation. For the Bell's Ansatz for the rate, this expression becomes

$$P_v^{\text{Bell}}(f|f_0) = \exp\left(-\frac{\omega(0)}{\alpha\kappa v}\left(e^{\alpha f} - e^{\alpha f_0}\right)\right).$$ (8.10)

The derivative of the survival probability taken with the minus sign, $-P'_v(f|f_0)$, represents the probability density of escape events at the force value f. Correspondingly, the most probable escape force, f_m, is the one which maximizes this quantity. Differentiating Eq. (8.8) with respect to the force and setting the derivative to zero, we find that the most probable rupture force is given by the equation

$$v = \frac{\omega^2(f_m)}{\kappa\omega'(f_m)},$$ (8.11)

which has to be solved with respect to f_m. Taking into account Eqs. (8.8) and (8.9), one can show that if the pulling velocity is smaller than some critical value, $v < v_c$, the probability of rupture/jump, $-P'_v(f|f_0)$, is a monotonically decreasing function of the acting force at $f > f_0$, because the solution of Eq. (8.11) is smaller than the initial force, $f_m < f_0$. In contrast, at fast pulling, $v > v_c$, the maximum force exceeds the initial force, $f_m > f_0$. To find the critical velocity, we note that at $v = v_c$ the initial force and the most probable force coincide. Therefore, the expression for v_c is

$$v_c = \frac{\omega^2(f_0)}{\kappa\omega'(f_0)}.$$ (8.12)

In particular, for the Bell's Ansatz we find $v_c^{\text{Bell}} = \omega(0)\exp(\alpha f_0)/(\alpha\kappa)$.

At $v > v_c$, the probability of a jump becomes a nonmonotonic function of the acting force f with a unique maximum determined by Eq. (8.11). The solution of Eq. (8.11) in the general case, Eq. (8.6), reads:

$$\frac{f_m}{f_c} = 1 - \left\{\frac{\gamma-1}{\gamma}\frac{k_B T}{\Delta E_0} W\left(\frac{\gamma}{\gamma-1}\left(\frac{\Omega f_c}{\gamma\kappa v}\right)^{\frac{\gamma}{\gamma-1}}\left(\frac{k_B T}{\Delta E_0}\right)^{\frac{1}{\gamma-1}}\right)\right\}^{1/\gamma},$$ (8.13)

where $W(x)$ is the Lambert function defined implicitly by

$$W(x)e^{W(x)} = x.\tag{8.14}$$

This function can be evaluated numerically using a very efficient scheme, for example, from [77].

For the Bell's approximation for the rate, the expression for the most probable escape force simplifies considerably to

$$f_m^{Bell} = \frac{1}{\alpha} \ln \frac{\alpha \kappa v}{\omega(0)}.\tag{8.15}$$

One can see that the parameters $\omega(0)$ and α can be determined by conducting experiments with several different pulling rates $v > v_c$. For each of them, the most probable rupture force f_m can be estimated, and the resulting f_m values are plotted versus $\ln v$. Finally, a linear fit to this graph yields an estimate for $\omega(0)$ and α. This procedure is henceforth referred to as the standard method of data analysis.

Apart from the fact that the Bell's Ansatz, Eq. (8.7), contains only two parameters – $\omega(0)$ and α – whereas the more general formula (8.6) has four parameters – Ω, ΔE_0, f_c, and γ – there is a more important reason for the fact that the Bell's approximation has been much more popular in interpreting experimental results than the more general expression. Namely, it is usually the case that the deviations of the experimentally observed force spectrum, that is, the plot f_m vs. $\ln v$, from the straight line predicted by the Bell's relation (8.15) are within the error bars of the experiment. This means that the exponent γ, which controls the curvature of the force spectrum, usually cannot be determined sufficiently accurately from the experimental data. Indeed, the fits to the experimental nanofriction data using the values $\gamma = 3/2$ and $\gamma = 1$ [61] were found to agree with the experimental data almost equally well. Moreover, it has recently been demonstrated [78] that accurate determination of the exponent γ is problematic even when one fits the distribution of jump/rupture forces, rather than the most probable force vs. velocity.

The inability to accurately distinguish the experimental force spectra with the exponent $\gamma \neq 1$ from the linear force spectra with $\gamma = 1$ implies that the force-free activation energy ΔE_0 and the prefactor Ω cannot be determined simultaneously. This is so because in the linear case, $\gamma = 1$ in Eq. (8.6), they are coupled in such a way that multiplication of the parameter Ω by an arbitrary constant and simultaneous addition of $k_B T$ times the logarithm of that constant to the activation energy ΔE_0 will result in the same rate $\omega(f)$. Therefore, it is usually the case that, based on the experimental force spectra, only Bell's parameters, α and $\omega(0) \equiv \Omega e^{-\Delta E_0/k_B T}$, can be determined with sufficient reliability.

The theory presented above is based on the one-step rate equation (8.2) and is the simplest method allowing one to make quantitative predictions about the system studied. Referring to the DFS measurements, important extensions of the single-step approach have been made to include the possibility of intermediate states in the bond dissociation path [31, 35, 36, 43], to account for a possible coexistence of several parallel bond rupture channels [40, 79], and the nonzero rebinding probability [38, 80]. Last but not least, multiple-state models were introduced to describe the catch-bond behavior [42, 44, 45].

8.2.3
DFS Pulling Experiments: Mean Rupture Force

As mentioned above, the method of determination of the rate parameters based on the most probable rupture force applies for velocities greater than the value given by Eq. (8.12). As an alternative to the most probable force at the moment of bond rupture, one may use the average force

$$\langle f \rangle = \int_{f_0}^{f_c} df\, f \left(-\frac{\partial P_v(f|f_0)}{\partial f} \right) = f_0 + \int_{f_0}^{f_c} df\, P_v(f|f_0), \qquad (8.16)$$

where integration by parts was employed to obtain the second part of this equality, and it is assumed that pulling proceeds sufficiently slowly, so that the system leaves the metastable state with overwhelming probability well before the critical force is reached, i.e. $P_v(f_c|f_0)$, is practically zero. If this is not the case, then the rate description is no longer valid.

For the survival probability corresponding to the Bell's Ansatz, Eq. (8.10), the average force for pulling experiments is given by [76, 81]

$$\langle f \rangle^{\text{Bell}} = f_0 + \frac{1}{\alpha} e^{\frac{\omega(0)\exp(\alpha f_0)}{\alpha \kappa v}} E_1 \left(\frac{\omega(0) e^{\alpha f_0}}{\alpha \kappa v} \right), \qquad (8.17)$$

where $E_1(u) := \int_1^{\infty} dx\, e^{-ux}/x$ is the exponential integral, which can be evaluated numerically using a standard algorithm [82]. In obtaining this relation, we have extended the upper limit of integration in Eq. (8.16) to infinity, which is admissible, as long as the survival probability drops to zero at forces much smaller in comparison to f_c. For small values of the argument (large loading rates), one can use the asymptotic properties of the exponential integral to simplify the expression (8.17) to

$$\langle f \rangle^{\text{Bell}} \cong f_0 + \frac{1}{\alpha} e^{\frac{\omega(0)\exp(\alpha f_0)}{\alpha \kappa v}} \left(\ln \frac{\alpha \kappa v}{\omega(0) e^{\alpha f_0}} - \gamma_{\text{EM}} \right), \qquad (8.18)$$

where $\gamma_{\text{EM}} = 0.5772156649\ldots$ is the Euler-Mascheroni constant [83].

For a more general rate expression (8.6), the most probable rupture force has been found approximately by Garg [64]:

$$1 - \frac{\langle f \rangle}{f_c} \cong \left(\frac{k_B T}{\Delta E_0} \ln X \right)^{1/\gamma} \left[1 + \frac{1}{\gamma \ln X} \left(\frac{1-\gamma}{\gamma} \ln \ln X + \gamma_{\text{EM}} \right) \right], \qquad (8.19)$$

where $X = \frac{f_c \Omega}{\kappa v \gamma} \left(\frac{k_B T}{\Delta E_0} \right)^{1/\gamma}$ and we have set the initial force f_0 to zero.

As in the case of the most probable force, f_m, the mean force $\langle f \rangle$ increases approximately logarithmically with velocity and can be used to establish the rate parameters. It has been argued [83] that, in comparison to the approach based on f_m, the reliability of Eqs. (8.18) and (8.19) as tools to process the experimental data is somewhat lower, because, apart from the thermal activation, other sources of randomness are operative in a real experimental situation. It is reasonable to expect that these additional sources of randomness (some of which are discussed in Sects. 8.4 and 8.5 of this chapter) affect the average force $\langle f \rangle$ to a greater degree than the most probable force f_m. On the other hand, they will also cause the observed distribution of rupture/jump forces to deviate from Eqs. (8.8) and (8.9) (see Sect.8. 4). Therefore, determination of the position of force maximum, f_m, usually involves fitting the experimental force distribution with a Gaussian curve, rather than with the expression following from Eqs. (8.8) and (8.9), and this may result in additional errors [84].

8.2.4
Atomic Friction: Average Force in the Stick-Slip Regime

While Eq. (8.11) involves the most probable slip force, in friction force spectroscopy experiments, see Fig. 8.3, it is more customary to measure the mean force of friction defined as the long-time limit

$$\bar{f} = \lim_{t \to \infty} \frac{1}{t} \int_0^t dt' f(t'). \tag{8.20}$$

In practice, this force is determined from the area of the experimental stick-slip curve (see Fig. 8.3, inset). To deduce the average force in the stick-slip regime of cantilever motion, one cannot directly transfer the results given by Eqs. (8.18) and (8.19) from the pulling experiments, because these results imply a unique value of the initial force f_0. In contrast, during the stick-slip motion, each stick phase begins with a different initial force. Hence, a more elaborate analysis is required.

For further convenience, we introduce the long-time limit of the probability distribution $W_v(f_0)$ that a given stick phase of motion begins with the force value in a small interval around f_0 for a given pulling velocity v. It obeys the following integral equation:

$$W_v(f_0) = \int_{-\infty}^{f_0 + \kappa a} df' \left(-\frac{\partial P_v(f_0 + \kappa a | f')}{\partial f_0} \right) W_v(f'). \tag{8.21}$$

The physical meaning of this equation is as follows. During each slip event, the cantilever tip gets displaced by one lattice constant a of the surface. Therefore, the elastic force drops by the same amount κa, and the probability $W_v(f_0)$ that a given stick phase starts around the force value f_0 equals the probability $-\partial P_v(f_0 + \kappa a | f')/\partial f_0$ that the previous stick phase ends at a value $f_0 + \kappa a$ averaged over all initial forces $f' < f_0 + \kappa a$ of the previous phase with the weighting function $W_v(f')$.

If we find the distribution of lower forces by solving Eq. (8.21), the time-averaged force follows immediately [85]:

$$\bar{f} = \int_{-\infty}^{\infty} df_0 \, f_0 W_v(f_0) + \frac{\kappa a}{2}. \tag{8.22}$$

This relation results from the linear character of force increase in each stick phase, and from the fact that the magnitude of force jump in each slip event is the same and equals κa. While the relation (8.22) between the time-averaged force and the first moment of the distribution $W_v(f_0)$ is simple, the iterative determination of the latter distribution according to Eq. (8.21) is only possible by means of a time-consuming numerical procedure. Therefore, our next goal is to obtain an approximate analytic relation between the average force $\bar{f}$ and velocity v without the knowledge of the distribution $W_v(f_0)$.

Multiplication of both sides of Eq. (8.21) by f_0 and integration yield

$$\kappa a = \int_{-\infty}^{\infty} df_0 \int_{-\infty}^{f_0} df' P_v(f_0|f')W_v(f') = \int_{-\infty}^{\infty} df_0 W_v(f_0) \int_{f_0}^{\infty} df' P_v(f'|f_0), \qquad (8.23)$$

where we used integration by parts with subsequent change of the variables of integration from f_0 to $f_0 + \kappa a$ to obtain the first equality, and interchanged the order of integration with subsequent interchange of the variables $f' \leftrightarrow f_0$ to obtain the second one.

The inner integral in the rightmost part of Eq. (8.23) represents the average force increment $\Delta F(f_0)$ during a given stick phase, provided that the initial force value is f_0, cf. Eq. (8.22). The double integral, therefore, is the force increment during the stick phase averaged over all initial forces. After the decay of transient processes, this quantity equals the force drop κa during the slip into the next potential well.

Next, we note that the behavior of the function

$$\Delta F(f_0) = \int_{f_0}^{\infty} df' P_v(f'|f_0) \qquad (8.24)$$

does not deviate strongly from linearity in that force interval around $\bar{f}_0$, where the distribution $W_v(f_0)$ is significantly different from zero. To validate this statement, let us consider two cases of high and low κ.

(1) At high spring constants, the probability of staying within the same well, Eq. (8.9), is close to one in a rather extended force interval above f_0 [because of a large factor κ in the denominator of the expression in the exponent in Eq. (8.9)]. This means that the statistics of jump events, and hence the average force at the moment of transition, is practically independent of the initial force. Correspondingly, the average force increment for a fixed initial force f_0 indeed behaves linearly with f_0, i.e., as $\Delta F(f_0) = \bar{f}_{\text{slip}} - f_0$ at high κ, where the average upper force at slip $\bar{f}_{\text{slip}}$ is practically independent of f_0.

(2) At low values of κ the magnitude of force fluctuations, which is of the order of κa, and hence the width of the distribution $W(f_0)$ is small. Then, the deviations of the function $\Delta F(f_0)$ from linearity can be neglected within the relevant force interval.

Since the function $\Delta F(f_0)$ is approximately linear in the physically important range of f_0 both at low and high κ, it can reasonably be expected that it is also almost linear in this range for intermediate stiffness. This allows us to replace the average value of this function in the inner integral of the rightmost expression (8.23) with the function evaluated at the average value of its argument, $\Delta F(\langle f_0 \rangle)$, resulting in the following implicit force-velocity relation [66, 67, 85]:

$$\int_{\bar{f}-\kappa a/2}^{\infty} df \exp\left(-\frac{1}{\kappa v}\int_{\bar{f}-\kappa a/2}^{f} df' \omega(f')\right) = \kappa a. \qquad (8.25)$$

Here, we wrote an explicit expression (8.9) for $P_v(f|\langle f_0 \rangle)$ and used the relation (8.22) between $\langle f_0 \rangle$ and $\bar{f}$.

For an analytic solution of Eq. (8.25), we need to further approximate the transition rate so as to be able to evaluate the integral in the left-hand side. According to Eq. (8.6), the rate depends exponentially on the force. On the other hand, the energy barrier $\Delta E(f)$ is a much weaker function, see Eq. (8.5). This observation suggests expanding the logarithm of the transition rate about some force value f_*, i.e. to take, in the spirit of Bell's Ansatz, Eq. (8.7),

$$\omega(f) \cong \omega(f_*) \exp\left(\alpha(f_*)(f - f_*)\right), \qquad (8.26)$$

$$\alpha(f) := \frac{\omega'(f)}{\omega(f)}. \qquad (8.27)$$

The next question is how to choose the force f_*, about which the expansion is performed. To answer this question, let us examine Eq. (8.25) more closely. Depending on the value of κ, the integrand, $\exp(\dots)$, may exhibit two kinds of behavior:

(1) At high κ, there is a rather wide region of forces between the average lower force, $\langle f_0 \rangle \equiv \bar{f} - \frac{\kappa a}{2}$, and the average upper force, $\langle f_{\text{slip}} \rangle \equiv \langle f_0 \rangle + \kappa a$, where the integrand has the value 1, followed by an abrupt drop to zero in the immediate vicinity of $\langle f_{\text{slip}} \rangle$. The nature of the approximation (8.26) is such that if we choose f_* to be in the region of the steepest descent of the integrand, we will correctly reproduce its behavior not only in this region, but also outside of it, where the integrand is very close to 0 (at higher forces) or 1 (at lower forces).
(2) At low κ, the integrand drops to zero in a rather narrow interval above $\bar{f} - \kappa a/2$. Therefore, we expect that the integral will not be very sensitive to the choice of f_*, provided that f_* belongs to that narrow region where the integrand is notably different from zero. This region extends from $\langle f_0 \rangle$ to a value slightly higher than the average force at the moment of transition, $\langle f_{\text{slip}} \rangle \equiv \langle f_0 \rangle + \kappa a$.

The choice of the expansion point in Eq. (8.26), which applies to both cases equally well, is simply

$$f_* = \langle f_{\text{slip}} \rangle = \bar{f} + \kappa a/2. \qquad (8.28)$$

Making a change of variables according to $x = e^{\alpha(\langle f_{\text{slip}} \rangle)(f - \langle f_{\text{slip}} \rangle)}$, we have from Eqs. (8.25), (8.26), (8.27), and (8.28):

$$\alpha\left(\langle f_{\mathrm{slip}}\rangle\right)\kappa a = e^{u\left(\langle f_{\mathrm{slip}}\rangle\right)}E_1\left(u\left(\langle f_{\mathrm{slip}}\rangle\right)\right),\tag{8.29}$$

$$u(f) := \frac{\omega(f)e^{-\alpha(f)\kappa a}}{\kappa v\alpha(f)}.\tag{8.30}$$

It follows from Eq. (8.29) that the sought relation between force and velocity has the form

$$v(\bar{f}) = a\omega(\bar{f} + \kappa a/2)Q\left(\frac{\omega'(\bar{f} + \kappa a/2)}{\omega(\bar{f} + \kappa a/2)}\kappa a\right),\tag{8.31}$$

where the function $Q(x)$ is defined implicitly by the relation

$$E_1\left(\left[xe^{xQ(x)}\right]^{-1}\right)\exp\left(\left[xe^{xQ(x)}\right]^{-1}\right) = x.\tag{8.32}$$

From the asymptotic properties of the exponential integral [86] it can be inferred that $Q(x)$ is a monotonically decreasing function with $Q(0) = 1$ and $Q(x) \propto e^{\gamma_{\mathrm{EM}}}x$ at $x \to \infty$. This function can be approximated by

$$Q(x) \cong \frac{1}{\sqrt{1 + (e^{\gamma_{\mathrm{EM}}}x)^2}}.\tag{8.33}$$

with a high accuracy [66].

In the analysis above, we have tacitly assumed that the jump probability of the cantilever in the direction against the force is much smaller than in the forward direction, and thus neglected the back-jumps in the rate Eq. (8.2). We note that inclusion of such transitions into consideration leads to the phenomenon of thermolubricity [87], i.e. vanishing friction at low velocities.

8.3
Alternative Method of Data Analysis

The standard approach to data analysis described in Sect. 8.2.2 has several obvious weak points.

First, for typical values of the rupture parameters in Bell's Ansatz, Eq. (8.7), for the rate (see Table 3 in Ref. [46]), $\omega(0) = 1\,\mathrm{s}^{-1}$, $\alpha = 0.1\,\mathrm{pN}^{-1}$, the critical loading rate in Eq. (8.12) for $f_0 = 0$ is $\kappa v_c = 10\,\mathrm{pN/s}$. Hence, out of the experimentally feasible loading rates between 10^{-1} and $10^5\,\mathrm{pN/s}$, the first two decades are useless and the fitting regime is restricted to the remaining three decades. Moreover, this critical velocity is not known a priori. Second, the usually employed Bell's Ansatz, Eq. (8.7), itself is an uncontrolled theoretical approximation. Third, one has the feeling that the method does not really make optimal use of the available data [78]: a reduction of the statistical uncertainty by means of a more sophisticated approach seems possible. Finally, to determine the most probable rupture force, one usually fits the rupture force distribution with a Gaussian, and this procedure causes a systematic overestimation of $\omega(0)$ by about 30% [84].

In the following, we briefly describe an alternative approach to data analysis, which allows one to verify, if the rate equation (8.2) is applicable in a given experimental situation, and, in the cases when its validity is confirmed, to deduce the rate $\omega(f)$ from the experimental data. The precursors of this method can be traced back to the works [25, 26], where the dynamic force spectroscopy of biocomplexes is analyzed, and to even earlier studies of Josephson junctions [88, 89].

It follows from the rate equation (8.8) in the force domain that the escape rate can be determined from the no-jump probability as

$$\omega(f) = -\kappa v \frac{P_v'(f|f_0)}{P_v(f|f_0)} = -\kappa v \frac{d \ln P_v(f|f_0)}{df}. \tag{8.34}$$

Experimentally, given a set of N_v rupture forces $f_1, f_2, \ldots, f_{N_v}$ at a fixed pulling velocity v, the best estimate for $P_v(f|f_0)$ that can be inferred from these data in the absence of any further a priori knowledge is

$$\tilde{P}_v(f|f_0) = \frac{1}{N} \sum_{n=1}^{N_v} \Theta(f - f_n), \tag{8.35}$$

where $\Theta(f) = \begin{cases} 0 \text{ for } f < 0 \\ 1 \text{ for } f \geq 0 \end{cases}$ is the Heaviside step function. Equivalently, one may

estimate $\tilde{P}_v(f|f_0)$ from the numbers $N_v(f)$ of the experimentally detected bond ruptures/slip events that occurred after the force value f:

$$\tilde{P}_v(f|f_0) = \frac{N_v(f)}{N_v(f_0)}. \tag{8.36}$$

Since the expression (8.34) involves differentiation of a piecewise-constant function, $\ln \tilde{P}_v(f|f_0)$, it is more convenient to deal with the quantity [25, 26, 30, 79, 84, 90, 101]:

$$\tilde{g}_v(f|f_0) = -\kappa v \ln \tilde{P}_v(f|f_0). \tag{8.37}$$

Note that the convergence of the experimental estimate (8.37) to the "true" value $g_v(f|f_0) = -\kappa v \ln P_v(f|f_0)$ at $N_v \to \infty$ is not uniform. Rather, for any fixed v, the majority of escape events f_n will sample a limited interval of forces around the most probable escape force. Only within this interval will an experimentally realistic finite number of trials admit a reliable estimate for the true function $g_v(f|f_0)$ in Eq. (8.37).

We now come to the central point of this Section, namely, the observation that, according to Eq. (8.34), the g-function from Eq. (8.37) is in fact independent of the pulling speed v. If the v-independence of the function $\tilde{g}_v(f|f_0)$ is verified in a given experiment, then the single-step rate equation (8.2) indeed holds, and the force-dependent bond rupture/slip rate is given by [cf. Eq. (8.34)]

$$\tilde{\omega}(f) = \frac{d\tilde{g}(f|f_0)}{df}. \tag{8.38}$$

By properly exploiting this universality, it should clearly be possible to reliably estimate $\tilde{g}(f|f_0)$ over a wide f-range by combining data for several different pulling speeds v. The technical details of how to do this in a way that makes optimal use of the information encapsulated in the available experimental data follow.

Consider an arbitrary but fixed $f > 0$. The reliability of the estimate (8.37) is quantified by the variance $\sigma^2[\tilde{g}_v(f|f_0)]$, whose explicit determination will be given shortly. With this amount of information at our disposition, according to the method of weighted averages [91], the best guess for the true $g_v(f|f_0)$ is represented by that argument x which minimizes the weighted sum of square deviations $\sum_v [x - \tilde{g}_v(f|f_0)]^2 /\sigma^2[\tilde{g}_v(f|f_0)]$, where summation is performed over all pulling velocities v. In other words, this best guess for $g_v(f|f_0)$ is given by the weighted average

$$\tilde{g}(f|f_0) = \sum_v c_v(f)\tilde{g}_v(f|f_0) \tag{8.39}$$

$$c_v(f) = \frac{1}{\sigma^2[\tilde{g}_v(f|f_0)]} \bigg/ \sum_{v'} \frac{1}{\sigma^2[\tilde{g}_{v'}(f|f_0)]}. \tag{8.40}$$

In order to determine the variances $\sigma^2[\tilde{g}_v(f|f_0)]$, we consider the number $N_v(f) = P_v(f|f_0)N_v$ of bonds/stick phases surviving up to the pulling force f. It follows that $N_v(f_0) = N_v$ and that for any fixed f, v, and N_v, the number $N_v(f)$ is distributed binomially according to

$$W[N_v(f)] = \frac{[P_v(f|f_0)]^{N_v(f)}[1 - P_v(f|f_0)]^{N_v-N_v(f)}N_v!}{N_v(f)![N_v - N_v(f)]!} \tag{8.41}$$

implying for the associated variance $\sigma^2[N_v(f)]$ the result

$$\sigma^2[N_v(f)] = N_v P_v(f|f_0)[1 - P_v(f|f_0)]. \tag{8.42}$$

An estimate $\tilde{\sigma}^2[N_v(f)]$ for the true $\sigma^2[N_v(f)]$ follows by replacing the true but unknown $P_v(f|f_0)$ in Eq. (8.42) with the approximation $\exp\left(-\frac{1}{\kappa v}\tilde{g}(f|f_0)\right)$. Then, by exploiting the error propagation law $\sigma^2[\tilde{g}_v(f|f_0)] = [d\tilde{g}_v(f|f_0)/dN_v(f)]^2 \sigma^2[N_v(f)]$ and Eq. (8.37), one finds for the coefficients $c_v(f)$ in Eq. (8.40) the result

$$c_v(f) = \frac{N_v \tilde{P}_v^2(f|f_0)e^{\tilde{g}(f|f_0)/(\kappa v)}}{(\kappa v)^2[1 - e^{-\tilde{g}(f|f_0)/(\kappa v)}]}\sigma^2[\tilde{g}(f|f_0)], \tag{8.43}$$

$$\sigma^2[\tilde{g}(f|f_0)] = \left(\sum_v \frac{N_v \tilde{P}_v^2(f|f_0)e^{\tilde{g}(f|f_0)/(\kappa v)}}{(\kappa v)^2[1 - e^{-\tilde{g}(f|f_0)/(\kappa v)}]}\right)^{-1}. \tag{8.44}$$

Finally, by taking into account Eq. (8.40), one readily verifies that $\sigma^2[\tilde{g}(f|f_0)]$ from Eq. (8.44) indeed coincides with the variance $\sum_v c_v^2\sigma^2[\tilde{g}_v(f|f_0)]$ describing the statistical uncertainty of $\tilde{g}(f|f_0)$ in Eq. (8.39).

In practice, this method boils down to the following two steps. First, the functions $\tilde{P}_v(f|f_0)$ and $\tilde{g}_v(f|f_0)$ are determined from the experimentally observed rupture forces f_n, $n = 1, \ldots, N_v$, for different pulling speeds v according to Eqs. (8.35) and (8.37). Second, upon verification that all those functions are sufficiently close to one another to make the conclusion about the applicability of the single-step rate equation in the given experiment, the weighted averages from Eqs. (8.39) and (8.40) are evaluated. Finally, the rate $\omega(f)$ is deduced based on Eq. (8.38).

Since the coefficients $c_v(f)$ in Eq. (8.40) themselves depend on the unknown quantity $\tilde{g}(f|f_0)$ according to Eq. (8.43), we are dealing with a transcendental equation for $\tilde{g}(f|f_0)$ for any fixed f-value. Among many other well-known methods to solve such an equation, one particularly simple way is to iteratively update the value of $\tilde{g}(f|f_0)$ on the basis of Eq. (8.43) until convergence is reached. The result is an estimate $\tilde{g}(f|f_0)$ for the true function $g(f|f_0)$ together with its statistical uncertainty given by Eq. (8.44).

8.4
Application to DFS Pulling Experiments

In Fig. 8.5 the g-function is evaluated for different pulling velocities v according to Eqs. (8.35), (8.36), and (8.37) for the same experimental system as in Fig. 8.2 (namely, rupture data obtained by dynamic AFM force spectroscopy for the DNA fragment *expE1/E5* and the regulatory protein ExpG; we refer to [92–94] for experimental details).

The salient feature of these experimental results is that, in contrast to Eq. (8.34), the functions $\tilde{g}_v(f|f_0)$ with the initial force $f_0 = 20\,\text{pN}$, evaluated from the experimental data using Eq. (8.36) at different values of v, do not collapse onto a single master curve. Rather, increasing the velocity results in an increased value of this function. In view of the very strong dependence of the experimental curves $\tilde{g}_v(f|f_0)$ on the pulling velocities v, we conclude that the experimental findings are incompatible with Eq. (8.34) and hence with the basic assumptions from Eqs. (8.2) and (8.3) of the standard theory [30, 79]. The same conclusion concerns other systems studied

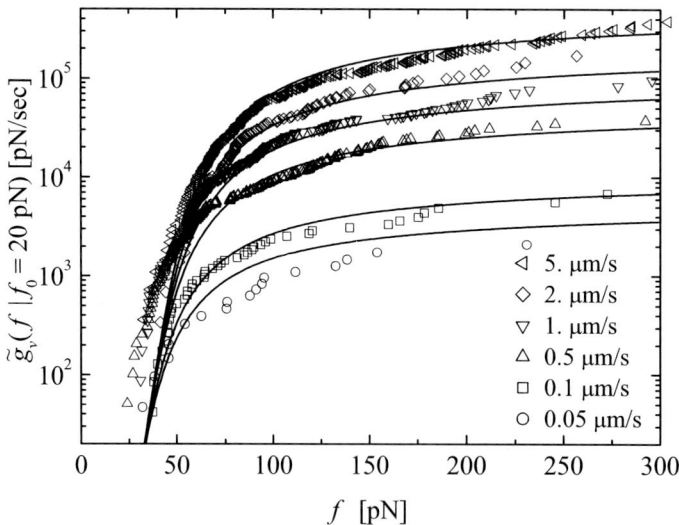

Fig. 8.5. *Symbols*: the experimental function $\tilde{g}_v(f|f_0)$ from Eq. (8.37) obtained for the bond formed by the DNA fragment *expE1/E5* and ExpG protein for the initial force $f_0 = 20\,\text{pN}$ and six different velocities, as detailed in the legend. *Solid lines*: fits based on the theory from Eqs. (8.45), (8.46), and (8.47)

experimentally, viz. a different DNA fragment and regulatory protein ExpG, a PhoB peptide (wild type) and the corresponding DNA target sequence, and a cationic guest molecule and a supramolecular calixaren host molecule [30].

We first note that only $f(t)$-curves surpassing $f_0 = 20\,\text{pN}$ have been taken into account in Fig. 8.5. Hence, rebinding after dissociation would require a huge and hence extremely unlikely random fluctuation and has indeed never been observed in the experiment at hand. Moreover, upon increasing f_0, no clear tendency towards a better data collapse than in Fig. 8.5 was observed, indicating that, indeed, rebinding events are negligible.

Concerning the accompanying equilibrium assumption implicit in Eq. (8.2), the most convincing possibility leading to its failure is the existence of several metastable (sub-)states of the bound complex with relatively slow transitions between them [31, 35, 36, 43] and possibly several different dissociation pathways [40]. As discussed in detail in [79], one indeed gets a spreading of the g-function for different velocities in this way. This spreading is, however, qualitatively quite different from that in Fig. 8.5 for a generic model with a few internal states. With more complex networks of internal states and dissociation channels – and a concomitant flurry of fit parameters in the form of transition rates between them – a satisfactory fit to the data in Fig. 8.5 may be possible, but their actual existence in all the different experimental systems seems quite difficult to justify.

For further unsuccessful attempts to quantitatively explain the noncollapse of the data to a single master curve in Fig. 8.5 see [79].

The most reasonable explanation of the experimental findings seems to be the heterogeneity of the chemical bonds. Basically, this means that Eqs. (8.2) and (8.3) remain valid except that the force-dependent dissociation rate $\omega(f)$ is subjected to random variations upon repeating the pulling trials. As a consequence, the experimentally determined $\tilde{P}_v(f|f_0)$ from Eqs. (8.35) and (8.36) should be compared not with the function $P_v(f|f_0)$ from Eq. (8.9), but rather with its average with respect to the probability distribution of the rates $\omega(f)$, henceforth denoted as $\overline{P}_v(f|f_0)$.

At a first glance, such an intrinsic randomness of the dissociation rate $\omega(f)$ might appear unlikely in view of the fact that, after all, it is always the same species of molecules which are dissociating. Yet, possible physical reasons for such random variations of the dissociation rate $\omega(f)$ might be:

- Random variations and fluctuations of the local molecular environment by ions, water and solvent molecules locally modulating ionic strength, pH and electric fields which may influence the dissociation process of the molecular complex [95].
- Structural fluctuations due to thermal activation may lead to different conformations of a (macro-) molecule.
- Orientation fluctuations of the molecular complex relative to the direction of the applied pulling force may amount to different dependences of the rate ω on f. In addition, the linker molecules may be attached to the complex at different positions, but also many other random geometrical variations may be possible [96, 97].
- In a number of dissociation events one is actually pulling apart not the specific molecular complex of interest but rather some different, unspecific chemical bond. In a small but not necessarily negligible number of such unspecific bond

ruptures, the force-extension curve may still look exactly like that in Fig. 8.2 and hence it is impossible to eliminate those events from the experimental data set.

We remark that not all those general reasons may be pertinent to the specific experimental data in Fig. 8.5 and that there may well exist additional sources of randomness which we have overlooked so far. Their detailed quantitative modeling is a daunting task beyond the scope of our present work and also beyond the present possibilities of experimental verification. Rather, we will resort to the ad hoc Ansatz that all those different sources of randomness approximately sum up to an effective distribution of the rate parameters. A brief quantitative description of the fit procedure follows (see [30] for a more complete account).

Our starting point will be the Bell's Ansatz for the rate, Eq. (8.7), which, apart from the pulling force f, depends on two fit parameters, $\omega(0)$ and α. These parameters are randomly distributed according to a certain probability density. In view of the exponential function in Eq. (8.7), we can expect that the randomness of α has a much stronger effect than that of $\omega(0)$. Hence, for simplicity, we assume that the force-free rupture rate has a constant value, and a Gaussian distribution of the parameter α:

$$\rho(\alpha) = N \, \exp\left(-\frac{(\alpha - \alpha_{\mathrm{m}})^2}{2\sigma^2}\right) \Theta(\alpha). \tag{8.45}$$

Negative α-values in Eq. (8.45) appear quite unphysical and hence are suppressed by the factor $\Theta(\alpha)$, while N is a normalization constant. The truncated Gaussian from Eq. (8.45) may be viewed as a poor man's guess in order to effectively take into account the many different possible sources of bond randomness mentioned above. The parameters α_{m} and σ approximate the mean and the dispersion of α, provided the relative dispersion $\sigma/\alpha_{\mathrm{m}}$ is sufficiently small. Otherwise, the actual mean value, $\bar{\alpha} = \int d\alpha \, \alpha \, \rho(\alpha)$, may exceed the most probable value α_{m} quite notably.

The experimentally observed bond survival probability should be compared with the expression (8.10) averaged with respect to the distribution from Eq. (8.45):

$$\overline{P}_v(f|f_0) = \int\limits_0^\infty d\alpha \, \rho(\alpha) \exp\left(-\frac{\omega(0)}{\alpha \kappa v} \left(e^{\alpha f} - e^{\alpha f_0}\right)\right). \tag{8.46}$$

Finally, the fit parameters α_{m} and σ are determined so that the resulting g-function

$$\overline{g}_v(f|f_0) = -\kappa v \ln \overline{P}_v(f|f_0) \tag{8.47}$$

reproduces the experimentally observed $\tilde{g}_v(f|f_0)$ as closely as possible. The resulting optimal parameters α_{m} and σ yield an estimate for the heterogeneity of the chemical bonds in the form of the probability distribution of the parameter α from Eq. (8.45). The excellent agreement of the fit with the experimental results is achieved with the following fit parameter values:

$$\omega(0) = 0.0033 \, \mathrm{s}^{-1}, \; \alpha_{\mathrm{m}} = 0.13 \, \mathrm{pN}^{-1}, \; \sigma = 0.07 \, \mathrm{pN}^{-1}. \tag{8.48}$$

The fit results are presented in Fig. 8.5 as solid lines. We note that two other functional forms for the distribution of α have been tried, namely, box and parabolic distribution, with similar fit results for the rate prefactor $\omega(0)$, the most probable value α_m, and distribution width σ [30].

8.5
Application to Atomic Friction

Although the logarithmic relation similar to Eqs. (8.15) and (8.31) between the pulling velocity and the friction force has been shown to agree well with the experimental data [63, 98], we argue that also in nanofriction experiments this success does not prove unambiguously the validity of the simple rate equation (8.2) for the description of the stick-slip process. Indeed, apart from thermal noise effects encapsulated in Eq. (8.2), other sources of randomness of jump events are conceivable in experiments. Some of them are as follows.

The tip–substrate contact is a complicated system involving many atoms. After each slip, the contact is formed anew, and there is no a priori reason to expect that all participating atoms will establish the same positions and interactions in the new stick phase as in the previous one. Furthermore, the cantilever tip may even acquire or lose surface atoms during a stick-slip cycle. Because of the exponential form of the rate from Eq. (8.3), even a small variation of contact properties may result in large variations of $\omega(f)$ in the rate Eq. (8.2). Finally, it has been suggested recently [99, 100] that the tip can form multiple contacts, whose behavior cannot be captured by a single-step rate Eq. (8.2), but rather requires a more complicated multiple-step/multiple-state model describing several possible bonds in the contact region.

Because the rate $\omega(f)$ determines the number of interstitial transitions per unit time, the logarithmic force-velocity relation reflects the exponential character of this function. Since in all cases mentioned above it is not necessary to abandon the rate description concept, but rather amend it, all of these possibilities preserve the logarithmic character of the force-velocity relation. However, when fitting the experimental force-velocity curve according to the simple rate Eq. (8.2) [63, 98], it remains unclear, what exactly the resulting fitting parameters characterize – a single tip–sample contact, the average effect of differently formed contacts in each stick phase, multicontact connection between the tip and the substrate, or possibly some other mechanism not mentioned above? To resolve this ambiguity, it is imperative to directly check the validity of the rate Eq. (8.2) itself in each experimental situation [90, 101].

In what follows, we apply the method of data analysis from Sect. 8.3 to the friction force experiments on highly oriented pyrolytic graphite (HOPG), which were conducted with a commercial AFM under the pressure of 2×10^{-10} mbar at room temperature. The experimental details can be found elsewhere [90, 101].

The stick-slip phenomenon is observed via the friction force contrast with atomic unit cell periodicity of the surface. A typical lateral force map is presented in Fig. 8.6a, and a friction signal for a single scan line is shown in Fig. 8.6b revealing the expected sawtooth-type behavior. The velocity dependence of friction forces was measured in the range from $v = 20$ nm/s to $v = 200$ nm/s.

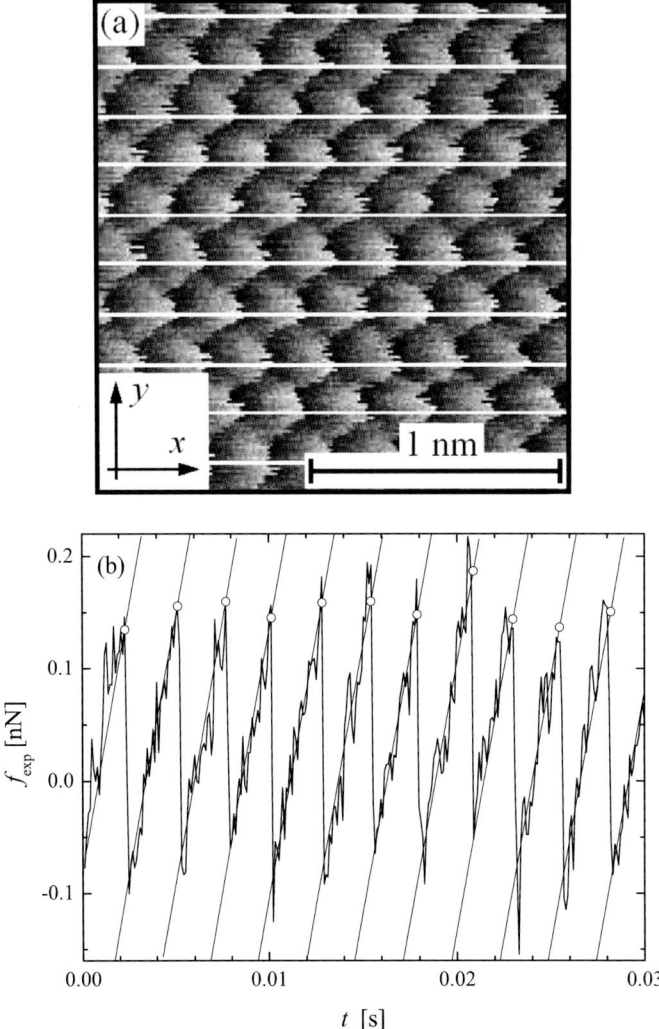

Fig. 8.6. a A representative lateral force map on HOPG obtained at a pulling velocity $v =$ 100 nm/s. The *horizontal lines* indicate those scan lines that were used for the collection of friction data. **b** The experimentally observed lateral force $f_{exp}(t)$ during one typical line scan from **a**. The *equidistant parallel lines* result from the fit of the stick segments with Eq. (8.49) as detailed in the text. The *circles* show the positions of the slip events

The individual carbon atoms form hexagonal rings arranged in a honeycomb structure [102]. During the scan the tip jumps in between the local energy minima located at the centers of the carbon rings ("hollow sites" [102] appearing as "humps" in Fig. 8.6a). The sample was oriented in such a way that the fast scan direction (x-direction) runs along the $(\overline{1}, 2, \overline{1}, 0)$ direction of the (0001) HOPG surface. This ensures that predominantly jumps along the fast scan direction are observed [59].

Since the surface potential varies along the y-axis even within one hollow site, only the scan lines from the equivalent y-positions at the center of the hollow sites were analyzed. Those positions are indicated with the horizontal lines in Fig. 8.6a.

A typical stick-slip scan line contained 10–12 slip events, see Fig. 8.6b. While the moments t_n of the slip events can be readily identified as almost instantaneous force drops, the forces themselves require a more careful consideration. As seen in Fig. 8.6b, the experimentally observed force evolution $f_{exp}(t)$ is composed of random fluctuations due to thermal and instrumental noise, whose details strongly depend on the experimental resolution, and linearly increasing regular segments, to be identified with the forces f entering the theory of Sects. 8.2 and 8.3. Hence, the random forces have to be separated from the regular ones in the experimental data before comparison with the theory. To this end, the measured $f_{exp}(t)$ in Fig. 8.6b were fitted with a piecewise linear function of the form

$$f_n(t) = \kappa(vt - na) + \Delta f, \tag{8.49}$$

increasing its index n at every slip instant t_n. As a result, each scan line yields an estimate for the unknown fit parameters in Eq. (8.49), namely the effective spring constant κ and the lattice constant a. Moreover, evaluating the fitting function at the slip times t_n yields the force values at the end of every stick phase, indicated by the circles in Fig. 8.6b and required to evaluate the theoretical quantities in Eqs. (8.35) and (8.36).

By fitting a and κ for each scan line, the former was found to vary only insignificantly around the mean value $a \approx 0.26\,\mathrm{nm}$, while the latter fluctuated by about 8% (standard deviation) around the mean value $\kappa \approx 0.93\,\mathrm{N/m}$. These variations can be understood as yet another signature of the abovementioned quite complicated behavior of the tip–substrate contact, considering that the tip apex is particularly prone to changes at the very beginning of every scan line and that the tip–substrate contact significantly influences the relevant effective spring constant κ [103, 104].

Our goal is to check by means of Eq. (8.37) whether the individual slip events can be viewed as single-step transitions described by Eq. (8.2). Note that the force dependence of the rate $\omega(f)$ in Eq. (8.3) is quite sensitive to changes of the effective spring constant κ and in Eq. (8.37) only one fixed κ-value is tacitly taken for granted. Because of the abovementioned variability of κ we thus cannot use all stick-slip data available, but rather have to restrict ourselves to a subset for which κ is confined to a narrow window around some specific value. We have processed the data corresponding to several such values of κ and found practically the same results for all of them. In the following, we report our findings for $0.92\,\mathrm{N/m} \leq \kappa \leq 0.94\,\mathrm{N/m}$.

The experimental estimate for the function $\tilde{g}_v(f|f_0 = 0)$ from Eq. (8.37), for several pulling velocities is presented in Fig. 8.7.

As shown in Fig. 8.7a, the measured functions $\tilde{g}_v(f|f_0 = 0)$ indeed collapse onto a single master curve, provided that the pulling velocities are sufficiently high, $v > 90\,\mathrm{nm/s}$. On the other hand, no such collapse is observed in the low-velocity range, see Fig. 8.7b, where the experimental g-function increases with increasing pulling velocity. This result indicates the inapplicability of the rate theory, Eq. (8.2), to the description of the stick-slip process at slow pulling. In the following, we amend the model from Eq. (8.2) to adequately describe the experimental results for all pulling velocities.

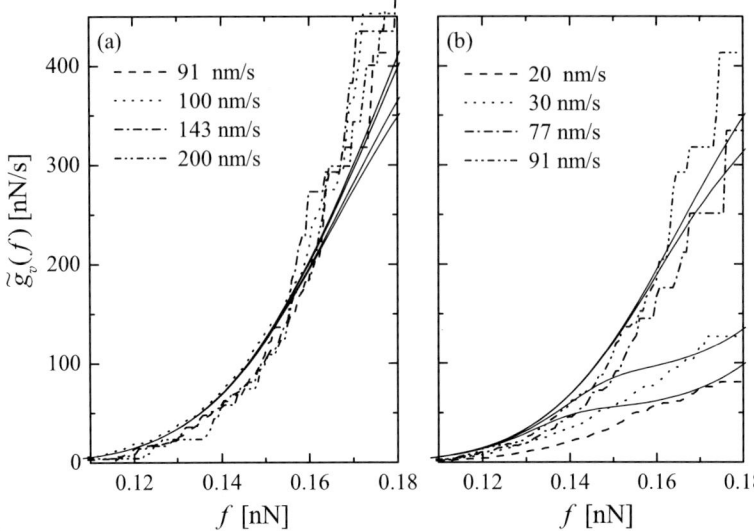

Fig. 8.7. The experimentally measured functions $\tilde{g}_v(f|f_0 = 0)$ from Eq. (8.37) for different pulling velocities, v, as specified in the legends, for zero initial force, and for the effective spring constants $0.92\,\text{N/m} \le \kappa \le 0.94\,\text{N/m}$. The smooth *solid lines* are theoretical fits using the model described in Sect. 8.5 [101]

The lack of universal behavior of the g-function for slow velocities, Fig. 8.7b, suggests that some other process besides thermal activation plays a significant role in this regime, while at fast pulling, thermal activation takes over. In the spirit of [99, 100], it has been hypothesized [90] that a possible candidate for such a process is multiple bond formation of the tip–substrate contact. More precisely, formation of additional bonds takes finite time, so that at fast pulling, new bonds do not have sufficient time to develop during a single stick phase. This leaves in effect a single tip–substrate contact, whose rupture is well described by Eq. (8.2), resulting in the collapse of the measured functions $\tilde{g}_v(f|0)$ onto a single master curve, Fig. 8.7a.

To quantify these ideas, we introduce the following minimalistic model [101]. Within this model, it is assumed that in a given stick phase, the tip–sample contact may either break, resulting in a slip event, or strengthen itself by means of forming new bonds; the latter possibility can be termed "contact ageing."

To account for this effect, we assume for simplicity that the contact can exist in only two different states, a strongly and a weakly bound one, characterized by two different off-rates $\omega_i(f)$, and the respective occupation probabilities $p_i(f)$, $i = 0, 1$. Here, the value of the subscript $i = 0$ refers to the initial weakly bound state of the contact, and $i = 1$ to the strongly bound state. We further assume that the contact can enter the strongly bound state 1 from the originally formed weakly bound state 0 at a rate λ independent of the value of the force. At the same time, the back-transitions $1 \to 0$ are neglected. The model is schematically illustrated in Fig. 8.8.

Fig. 8.8. Schematic representation of the proposed model of the tip–sample contact

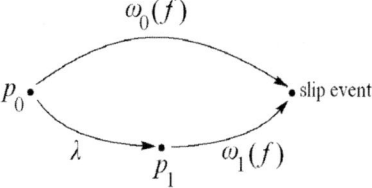

Within this simplified picture, the single-step rate Eq. (8.2) describing the state of the contact should be replaced by two rate equations for the probabilities p_0, p_1:

$$\dot{p}_0(t) = -(\lambda + \omega_0[f(t)])p_0(t), \tag{8.50a}$$

$$\dot{p}_1(t) = \lambda p_0(t) - \omega_1[f(t)])p_1(t), \tag{8.50b}$$

with the initial condition $p_i(0) = \delta_{i0}$.

In each stick phase, the force increases linearly [cf. Eq. (8.49)]

$$f(t) = \kappa v t + f(0), \tag{8.51}$$

allowing one to go from the time to the force domain

$$p_0'(f) = -\frac{1}{\kappa v}(\lambda + \omega_0(f))p_0(f), \tag{8.52a}$$

$$p_1'(f) = \frac{1}{\kappa v}[\lambda p_0(f) - \omega_1(f)p_1(f)]. \tag{8.52b}$$

From these equations, one obtains for the respective probabilities to find the contact in the states 0 and 1 at the force f explicit results:

$$p_0(f) = \exp\left\{-\frac{1}{\kappa v}\left[\lambda(f - f_0) + \int_{f_0}^{f} df'\,\omega_0(f')\right]\right\}, \tag{8.53a}$$

$$p_1(f) = \frac{\lambda}{\kappa v}\int_{f_0}^{f} df'\,p_0(f')\exp\left[-\frac{1}{\kappa v}\int_{f'}^{f} df''\,\omega_1(f'')\right]. \tag{8.53b}$$

The off-rates for the two states are assumed to be given by [cf. Eq. (8.6)]

$$\omega_i(f) = \Omega_i \exp\left[-\frac{\Delta E_i}{k_B T}\left(1 - \frac{f}{F_i}\right)^{\gamma_i}\right]. \tag{8.54}$$

Since the probability to find the tip in any bound state 0 or 1 is $p_0 + p_1$, the experimental g-function should be compared with

$$g_v(f|f_0) = -\kappa v \ln[p_0(f) + p_1(f)]. \tag{8.55}$$

In the limit of pulling velocities so high that new bonds do not have time to form in each stick phase, the tip remains in the state 0 throughout the whole stick-phase with overwhelming probability, so that the theoretical g-function from Eq. (8.55) simplifies to a velocity-independent expression:

$$g_\infty(f) = -\kappa v \ln p_0(f) = \int\limits_{f_0}^{f} df' \omega_0(f') =$$

$$\frac{\Omega_0 F_0}{\gamma_0} \left(\frac{k_B T}{\Delta E_0}\right)^{1/\gamma_0} \left\{\Gamma\left(\frac{1}{\gamma_0}, \frac{\Delta E_0}{k_B T}\left[1 - \frac{f_0}{F_0}\right]^{\gamma_0}\right) - \Gamma\left(\frac{1}{\gamma_0}, \frac{\Delta E_0}{k_B T}\left[1 - \frac{f}{F_0}\right]^{\gamma_0}\right)\right\}$$

$$(8.56)$$

where in the second line, the integral is evaluated explicitly for the rate Ansatz (8.54). Here, $\Gamma(a, x) = \int_0^x dy\, y^{a-1}\, e^{-y}$ is the incomplete gamma-function; it can be calculated numerically using an efficient algorithm, for example from [82].

The theory from the previous section contains nine fit parameters, eight characterizing the rates $\omega_{0,1}(f)$ (namely, Ω_i, ΔE_i, F_i, and γ_i, with $i = 0$, 1), and the $0 \to 1$ transition rate λ. Their determination proceeds consecutively in the following steps.

First, using Eqs. (8.39), (8.43), and (8.44), we have combined the g-curves in the high-velocity range, Fig. 8.7a, into a single master curve. The plot of the g-function together with the error bars is presented in Fig. 8.9. The combined g-function from Fig. 8.9 is fitted using the high-velocity asymptotic formula, Eq. (8.56), with the following values of the resulting fit parameters:

$$\Delta E_0 = 24\, k_B T = 97.\, 2\, \text{pN nm}; \quad F_0 = 0.\, 19\, \text{pN}; \quad \gamma_0 = 2.\, 4; \quad \Omega_0 = 12000\, \text{s}^{-1}.$$

$$(8.57)$$

The fitting curve is presented in Fig. 8.9 as a solid line.

It remains to determine five more values, namely, Ω_1, ΔE_1, F_1, γ_1, and λ. To reduce their number, we make the following additional assumptions: (1) We assume that the overall geometry of the tip energy landscape in contact with the surface is the same in both states. (2) The critical force is directly proportional to the barrier height

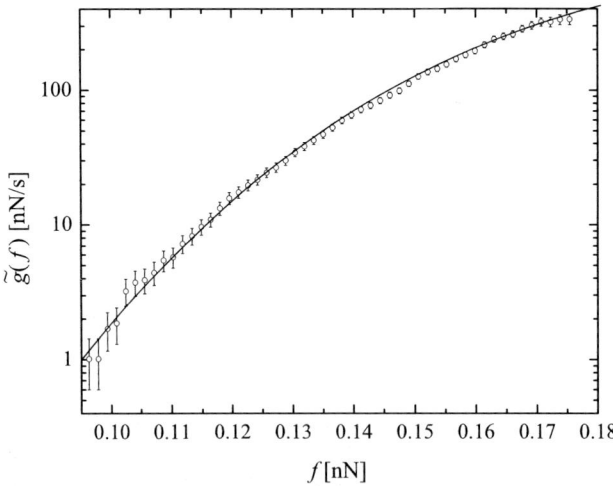

Fig. 8.9. *Circles*: the combined function $\tilde{g}(f|0)$ based on the experimental g-curves for $90\, \text{nm/s} < v \leq 200\, \text{nm/s}$. *Solid line*: theoretical fit using Eq. (8.56)

with the same proportionality coefficient in both states. (3) As in the Kramers' theory of thermally activated escape [12], we assume that the prefactor Ω_i is proportional to the value $\sqrt{U''_{min}|U''_{max}|}$, where $U''_{min,max}$ denote the curvatures of the tip energy landscape at the two extrema. These three assumptions can be expressed mathematically as

$$\gamma_1 = \gamma_0; \ F_1 = \frac{\Delta E_1}{\Delta E_0} F_0; \ \Omega_1 = \frac{\Delta E_1}{\Delta E_0} \Omega_0, \tag{8.58}$$

leaving only two unknown fit parameters, ΔE_1 and λ. Their values are determined from the $\tilde{g}_v$-curves in the low-velocity range, Fig. 8.7b:

$$\Delta E_1 = 32\,k_B T = 129.6\,\text{pN nm}; \ \lambda = 10\,\text{s}^{-1}. \tag{8.59}$$

The reasonable agreement of the fitting curves from Eqs. (8.53), (8.54), and (8.55) with the parameters from Eqs. (8.57), (8.58), and (8.59) with the experimental results at low velocities is clear from Fig. 8.7. We conclude that the typical time for the formation of new bonds is of the order of $\lambda^{-1} = 0.1\,\text{s}$. This number can be compared with the duration of the stick phase $av^{-1} \approx 3\,\text{ms}$ at the slowest pulling velocity $v = 91\,\text{nm/s}$, for which the validity of the rate Eq. (8.2) has been verified, see Fig. 8.7.

Although the theoretical and experimental g-curves in Fig. 8.7 are in reasonable agreement with each other, the discrepancy between the two sets of data is quite noticeable. This is because the theory presented above together with the simplifying assumptions from Eq. (8.58) provides only a crude description of the contact ageing process, and captures only its most salient feature: strengthening of the tip–sample contact. This crude description can be refined in several ways.

While our model uses only two contact states of the cantilever, see Fig. 8.8, one can generalize this approach and introduce additional contact states, all characterized by different occupation probabilities $p_i(f)$ and off-rates $\omega_i(f)$. In addition, the rate parameters for each such state can be considered as random variables, with respect to which the final no-jump probability $\Sigma_i\,p_i(f)$ has to be averaged, as shown in Sect. 8.4.

Unfortunately, these amendments to the model from Eq. (8.52) will make it impossible to solve the resulting rate equations analytically, while within our simplified description, such an analytic solution is still possible, see Eq. (8.53). Even worse, making the abovementioned amendments to Eq. (8.52) means introducing a number of new fit parameters, whose reasonably accurate determination based on the experimental data may be questionable.

Despite its crude simplifications, our analysis shows clearly that contact ageing plays an important role in realistic friction force experiments. The presented simple model, which can still be solved analytically, captures the essential features of this process.

8.6
Conclusions

In this chapter, we have considered two research applications of AFM, one aimed at the determination of the binding strength of various biological complexes, and the other at the studies of friction at the nanoscale. The common feature of the two research fields is that both forced bond rupture and interstitial tip transitions are thermally activated events and can be described in a first approximation by a single-step rate equation. The off-rate entering that equation, i.e. the probability to perform a thermally activated escape out of a metastable state per unit time, depends on the acting force, and one can use this dependence to characterize the properties of the biocomplex bond and the tip–sample interaction potential. The standard approach to data analysis uses the relation between the velocity and the most probable force at which the transition takes place.

A careful examination of the whole distribution of the escape events allows one to gain further insight into the properties of the system studied. In particular, we have described an alternative approach to the analysis of the experimental data, which allows one to check the validity of the single-step rate equation, and, upon its verification, to accurately deduce the force-dependent off-rate. The main idea of the method is that, if the single-step rate equation applies, then the logarithm of no-jump probability times the pulling velocity is a velocity-independent function of the acting force, whose derivative with respect to the force is the off-rate.

Application of this method to the real experimental data has indicated that in both forced bond rupture experiments and nanofriction experiments, the single-step rate equation alone cannot describe adequately the statistics of escape events. The deviations of the jump statistics in the former experimental situation from the predictions of the single-step rate equation has been explained as resulting from the heterogeneity of the chemical bonds involved. On the other hand, such deviations in nanofriction measurements have been attributed to the process of contact ageing, i.e. gradual change of contact properties in a stick phase. By amending the rate equation so as to correctly reproduce the observed statistics of bond ruptures and slip events, we were able to quantitatively characterize both heterogeneity of chemical bonds and contact ageing process.

Acknowledgments. The author is grateful to Alexander von Humboldt-Stiftung, Deutsche Forschungsgemeinschaft (SFB613 and RE 1344/3–2), and European Science Foundation (collaborative research project Nanorama-07-FANAS-FP-009) for financial support, to Peter Reimann for many fruitful discussions, and to Sebastian Getfert, André Schirmeisen, and Ralf Eichhorn for critical reading of this chapter.

References

1. Binnig G, Quate CF, Gerber C (1986) Phys Rev Lett 56:930
2. Meyer G, Amer NM (1988) Appl Phys Lett 53:1045
3. Alexander S, Hellemans L, Marti O, Schneir J, Elings V., Hansma PK, Longmire M, Gurley J (1989) J Appl Phys 65:164
4. Grubmüller H, Heymann B, Tavan P (1996) Science 271:997
5. Izrailev S, Stepaniants S, Balsera M, Oono Y, Schulten K (1997) Biophis J 72:1568
6. Sorensen MR, Jacobsen KW, Stoltze P (1996) Phys Rev B 53:2101
7. Livshits AI, Shluger AL (1997) Phys Rev B 56:12482
8. Heymann B, Grubmüller H (2001) Biophys J 81:1295
9. Grabert H (1992) Projection operator techniques in nonequilibrium statistical mechanics. Springer, Berlin

10. Hänggi P, Thomas H (1982) Phys Rep 88:207
11. Risken H (1984) The Fokker-Planck equation, Springer, Berlin
12. Hänggi P, Talkner P, Borkovec M (1990) Rev Mod Phys 62:251
13. Reimann P (2002) Phys Rep 361:57
14. Florin E-L, Moy VT, Gaub HE (1994) Science 264:415
15. Lee GU, Kidwell AD, Colton RJ (1994) Langmuir 94:354
16. Moy VT, Florin E-L, Gaub HE (1994) Science 266:257
17. Chilcotti A, Boland T, Ratner BD, Stayton PS (1995) Biophys J 69:2125
18. Lo Y-S, Zhu Y-J, Beebe Jr TB (1995) Langmuir 17:3741
19. Dammer U, Hegner M, Anselmetti D, Wagner P, Dreier M, Huber W, Güntherodt H-J (1996) Biophys J 70:2437
20. Hinterdorfer P, Baumgartner W, Gruber HJ, Schilcher K, Schindler H (1996) Proc Natl Acad Sci USA 93:3477
21. Allen S, Chen X, Davies J, Davies MC, Dawkes AC, Edwards JC, Roberts CJ, Sefton J, Tendler SJB, Williams PM (1997) Biochemistry 36:7457
22. Schwesinger F, Ros R, Strunz T, Anselmetti D, Güntherodt H-J, Honegger A, Jermutus L, Tiefenauer L, Plückthun A (2000) Proc Natl Acad Sci USA 97:9972
23. Strunz T, Oroszlan K, Schäfer R, Günterodt H-J (1999) Proc Natl Acad Sci USA 96:11277
24. Green NH, Williams PM, Wahab O, Davies MC, Roberts CJ, Tendler SJB, Allen S (2004) Biophys J 86:3811
25. Liphardt J, Onoa B, Smith SB, Tinoco Jr I, Bustamante C (2001) Science 292:733
26. Imparato A, Peliti L (2004) Eur Phys J B 39:357
27. Hukkanen EJ, Wieland JA, Gewirth A, Leckband DE, Braatz RD (2005) Biophys J 89:3434
28. Wieland JA, Gewirth AA, Leckband DE (2005) J Biol Chem 280:41037
29. Schlierf M, Li H, Fernandez JM (2004) Proc Natl Acad Sci USA 101:7299
30. Raible M, Evstigneev M, Bartels FW, Eckel R, Nguyen-Duong M, Merkel R, Ros R, Anselmetti D, Reimann P (2006) Biophys J 90:3851
31. Evans E, Ritchie K (1997) Biophys J 72:1541
32. Bell GI (1978) Science 200:618
33. Rief M, Fernandez JM, Gaub HE (1998) Phys Rev Lett 81:4764
34. Shillcock J, Seifert U (1998) Phys Rev E 57:7301
35. Merkel R, Nassoy P, Leung A, Ritchie K, Evans E (1999) Nature 397:50
36. Strunz T, Oroszlan K, Schumakovitch I, Güntherodt H-J, Hegner M (2000) Biophys J 79:1206
37. Heymann B, Grubmüller H (2000) Phys Rev Lett 84:6126
38. Seifert U (2000) Phys Rev Lett 84:2750
39. Evans E (2001) Annu Rev Biomol Struct 30:105
40. Bartolo D, Derényi I, Ajdari A (2002) Phys Rev E.65:051910
41. Nguyen-Duong N, Koch KW, Merkel R (2003) Europhys Lett 61:845
42. Evans E, Leung A, Heinrich V, Zhu C (2004) Proc Natl Acad Sci USA 101:11281
43. Derényi I, Bartolo D, Ajdari A (2004) Biophys J 86:1263
44. Barsegov V, Thirumalai D (2005) Proc Natl Acad Sci USA 102:1835
45. Barsegov V, Thirumalai D (2006) J Phys Chem B 110:26403
46. Merkel R (2001) Phys Rep 346:343
47. Kramers HA (1940) Physica (Utrecht) 7:284
48. Fleming GR, Hänggi P (1993) (eds) Activated Barrier Crossing. World Scientific, Singapore
49. Talkner P, Hänggi P (1995) (eds) New Trends in Kramers Reaction Rate Theory. Kluwer, Dordrecht
50. Persson BNJ (1999) Surf Sci Rep 33:83
51. Persson BNJ (2000) Sliding Friction, Springer, Berlin
52. Urbakh M, Klafter J, Gourdon D, Israelachvili J (2004) Nature 430:525

53. Mosey NJ, Müser MH, Woo TK (2005) Science 307:1612
54. Mate VM, McClelland GM, Erlandsson R, Chiang S (1987) Phys Rev Lett 59:1942
55. Dedkov GV (2000) Physics Uspekhi 43:541
56. Gnecco E, Bennewitz R, Gyalog T, Meyer E (2001) J Phys: Condens Matter 13:R619
57. Braun OM, Naumovets AG (2006) Surf Sci Rep 60:79
58. Gnecco E, Meyer E (2007) (eds) Fundamentals of Friction and Wear on the Nanoscale. Springer, Berlin Heidelberg New York
59. Schirmeisen A, Jansen L, Fuchs H (2005) Phys Rev B 71:245403
60. Fujisawa S (1998) Phys Rev B 58:4909
61. Sills S, Overney RM (2003) Phys Rev Lett 91:095501
62. Schirmeisen A, Jansen L, Hölscher H, Fuchs H (2006) Appl Phys Lett 88:123108
63. Riedo E, Gnecco E, Bennewitz R, Meyer E, Brune H (2003) Phys Rev Lett 91:084502
64. Garg A (1995) Phys Rev B 51:15592
65. Dudko OK, Hummer G, Szabo A (2006) Phys Rev Lett 96:108101
66. Evstigneev M, Reimann P (2005) Phys Rev E 71:056119
67. Evstigneev M, Reimann P (2005) Phys Rev B 73:113401
68. Yukalov V I, Gluzman S, Sornette D (2003) Physica A 328:409
69. Gluzman S, Yukalov VI, Sornette D (2003) Phys Rev E 67:026109
70. Dembo M, Tourney DC, Saxman K, Hammer D (1988) Proc R Soc Lond B 234:55
71. Marshall BT, Long M, Piper JW, Yago T, McEver RP, Zhu C (2003) Nature 423:190
72. Sang Y, Dubé M, Grant M (2001) Phys Rev Lett 87:174301
73. Dudko OK, Filippov AE, Klafter J, Urbakh M (2002) Chem Phys Lett 352:499
74. Dudko OK, Filippov AE, Klafter J, Urbakh M (2003) Proc Natl Acad Sci USA 100:11378
75. Sheng Y-J, Jiang S, Tsao H-K (2005) J Chem Phys 123:091102
76. Hummer G, Szabo A (2003) Biophys J 85:5
77. Corless RM, Gonnet GH, Hare DEG, Jeffrey DJ, Knuth DE (1996) Adv Comp Math 5:329
78. Getfert S, Reimann P (2007) Phys Rev E 76:052901
79. Raible M, Evstigneev M, Reimann P, Bartels FW, Ros R (2004) J Biotech 112:13
80. Seifert U (2002) Europhys Lett 58:792
81. Gergely G, Voegel J-C, Schaaf P, Senger B, Maaloum M, Hörber JKH, Hemmerlé J (2000) Proc Natl Acad Sci USA 97:10802
82. Press WH, Teukolsky SA, Vetterling WT, Flannery BP (1999) Numerical recipes in C, Cambridge University Press
83. Williams PM (2003) Analytica Chimica Acta 479:107
84. Evstigneev M, Reimann P (2003) Phys Rev E 68:045103
85. Evstigneev M, Reimann P (2004) Europhys Lett 67:907
86. Abramowitz M, Stegun I (1965) (eds) Handbook of mathematical functions, Dover, New York
87. Krylov SY, Jinesh KB, Valk H, Dienwiebel M, Frenken JWM (2005) Phys Rev E 71:65101
88. Kurkijärvi J (1972) Phys Rev B 6:832
89. Fulto TA, Dunkleberger LN (1974) Phys Rev B 9:4760
90. Evstigneev M, Schirmeisen A, Jansen L, Fuchs H, Reimann P (2006) Phys Rev Lett 97:240601
91. Taylor JR (1982) An Introduction to Error Analysis, University Science Books, Mill Valley, CA
92. Bartels FW, Baumgarth B, Anselmetti D, Ros R, Becker A (2003) J Struct Biol 143:145
93. Eckel R, Wilking S-D, Becker A, Sewald N, Ros R, Anselmetti D (2005) Angew Chem Int Ed Engl 44:3921
94. Eckel R, Ros R, DeckerB, Mattay J, Anselmetti D (2005) Angew Chem Int Ed Engl 44:484
95. Vijayendran RA, Leckband DE (2001) Anal Chem 73:471
96. Simson D A, Strigl M, Hohenadl M, Merkel R (1999) Phys Rev Lett 83:652

97. Strigl M, Simson DA, Kacher CM, Merkel R (1999) Langmuir 15:7316
98. Gnecco E, Bennewitz R, Gyalog T, Loppacher C, Bammerlin M, Meyer E, Güntherodt H-J (2000) Phys Rev Lett 84:1172
99. Maier S, Sang Y, Filleter T, Grant M, Bennewitz R, Gnecco E, Meyer E (2005) Phys Rev B **72**:245418
100. Filippov AE, Klafter J, Urbakh M (2004) Phys Rev Lett **92**:135503
101. Evstigneev M, Schirmeisen A, Jansen L, Fuchs H, Reimann P (2008) J Phys: Condens Matt 20:35400
102. Hölscher H, Schwarz UD, Zwörner O, WiesendangerR (1998) Phys Rev B 57:2477
103. Socoliuc A, Bennewitz R, Gnecco E, Meyer E (2004) Phys Rev Lett 92:134301
104. Reimann P, Evstigneev M (2005) New J Phys **7**:25

Subject Index

Printing: Krips bv, Meppel, The Netherlands
Binding: Stürtz, Würzburg, Germany